Exactly how modern humans evolved is a subject of intense debate. This book deals with the evolution of modern humans from an archaic ancestor and the differentiation of modern populations from each other. The first section of the book investigates whether modern populations arose from regional archaic hominid groups that were already different from each other, and argues that in fact, most lines of evidence support a single, recent origin of modern humans in Africa. Dr Lahr then goes on to examine ways in which this diversification could have occurred, given what we know from fossils, archaeological remains and the relationships of existing populations today. This book will be a must for all those interested in human evolution.

Cambridge Studies in Biological Anthropology 18

The evolution of modern human diversity

Cambridge Studies in Biological Anthropology

Series Editors

Also in the series

G.W. Lasker *Surnames and Genetic Structure*
C.G.N. Mascie-Taylor and G.W. Lasker (editors) *Biological Aspects of Human Migration*
Barry Bogin *Patterns of Human Growth*
Julius A. Kieser *Human Adult Odontometrics – The study of variation in adult tooth size*
J.E. Lindsay Carter and Barbara Honeyman Heath *Somatotyping – Development and applications*
Roy J. Shephard *Body Composition in Biological Anthropology*
Ashley H. Robins *Biological Perspectives on Human Pigmentation*
C.G.N. Mascie-Taylor and G.W. Lasker (editors) *Applications of Biological Anthropology to Human Affairs*
Alex F. Roche *Growth, Maturation, and Body Composition – The Fels Longitudinal Study 1929–1991*
Eric J. Devor (editor) *Molecular Applications in Biological Anthropology*
Kenneth M. Weiss *The Genetic Causes of Human Disease – Principles and evolutionary approaches*
Duane Quiatt and Vernon Reynolds *Primate Behaviour – Information, social knowledge, and the evolution of culture*
G.W. Lasker and C.G.N. Mascie-Taylor (editors) *Research Strategies in Biological Anthropology – Field and survey studies*
S.J. Ulijaszek and C.G.N. Mascie-Taylor (editors) *Anthropometry: the individual and the population*
C.G.N. Mascie-Taylor and B. Bogin *Human Variability and Plasticity*
S.J. Uligaszek *Human Energetics in Biological Anthropology*
R.J. Shephard and A. Rode *The Health Consequences of 'Modernisation'*

The evolution of modern human diversity

a study of cranial variation

MARTA MIRAZÓN LAHR
Department of Biological Anthropology
University of Cambridge

CAMBRIDGE UNIVERSITY PRESS
Cambridge, New York, Melbourne, Madrid, Cape Town, Singapore, São Paulo

Cambridge University Press
The Edinburgh Building, Cambridge CB2 2RU, UK

Published in the United States of America by Cambridge University Press, New York

www.cambridge.org
Information on this title: www.cambridge.org/9780521473934

First published 1996
This digitally printed first paperback version 2005

A catalogue record for this publication is available from the British Library

Library of Congress Cataloguing in Publication data

Lahr, Marta Mirazón
The evolution of modern human diversity: a study of cranial
variation / Marta Mirazón Lahr.
p. cm. – (Cambridge studies in biological anthropology; 18)
Includes bibliographical references and index.
ISBN 0–521–47393–4 (hc)
1. Craniology. 2. Craniometry. 3. Skull–Evolution.
4. Human evolution. 5. Human anatomy–Variation.
I. Title. II. Series.
GN71.L35 1996
573′.7–dc20 95–40250 CIP

ISBN-13 978-0-521-47393-4 hardback
ISBN-10 0-521-47393-4 hardback

ISBN-13 978-0-521-02031-2 paperback
ISBN-10 0-521-02031-X paperback

To Fabio, Phillip and Christopher

Contents

Foreword

When Anders Retzius, a century and a half ago, invented the cranial index, he gave us an answer for which there was no question. Nonetheless this, and other indices for face, nose and so on, were useful in description, and there ensued a long period of setting up and standardising cranial measurements, and the gathering and publication of the same – literally in volumes. Results were not commensurate with the effort: they remained largely descriptive, except perhaps as used in the discernment of races of dubious actuality. Eventually English and Indian biometricians (Pearson, Hobgen, Fisher, Mahalanobis, Rao) addressed the problems of population parameters, generalised distances and factors of shape – a turning point because redundancy of information could now be controlled, and significant elements of shape could be elicited. But it was only in the computer age that various workers of the last few decades were able to build on this, largely using intergroup distances and clustering techniques. These have given much significant information, although actual results have differed among workers because of differing data, exact form of analysis or suitability of method to problem.

Now Marta Mirazón Lahr, in this book, goes us all one better, tailoring measurement and method to the precise problems perceived, in a well integrated study. She attacks the main question of the moment, that of modern human origins. She sharply defines what should be looked for in the existing evidence. She selects or creates measurements and graded morphological traits, clearly defined and tested for reliability. She analyses these over regional and other samples, simultaneously handling the fossil data, the context of time, and various probable processes like loss of robusticity or modes of population dispersal. She has looked under a lot of old stones. She ends with a broad and suggestive reconstruction of probable events, without being dogmatic or overly specific, and thus getting into side issues.

And all this is eminently readable – clarity of reasoning makes for clarity of prose. Her main conclusion supports a recent evolution of moderns and a stepwise dispersal from Africa, as against the alternative, the Multiregional Model of separate development. Whether or not some remain unpersuaded, she has made a solid contribution to the discussion. I for one can only contemplate in some awe the formidable amount of exacting work she has accomplished.

W.W. Howells
Maine, May, 1995

Preface

This book is the outcome of my years of study at Cambridge. Cambridge has one of the world's great skeletal collections, collections that remain a major source of information on our history and biology. The study of human skeletal, and particularly cranial, morphology was one of the main interests of physical anthropologists for a long period of time, until it fell into disregard in the middle of this century. The fall of *craniology* had many reasons, some historical, some political and some scientific. Nevertheless, the study of recent human cranial variation in the past 40 years continued, largely through the works of a number of people working with the applications of multivariate statistics to anthropology. These researchers have developed a consistent framework of relationships between modern populations, and my work rests heavily on their studies, especially the works of W.W. Howells, L. Brace, M. Pietrusewsky and G. van Vark. My own research, however, does not aim at disclosing close population relationships in terms of statistical distances. My greatest interest is in terms of mechanisms that can disclose the evolutionary pathways of diversification, be these social, ecological or morphological. I believe that all these played a role, varying in importance in different circumstances. This book represents an attempt at examining some of these processes, and suggesting a possible pattern of the evolution of modern humans.

The first and second parts of this book represent the research I carried out for my PhD, under the supervision of Rob Foley. Nick Macie-Taylor advised on many of the statistical analyses used, and I gained from his knowledge, dedication and enormous patience. Chris Stringer gave many suggestions and ideas, and I thank him for his much appreciated advice. The third part of the book reflects the research I carried out in the past three years, with a post-doctoral Fellowship from Clare College, Cambridge. I have had the pleasure and benefit of discussing my work with L. Aiello, Baruch Arensburg, J. Bowman, C. Duhig, H. Eeley, M. Gomendio, L. Humphrey, R. Kruszinski, P. O'Connor, K. Robson Brown, E. Roldán, V. da Silva and Nan Stevens.

My recent work has been partly done in collaboration with Richard Wright from Australia, Rebeca Haydenblit from Mexico (currently in

Cambridge), and especially with Rob Foley. I have gained from Richard's vast statistical knowledge and concerns, Rebeca's practicality and realistic approach, and Rob's pull towards theory, and literally, millions of ideas. With collaborations like these, ideas are shared to the extent in which they no longer come from one or the other. So I am sure that they will all recognise many of their own thoughts in my writings – and I gratefully acknowledge their very different contributions to my own thinking. My most sincere thanks to Bill Howells, who patiently read and commented on the entire manuscript, and whose knowledge and generosity I so much admire. He has been a most important source of support and encouragement. My greatest intellectual debt is to Rob Foley, who throughout the years has been teacher, supervisor, colleague and friend. Discussing and arguing about the evolution of modern humans with him is not only always interesting but enormous fun, and it only gets better when Baruch Arensburg is in the room. To both of them, who are very dear friends, my greatest thanks.

I thank the following people for allowing access to skeletal material under their care and much appreciated hospitality: B. Arensburg, M. Berner, S. Condemi, G. D'Amore, R. Foley, D. Hervet-Grimaud, R. Kritscher, R. Kruszinski, H. de Lumley, J. Moggi-Cecchi, M. Piccardi, C. Stringer and J. Zias. Cambridge Colleges are a real experience (especially to a South American), and through Clare and the generosity of its Fellows, I discovered the important role they play in nurturing and promoting research. I received several small grants from Clare, as well as from the Leakey Trust, the CARE Foundation, and a full PhD scholarship from CNPq-Brasil. Many thanks also to the editorial staff of CUP, especially Tracey Sanderson and Rita Owen, for their able assistance.

My most sincere thanks to my parents, Luis and Martha Mirazón, who helped me in every way, and to my sisters Gabi, Marina and Nora.

I dedicate this work to Fabio and our children, Phil and Chris. Those who know us, know how much Fa's care, help, support and interest mean. Without him, nothing of this would be. Phil and Chris are the greatest joy of my life.

Marta Mirazón Lahr
Cambridge, June 1995

I The problem of modern human origins

1 *Introduction*

Modern human diversity is a subject that interests anthropologists and non-anthropologists alike, and most people have a formed opinion as to the scale of diversity existing today. A poll of these opinions among any group of people would most probably reveal greatly disparate views, from those that consider modern human diversity as vast, to those who think of it as relatively small. These different perceptions are explained by the fact that the scale of diversity varies according to the point of reference. Culturally and socially, modern human diversity is indeed vast. The number of social systems, languages, religions, means of subsistence and artistic expression, is so large that a view of a greatly diverse human species is justified. In biological terms, however, modern humans are a relatively homogenous group, as shown by comparisons to closely related species such as chimpanzees. The amount of genetic diversity within a single population of chimpanzees can be greater than that between human populations. This socio-biological paradox has been an important component of socio-historical misunderstandings of the degree of differences between human populations.

Sources of diversity

In order to obtain a measure of the scale of past human diversity and its development different approaches can be used. One of these uses a measure of present diversity, either genetic, morphological, social or linguistic, to establish the degree of relationships between present populations, and through these relationships infer evolutionary patterns. This way of studying the evolution of diversity is essential to provide the framework of population relationships for which an explanation is needed. However, by definition, this method can only work with those populations that exist today or in the very recent past, and even historical records show that a large number of human groups have become extinct. This is also evident when the affinities of many individual modern fossils are assessed and no close relationship to any present modern populations is found. Therefore, both a framework from recent relationships and degrees of differences and a direct assessment of past diversity through the palaeoanthropological record, human fossils, their material culture and their palaeoenvironmental context, are needed. This shift from real present day parameters to inferential ones requires very strict assumptions to be made, so that human fossils can be taken to represent biological parameters, stone tools and associated assemblages taken to represent socio-cultural context, and palaeoclimatic and ecological reconstructions over broad periods of time taken to represent the sum of the small nuances in environment that influence an individual's life-time. Such assumptions are made by anthropologists and archaeologists every day, and are imbedded in each conclusion and each hypothesis proposed.

However, this exercise is less straightforward than this. All those familiar with palaeoanthropological debates know that many of these discussions are fuelled by researchers speaking past each other in what seem to be different languages. And the analogy to different languages is correct. The moment that anyone interprets the prehistoric record, assumptions are made as to what that record represents. Fossils are interpreted in biological terms, but the biological level may not be the same. One may assume that the differences between two fossil groups represent real genetic differences and thus two taxa, or that they represent adaptations to specific environments within a single taxon, or even that they represent individual variations within a single population. The interpretation of archaeological assemblages is even more difficult, for it encompasses biological and cultural aspects. Elements within an archaeological assemblage may be taken to represent the biological identity of a group, and differences in the record through space and time may be taken to represent population differences and/or migratory movements. Alternatively, archaeological remains may be taken to reflect only the utilisation of resources and environment, and thus differences in the record through space and time would reflect different environmental conditions, social connections, activities, trade routes and raw material. A related problem is whether the differences in material culture should be taken to represent the level of cognition of a population or just the exploitation of available materials within a specific context.

Although in practical terms, the sources of information on the evolution of human diversity can be divided into those that reflect reconstructions from present diversity, for which there is clear evidence and statistically correct sample sizes, and those that reflect past diversity, for which there is patchy but direct evidence, the assumptions necessary to interpret the past would suggest a different division of information. In terms of the inferences necessary, the fossil data represent biological parameters similar to those of the present genetic and morphological data, and are thus analysed in similar terms. On the other hand, the archaeological data represent socio-cultural information, and with their characteristic horizontal as well as vertical transmission, are thus paralleled by the analytical problems posed by the present distribution of languages and social systems.

Paradigms: continuity and discontinuity

All these assumptions necessary to interpret the prehistoric record are at the root of the debates in palaeoanthropology. They represent the theoretical paradigms within which each researcher works. Therefore, some researchers do not find evidence for biological identity and distinctions in either the fossil or archaeological records, and thus interpret regional records as the result of the continuous temporal change of a single population. Also, there are those researchers that identify different populations in the morphological and archaeological differences they observe, and thus interpret regional records as discontinuous and characterised by population turn-over. These differences in paradigms in palaeoanthropological research are not unique to our subject, and reflect a more general debate on the gradualism of the evolutionary process.

However, besides the issue of gradual versus punctuated change in evolution, there is the question of universality of mechanism. It is likely that all possible different interpretations mentioned above (and many more!) probably apply to different hominid groups at different times in different contexts. There is no reason to believe that the same conceptual biological and social parameters, i.e. a single paradigm, apply to different hominids or even to different populations of a hominid species. Within the dynamics of the evolution of populations, the demographic and environmental conditions will determine the character of the record, and define the changes in a population as continuous or discontinuous, local or migratory, resulting in speciation or subspeciation events.

The issue of modern origins and origins of diversity

The research presented in this book arises from this problem of gradualism and universality in the interpretation of the evolution of modern populations. It tackles two main questions, that of the origins of modern humans from an archaic ancestor and the origins of the differences between modern populations. Depending on the evolutionary perspective, these issues may or may not be coupled together.

If a gradual and continuous view of modern human origins is assumed, then the regional differences between populations are related to the first establishment of widely dispersed regional hominid groups, whether modern or archaic, for it is assumed that regional continuity of occupation and form occurred. This view has been formulated into an evolutionary hypothesis called the Multiregional Model of modern human origins. It argues that there was no event associated with the appearance of a modern human form, but rather the development of regional variants of a single species which followed similar evolutionary trends because of extensive inter-regional gene flow. The establishment of regional variants would have occurred approximately one million years ago, when *Homo erectus* expanded out of Africa towards Southeast Asia, Eastern Asia, Central Asia and Europe, while the transition from archaic to modern would have occurred in parallel in all regions because of genetic diffusion. According to this view, a modern human morphology is secondary to regional diversity, and thus, the origin of human diversity precedes the origin of modern humans. The first section of this book explores a multiregional perspective of origins, both of humans and human diversity.

An alternative view would see the record as discontinuous and anachronic, and interprets the appearance of modern humans as a distinct localised event followed by replacement of archaic hominids. This view has also been formalised into an evolutionary model of origins, called the Out of Africa hypothesis. It argues that modern humans appeared in Africa in the late Middle to early Upper Pleistocene, and subsequently expanded to form regional populations which diversified from an already modern ancestral form. According to this view, modern human regional diversity is secondary to a modern morphology, and thus, the origin of modern humans precedes the origin of modern diversity. This view is explored in the second section of this book, in which evidence for a distinct event leading to the appearance of modern humans in Africa, significantly earlier than in other regions of the world, is discussed. In interpreting the apparent discontinuity in the record as supporting a single and recent origin of modern humans, the temporally and spatially variable character of the later modern human regional record is developed further into a model of multiple events starting from a single ancestral source and multiple evolutionary mechanisms that may account for the non-universality of process.

The key to this argument lies in the role of gene flow between regional populations of archaic and modern hominids. Besides temporal regional continuity, a multiregional model of Pleistocene hominid evolution requires that large amounts of spatial continuity between regions took place. This large amount of genetic exchange between widely separated areas like Java, East Africa, Europe and China, is necessary in order to maintain the temporal overall similarities and prevent local speciation occuring. Again, past gene flow cannot be directly measured, and traditionally two sources of inferential information have been used. The first of these is inferred from current genetic exchange between foraging groups like the Inuits, and transposed into the past. This method gives a good measure of possible gene flow in hunter–gathering groups, but does not take into account past demography and the opening and closing of routes in the past million years. Whether the world population of *H. erectus* had the necessary critical size to maintain gene flow patterns similar to any modern group is again debatable, and current interpretations of mtDNA data suggest not. The second source of evidence for inter-regional gene flow comes from the appearance of fossils and/or archaeological remains in one area which have strong affinities with another region of the world. These new forms are interpreted as resulting from gene flow from another regional hominid population, which acts to maintain the gradual multiregional change through time. However, this interpretation is again part of the paradigm of gradualism and continuity, and an alternative punctuated event leading to the migration of new forms into an area and eventual replacement of the original local inhabitants is also possible.

The question of replacement

The issue of whether gradual and universal mechanisms characterise the evolution of modern populations also has a socio-political dimension. It has been argued that the view of punctuated and rapid events creating discontinuity of populations (i.e. population extinction) is a 'racist' view of prehistory, one that highlights the supremacy of one form over the other (be it modern humans over Neanderthals, or one modern population over a previously existing one). However, the question of evolutionary advantages of one group in relation to other closely related groups or even individuals is the whole basis of Darwinian thought. That the observed succession of forms in the fossil record represents the advantage of one form over another is the reality of evolution. This advantage may be biological, social, demographic, or even incidental, in the sense that a population may out-compete another only for the particular circumstances (environmental or demographic) of the moment in which they are competing for resources rather than for an absolute biological or social advantage. In such cases, the replacement mechanism may be reversed as circumstances change. A good example of such a mechanism is provided by the alternation of modern and Neanderthal forms in the Middle East during the Upper Pleistocene. It is probable that early modern humans had neither a biological nor a technological advantage over early Neanderthals, as shown by the fact that early modern humans did not expand into Europe during the last interglacial. Therefore, the alternate occupation of the Middle East (an area that suffers strong climatic changes during interglacial and early glacial conditions through the northward and southward displacements of African and Asian climatic regimes) by early moderns and Neanderthals may only reflect circumstantial environmental conditions favouring one or other subsistence adaptation, rather than a real bio-social advantage of one group over the other. However, even in these circumstantial cases, replacement does take place. Evolution results in population replacement through time, as new forms pass to occupy the area previously inhabited by another group of similar niche when biological, social or environmental conditions change.

However, these replacements as seen on a geological time-scale may be real or an artefact of the record. It is possible that a change of fauna in the history of a region represents a real replacement event, by which a new form (locally evolved or migrant) will out-compete a resident population. The only clue for such an event would be evidence of a four-stage sequence: (1) the presence of a single population in an area, or one that forms the vast majority; (2) the appearance of a new form, or the increase in numbers of a minor group; (3) a period of contemporaneity of the two groups; and (4) the disappearance or strong reduction of the first population. However, incomplete replacements, i.e. when the new population becomes the great majority but not the only group in the area, imply the formation of small isolated pockets where evolutionary relics of the earlier form can survive. Such pockets result from either geographical refugia (i.e. areas where the new group does not reach) or ecological refugia (i.e. areas where the conditions are so stable that the existing adaptations of the early form are not out-competed by the expanding group). The first relates to geographical barriers, while the second to stability of the niche. Again, both these situations are realities of animal evolution, and examples of the persistence of evolutionary relics, in isolation or exploiting specific ecological niches, abound. The existence of these relics, for whatever length of time, are evidence of the non-universality of events. However, there is yet another mode of change of population through time that may seem to take the form of population replacement as in the first case, but is not. That is the case of extinction of regional populations, for reasons not related to competition or contact with closely related groups, followed by later occupation of that area by a new related form. In this case, actual replacement of one population by another did not occur, but only as an artefact of the geological scale of palaeontology.

In the history of modern humans, evidence for all these modes of replacement can be found. The turn-over from a Neanderthal to a modern population in Europe between 40 and 30 ka (thousand years) ago is an example of the first mode. The Neanderthals were the regional population of Europe until the first modern forms appear in the record just before 40 ka. A period of contemporaneity follows, and later only modern humans are found in the area. Geographical relics of Neanderthals existed in certain areas, like southern Spain, for some time. This replacement process does not rule out population admixture, that may have occurred to greater or lesser degrees, but never to the extent of creating a population of equal Neanderthal and modern affinities. Replacement events with the formation of geographical and/or ecological relics abound in the history of modern populations, and the recent expansions of agricultural groups, with the persistence of foraging communities in small numbers, is only one example. The third mode of replacement, the artefactual one, may be represented in Southeast Asia. It has been traditionally assumed that evolutionary continuity from Javanese late *Homo erectus* to Southeast Asian and Australian modern humans took place, and suggestions to the contrary imply replacement of these late archaic Javanese forms. However, the age of the last *H. erectus* in Southeast Asia is uncertain, possibly last interglacial but possibly earlier. There is no fossil or archaeological record in the area to suggest that this population was still there 50 thousand years later (at least) to be replaced by early modern humans. In this case, the occupation of the area by modern people would seem to have replaced an earlier archaic group, but separated by a period of time during which the area may have been uninhabited.

Continuity of occupation

This brings us to the question of continuity of occupation. Hominids have had an almost world-wide geographical range for at least a million years (excluding Australia, the Americas, high altitudes and the Siberian plains). The capacity of exploration, exploitation and expansion of *Homo erectus,* evidenced by this world expansion, is one of the reasons for assuming that, once it had been colonised, an area would be continuously occupied by hominids. This is an assumption that is difficult to test with the palaeoanthropological record. In some areas the archaeological record is sufficiently rich as to allow some measure of stability of occupation, although the scale of measurement in terms of hundreds, thousands or tens of thousands of years defies assertions of continuity in the generational scale needed for evolutionary continuity to take place. However, other areas lack such archaeological record and have instead sporadic fossils to attest to hominid presence in the region. Such an area is Java, where stone tools are largely absent but from where hominid remains dated to before one million years, approximately 700 ka and 100 ka ago, have been interpreted as reflecting continuous occupation. The continuity of occupation of Java is very possible, but certainly not proven by the available record.

The question lies in the stability of hominid populations and the interaction with unstable environmental conditions. That most hominids identified in the Pleistocene fossil record were at one point successfully adapted to their particular region of the world there is little doubt. However, biological and non-biological components of all environments are continuously changing, and with a wider spectrum during the climatic fluctuations of glacial cycles. Such drastically changing conditions that brought about extremely arid and cold phases, interrupted by short episodes of wet and warm conditions, must have affected the availability of resources within any one area. Furthermore, the scale of climatic change in the Pleistocene formed and destroyed geographical barriers, intermittently bringing normally allopatric animal and human populations into contact with each other. These changing conditions must have implied changing strategies in subsistence foraging, resulting in either more densely and localised or more thinly and widely distributed populations. These demographic fluctuations in terms of population numbers, density and range, are at the root of the problem of continuity of occupation.

Although it is not possible to obtain realistic measures of palaeodemography in many situations (like the case of Java mentioned above), the comparatively rich Upper Pleistocene record shows that such fluctuations were a fact of human history. The reconstruction of the late archaeological history of certain areas with rich records, like Europe, Australia and South Africa, clearly show that during specific periods of time parts of these areas were virtually uninhabited. In the case of Europe, the northern plains were depopulated for a short interval during the last glacial maximum, giving rise to very dense occupation of southwestern Europe and temporal discontinuity with Italy and central Europe. In Australia, the extremely arid conditions of the last glacial maximum also had a demographic effect, giving rise to a period of population refugia during which some groups could have become extinct. In South Africa, as in pleniglacial northern Europe, the archaeological record also points to depopulation, but this seems to have preceded the glacial maximum by some 20 ka.

These few examples of late Pleistocene fluctuations in demographic patterns show that not only the process of evolution of populations is a dynamic one, but that it is very circumstantial in nature. Macroclimatic events seem to have affected regional populations in different ways and to a different extent, while the same regional population seems to have responded differently at different times to events of apparent similar scale and extent. The circumstantial determinants of the evolutionary process that make a population expand, contract or change in response to a particular environment at a particular time, are the source of the anachronical spatial patterns in the palaeoanthropological record.

Multiple events and the universality of process

In the course of the last few decades, palaeoanthropologists have come to see major events in hominid evolution as independent – bipedalism, tool use, brain expansion, life-history, language and many more. It is even accepted that the earliest hominids share with African apes important aspects of biology, such as size, life-history and probably cognition, although their habitual posture allies them with our own family. The same principle of independence of events should be applied to the late Pleistocene. It is necessary to decouple events in order to establish the patterns and allow the interpretation of process. In the study of modern human origins and diversity, the events relate to biological and social spheres. I started this introduction by saying that the extent of modern diversity in one sphere is far greater than the other. To this we should add that the temporal changes in any one of these spheres may not have been universal or synchronous across the range of modern human populations in the past or even the present.

Disclosing human diversity requires the establishment of the evolutionary patterns and processes, while understanding diversity requires the investigation of biosocial adaptations and demography. This book is mainly concerned with disclosing the evolutionary patterns and processes that gave rise to modern human diversity, although genetico-demographical parameters are briefly discussed in order to establish regional mechanisms. By proposing a model for the evolution of modern human diversity from an ancestral source. this book offers a theoretical framework with which to work towards an understanding of the biosocial parameters of each past population that created the stability or instability of past human occupation, and thus the character of the evolutionary patterns and processes that gave rise to modern human diversity.

2 *The modern human origins debate*

In the past 10 years there has been a revived interest in the origins of modern humans, evidenced by the number of recent books dealing with the subject (Bräuer & Smith, 1992; Hublin & Tillier, 1992b; Mellars & Stringer, 1989; Smith & Spencer, 1984; Trinkaus, 1989a). The previous three decades of palaeoanthropological research saw the establishment of the Australopithecines as our early ancestors, and most studies reflected this interest. In terms of taxonomy, this period also reflects the preoccupation with the number and degree of differentiation of hominid taxa, followed by a tendency towards reducing the accepted number of hominid species recognised. Consequently, the list of ancestral hominids passed from 29 genera and 100 species in the 1940s, to one genus and three species by the 1950s. In the late 1970s, the controversial hominid taxonomy stabilised at two genera (*Australopithecus* and *Homo*) and seven species (*afarensis, africanus, boisei, robustus, habilis, erectus, sapiens*). Modern humans, descendants from a *Homo erectus* ancestor, came to be widely considered as only minorly different from Neanderthals, separated at the subspecific level. However, the last 15 years witnessed the development of new dating techniques applicable to the period 150–50 ka and the advent of studies of modern human mtDNA diversity, which jointly shifted an important segment of palaeoanthropological research interests towards events in the later Pleistocene. This new research has transformed our view of late human evolution and has renewed the controversy about morphological diversity and number of late hominid species.

In the light of the controversies created by the new chronological and genetic evidence, and the amount of new archaeological and palaeontological information available, the aim of this chapter is to review the current state of palaeoanthropological research into the origins of modern humans. In order to do so, four aspects of this research will be discussed. Firstly, the theoretical basis of the competing explanations; secondly, the most important new dates; thirdly, the fossil and archaeological evidence of the late Middle to Upper Pleistocene; and fourthly, the implications drawn from genetic and linguistic studies.

Recent theories of modern human origins

Multiregional model

In 1939–43, Weidenreich proposed a polycentric theory of modern human origins. This approach was further developed by several workers based on the data from four main regions of the world: Australasia (Jelinek, 1982a; Thorne & Wolpoff, 1981; Weidenreich, 1946), Northeast Asia (Weidenreich, 1943a; Wolpoff, Wu & Thorne, 1984; Wolpoff, 1980a, 1989a,b, 1992a), North Africa (Ferembach, 1979; Jelinek, 1980, 1985), and Central Europe (Jelinek, 1969, 1976, 1978, 1985; Smith, 1982, 1984, 1985, 1992; Wolpoff, 1982).

In this model, the evolution of anatomically modern forms is seen as a continuous gradual process resulting from the late Lower to Middle Pleistocene radiation of *Homo erectus* out of Africa, the establishment of local populations through the appearance of local morphological features, and the successive evolution through a series of evolutionary grades into modern humans. In contrast to Coon's (1962) theory of independent evolution of various regional lineages, the model proposed today emphasises the role of gene flow in maintaining grade similarities throughout the world and preventing speciation, with local selection acting on the persistence of regional features in morphology (Wolpoff, 1989a,b). Thorne (1977) proposed 'The Centre and Edge' hypothesis, in which he suggests that there is less polymorphism at the borders of a species range because of higher levels of selection, with peripheral populations frequently being monomorphic because of greater multidirectional gene flow at the centre of the species range. Therefore, the levels of gene flow at the peripheries would be such as to allow both differentiation and stabilisation of morphological patterns, and prevent speciation events (Figure 2.1). This point is taken further by Wolpoff *et al.* (1984), who do not recognise the occurrence of any speciation event in the last 1.5 Ma, suggesting that the remains of *H. erectus* should be allocated to the species *H. sapiens*. According to the regional continuity model, transitional 'mosaic' forms should be regularly found throughout the main geographical regions of the world linking archaic groups and modern humans.

Wolpoff (1986a, 1989a,b, 1992a) suggests four reasons for advocating a Multiregional Model of modern human origins: firstly, that the evidence for earlier dates of modern humans in Africa and the Middle East is equivocal; secondly, that the earliest modern fossils from each region lack any distinctly African morphological features; thirdly, that most traditional descriptions and classifications of modern morphological features do not apply to all the different geographical populations; and fourthly, that clade

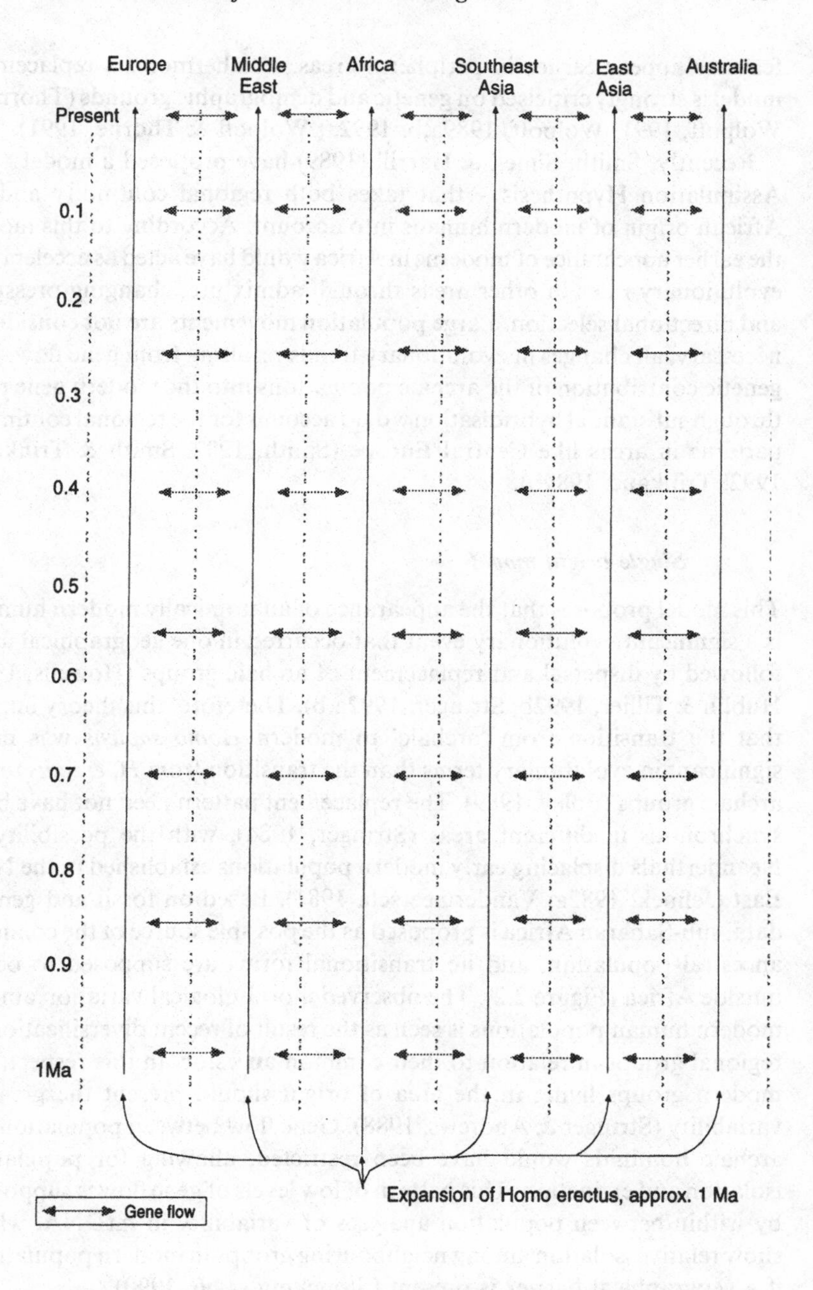

Figure 2.1 Schematic representation of the Multiregional Model of modern human origins showing continuity of regional occupation after the expansion of *Homo erectus* approximately 1 Ma. Transversal arrows indicate gene flow.

features appear earlier in peripheral areas. Furthermore, a replacement model is strongly criticised on genetic and demographic grounds (Thorne & Wolpoff, 1991; Wolpoff, 1989a,b, 1992a; Wolpoff & Thorne, 1991).

Recently, Smith, Simek & Harrill (1989) have proposed a model – the Assimilation Hypothesis – that takes both regional continuity and an African origin of modern humans into account. According to this model, the earlier appearance of moderns in Africa would have acted as accelerating evolutionary rates in other areas through admixture, changing pressures and directional selection. Large population movements are not considered necessary, all changes in evolutionary trends resulting from gene flow. The genetic contribution of the archaic populations into the modern gene pool through substantial hybridisation would account for the regional continuity patterns in areas like Central Europe (Smith, 1992; Smith & Trinkaus, 1992; Trinkaus, 1989b).

Single origin model

This model proposes that the appearance of anatomically modern humans is a significant evolutionary event that occurred in one geographical area, followed by dispersal and replacement of archaic groups (Howells, 1976; Hublin & Tillier, 1992b; Stringer, 1992a,b). Therefore, this theory implies that the transition from 'archaic' to modern *Homo sapiens* was more significant in evolutionary terms than the transition from *H. erectus* to the archaic groups (Foley, 1989). The replacement pattern need not have been synchronous in different areas (Stringer, 1984), with the possibility of Neanderthals displacing early modern populations established in the Near East (Jelinek, 1982a; Vandermeersch, 1981). Based on fossil and genetic data, sub-Saharan Africa is proposed as the possible source of the common ancestral population, and no transitional forms are supposed to occur outside Africa (Figure 2.2). The observed morphological variation among modern human populations is seen as the result of recent diversification of regional groups in relation to their common ancestor. In this respect, the modern groups living in the area of origin should present the greatest variability (Stringer & Andrews, 1988). Gene flow between populations of archaic hominids would have been restricted, allowing for population isolation and extinction. This pattern of low levels of gene flow is supported by within/between population analyses of variability in mtDNA, which show relative isolation among neighbouring groups in modern populations if a geographical barrier is present (Stoneking *et al.*, 1990).

A less extreme scenario is that originally proposed by Bräuer (1984a,b, 1989, 1992a,b) – the Afro-European *sapiens* Hypothesis or the Hybridisation

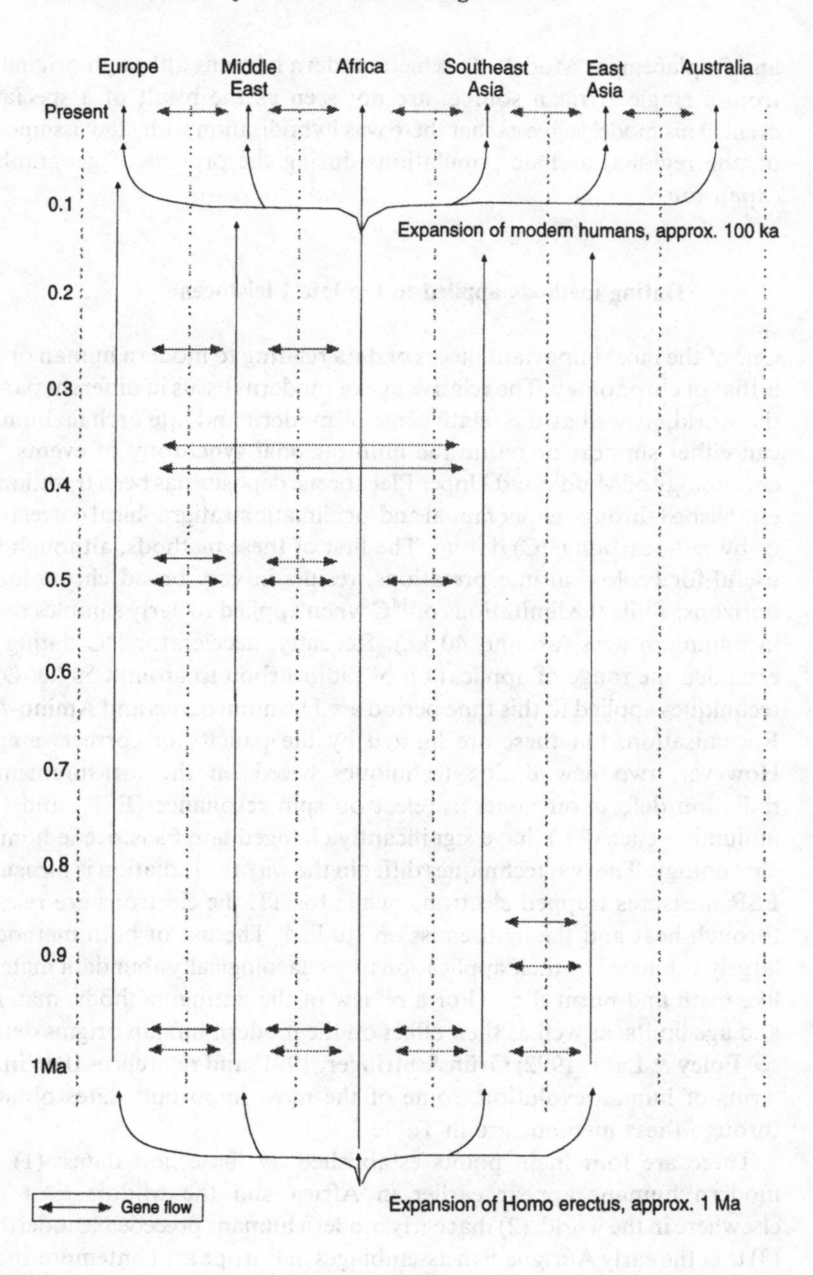

Figure 2.2 Schematic representation of the Single Origin Model of modern human origins showing discontinuity of occupation in all regions but Africa over the last million years. Transversal arrows indicate gene flow.

and Replacement Model – in which modern humans although originating from a single African source, are not seen as the result of a speciation event. This model suggests that there was hybridisation with, and assimilation of, the regional archaic populations during the process of geographical expansion.

Dating methods applied to the late Pleistocene

One of the most important pieces of data relating to modern human origins is that of chronology. The relative age of modern fossils in different parts of the world, as well as the relative age of modern and late archaic humans, can either support or refute the multiregional synchrony of events. The chronology of Middle and Upper Pleistocene deposits has been traditionally established through either faunal and/or climatic stratigraphical correlations or by radiocarbon (^{14}C) dating. The first of these methods, although very useful for ecological interpretations, results in very broad chronological horizons, while the limitations of ^{14}C when applied to early samples results in minimum ages (around 40 ka). Recently, accelerator ^{14}C dating has extended the range of application of radiocarbon to around 50 ka. Other techniques applied to this time period are Uranium Series and Amino-Acid Racemisation, but these are limited by the paucity of correct samples. However, two new dating techniques based on the measurement of radiation defects on minerals, electron spin resonance (ESR) and thermoluminesence (TL), have significantly changed late Pleistocene hominid chronology. The two techniques differ in the way the radiation is measured: ESR measures trapped electrons, while for TL the electrons are released through heat and the light emission studied. The use of both methods is largely enhanced by their application to archaeologically abundant material, like teeth and burnt flint. (For a review of the dating methods, materials and age limits, as well as their effect on the modern human origins debate, see Foley & Lahr, 1992; Grün & Stringer, 1991, and references therein.) In terms of human evolution, some of the most important dates obtained through these methods are in Table 2.1.

There are four main points established by these new dates: (1) that modern humans appear earlier in Africa and the Middle East than elsewhere in the world; (2) that early modern humans precede Neanderthals; (3) that the early Aurignacian assemblages in Europe are contemporaneous with late Neanderthal deposits; and (4) that the first occupation of Australia precedes the disappearance of Neanderthals from Europe. It is clear that the chronology of the last 150 ka shows a very marked

Table 2.1. *Some important recent dates in late human evolution*

Amud, Israel – site with Neanderthal remains.	ESR – EU: 42 ± 3/ 41 ± 3 ka, LU: 49 ± 4/ 50 ± 4 ka (Grün & Stringer, 1991).
Border Cave, South Africa – site containing fully modern skeletons previously considered 40 ka.	TL: 70-80 ka (Grün *et al.*,1990).
Djebel Irhoud, Morocco – late archaic *Homo sapiens*/early robust modern humans.	Preliminary ESR dates: 190–90 ka (Grün & Stringer, 1991).
El Castillo, Spain – early Aurignacian levels.	AMS 40 ± 2.1 ka (Cabrera & Bischoff, 1989)
Grotta Guattari, Italy – site with a Neanderthal cranium.	U-series: 51 ± 3 ka, ESR – EU: 44 ± 5 ka, LU: 54 ± 4 to 62 ± 6 ka (Schwarcz *et al.*, 1991).
Kebara, Israel – site with an almost complete Neanderthal skeleton.	TL: 60–48 ka (Valladas *et al.*, 1987).
La Chapelle-aux-Saints, France – site with a Neanderthal skeleton.	ESR – EU: 47 ± 3 ka, LU: 56 ± 4 ka (Grün & Stringer, 1991).
Le Moustier, France – Neanderthal skeleton.	TL: <41 ka (Valladas *et al.*, 1987).
Malakunanja, Australia – archaeological site.	OSL: 61 ±10 ka(Roberts & Spooner, 1992).
Nauwalabila I, Deaf Adder Gorge, Australia – archaeological site.	OSL: 55 ± 11 ka (Roberts & Spooner, 1992).
Qafzeh, Israel – site containing robust modern humans associated with Mousterian tools.	TL: 92 ± 5 ka (Valladas *et al.*, 1988); ESR – EU: 96 ± 13 ka, LU: 115 ± 15 ka (Schwarcz *et al.*, 1988).
Skhūl, Israel – site containing robust modern humans associated with Mousterian tools, previously considered ca. 40 ka.	ESR – EU: 81 ± 15 ka, LU: 101 ± 12 ka (Stringer *et al.*,1989).
St. Cesaire, France – site with a Neanderthal skeleton associated with Chatelperronian tools.	36 ± 3 ka (Mercier *et al.*, 1991).
Trou Magrite, Belgium – early Aurignacian site.	AMS 41.3 ± 1.7 ka (Strauss, 1994).
Zafarraya, Spain – late Mousterian site with Neanderthal fossil remains.	<35 ka (Hublin, 1994).
Zhoukoudian Upper Cave, China	AMS 29–24 ka (Hedges *et al.*, 1988, 1992).

AMS: Accelerator mass spectrometer radiocarbon dating; ESR: Electron spin resonance; TL: Thermoluminescence; ka: thousands of years; U-series: Uranium series.

asynchrony of events. This point argues strongly against a Multiregional view of 'grade' similarities through time and space.

The fossil evidence

Late Pleistocene (350–10 ka) hominids can be separated into four distinct hominid groups: persistent populations of *Homo erectus* in East and

Southeast Asia; 'archaic' *H. sapiens* populations in Africa, Europe, the Middle East and East Asia; Neanderthals in Europe and the Middle East; and modern humans. The first two hominids, *H. erectus* and 'archaic' *H. sapiens*, are found in regional chronological sequence in Africa and Europe, but current evidence suggests they were contemporaneous in East Asia where the *H. erectus* fossil of Hexian is considerably later than the early 'archaic' *H. sapiens* of Jinniu Shan (Pope, 1992). The Neanderthals and early modern humans are contemporaneous forms.

In order to review the palaeontological evidence, six regions will be considered: Africa, Europe, Middle East, East Asia, Southeast Asia and Australia. This review focuses on the first appearance of modern humans within each region, while a more detailed discussion of subsequent modern fossils in each area will be discussed in chapter 11.

Africa

The sub-Saharan African fossils of the period concerned have been divided into three groups (Bräuer, 1984b; Rightmire, 1984; Smith 1985, 1992; Smith *et al.*, 1989):

(1) An early, very robust 'archaic' *H. sapiens* sample, with retained *H. erectus* primitive features, represented by Kabwe, Bodo, Elandsfontein, Baringo, Melka Kunture, Eyasi and Ndutu.
(2) A heterogeneous sample in the late Middle to early Upper Pleistocene, with affinities both towards the archaic and modern groups, represented by Omo 2, LH 18, Florisbad, and possibly Dire-Dawa and Broken Hill 2. This group of fossils is generally considered an archaic/modern transitional sample.
(3) Anatomically modern humans in the early Upper Pleistocene, represented by Klasies River Mouth (KRM), Border Cave, Singa, Kanjera, Mumba and Omo 1.

These three groups show a temporally continuous cranial morphological pattern (Bräuer, 1984b; Rightmire, 1984; Wolpoff, 1980a), represented by a trend towards increasing cranial height and cranial capacity, less flattened frontals, reduction of the supraorbital and occipital tori, with the appearance of an external occipital protuberance in the modern group and in some of the 'archaic' *H. sapiens* like Eyasi (Hublin, 1978a); a reduction in facial height and prognathism, with the development of a canine fossa in the modern sample, Florisbad and LH 18; a reduction in posterior dental size; and the development of a mandibular eminence (chin) in the modern sample, although some specimens in KRM have this latter feature poorly developed (Smith *et al.*, 1989).

In northern Africa, the fossils of Sidi Abderrahman, Thomas Quarry and Rabat, generally considered 'archaic' *H. sapiens*, are relatively early, about 400 ka (Biberson, 1961; Hublin, 1985; Jaeger, 1981), and the Salé cranium, with a cranial capacity of 860 cm^3, is generally considered *H. erectus* (Day & Stringer, 1982; Holloway, 1981; Howell, 1978; Howells, 1980; Hublin, 1985; Wolpoff, 1980a; but see Bräuer, 1984a). The Djebel Irhoud material differs from this early group, presenting a mosaic of modern and primitive features (Howell, 1984), with Djebel Irhoud 2 being considered anatomically modern by Stringer *et al.* (1984). The occipital bunning observed in the Irhoud material is interpreted by Smith (1992) as evidence of Neanderthal genetic admixture. However, although not possessing a canine fossa, Irhoud 1 shows general similarities of the facial region with LH 18, greatly diverging from the Neanderthal pattern (Hublin, 1992). Furthermore, multivariate analyses place this fossil close to Skhul 5, although it does not share with the latter a close relationship with Omo 1, Upper Palaeolithic and recent crania (Stringer, 1974). Other North African fossils in which Neanderthal similarities have been observed are the Haua Fteah mandible associated with Mousterian tools (Howell, 1984; McBurney, 1967), and Mugharet el'Aliya (Coon, 1962). The remains from Temara and Dar-es-Soltane are later and anatomically modern (Ferembach, 1976; Roche & Texier, 1976; Stringer *et al.*, 1984).

The African fossil record indicates that the transition from *H. erectus* to 'archaic' *H. sapiens* was a relatively slow and continuous process. The late Middle to early Upper Pleistocene fossils reveal considerably greater morphological variation within a relatively faster evolutionary process (Bräuer, 1984b; Delson, 1981; Rightmire, 1981). However, although the phenetic similarities between the transitional and modern groups are clear (Stringer, 1992c), there are no strictly clade features that characterise the transition (Habgood, 1989; Rightmire, 1986; Smith, 1992; Smith *et al.*, 1989), a point associated with the difficulty of defining modern humans.

Accordingly, criticisms of the early modern African evidence have been made. For example, it has been suggested that the KRM remains are not fully modern and that the Border Cave remains are 'too' modern to be part of the ancestral population (Wolpoff, 1992a). The remains from Klasies River Mouth are very fragmentary and show pronounced sexual dimorphism and variable chin development (Bräuer, 1984b, 1989; Caspari & Wolpoff, 1990; Deacon, 1988; Deacon & Shuurman, 1992; Rightmire, 1984, 1986, 1989; Smith, 1992; Wolpoff, 1986a, 1989a; 1992a; Wolpoff & Caspari, 1990) and their modernity has been disputed (Wolpoff & Thorne,

1991; but see Smith *et al.*, 1989). These arguments highlight the problems of the lack of a formal definition of modern human cranial morphology and its limits of variation. The Border Cave remains are fully modern, although their claimed affinities to recent southern African populations (Beaumont *et al.*, 1978; Rightmire, 1978, 1981) are not accepted by others (Ambergen & Schaafsma, 1984; De Villiers & Fatti, 1982; Morris, 1992; van Vark *et al*, 1989). The different studies of the affinities of the Border Cave remains have been criticised for lack of comparability in terms of measurements, statistical methods and reference samples employed (Fatti, 1986; Morris, 1992). In support of Wolpoff's argument, these fossils seem to be too derived to be considered an ancestral form to robust modern populations like the Australians. It is possible that Border Cave, at 70–80 ka, should be considered an already differentiated African form and excluded from the group of ancestral early moderns. We will see below that recent genetic analyses would support such a view.

Europe

As a group, the European Middle Pleistocene fossils show great variability, from forms resembling *Homo erectus* to others approaching the Neanderthal pattern. Although this variation has been interpreted as reflecting high levels of sexual dimorphism (Wolpoff, 1980b), most authors would consider that it represents temporal and regional differences within a gradual evolutionary process of 'archaic' *H. sapiens* into the Neanderthals (Condemi, 1985; Cook *et al.*, 1982; Hublin, 1978a,b; Stringer, 1985; Trinkaus, 1982, 1983a, 1989b; Vandermeersch, 1985). The fossil record shows a gradual change from generalised archaic forms, such as Mauer, Petralona, Arago, Vertesszöllos and Bilzingsleben, through a later Middle Pleistocene group with strong Neanderthal affinities, like Biache, La Chaise (Suard), Swanscombe, Atapuerca and possibly Steinheim, to the last interglacial forms that are clearly Neanderthal-related, such as La Chaise (Bourgeois-Delauney), Ehringsdorf, Krapina and Saccopastore, to the 'classical' Neanderthals of the last glaciation (Stringer *et al.*, 1984; Trinkaus, 1988).

European Neanderthals have been found in over 60 sites, with some 10 relatively complete specimens: Spy 1 and 2; La Quina, Le Moustier, La Ferrassie, La Chapelle-aux-Saints, Saint-Cesaire; Monte Circeo; and Devil's Tower (the Central European sample, although extensive, is very fragmentary). As such, it is the best represented fossil hominid population, and its morphological pattern has been intensely studied (Condemi, 1985; Franciscus, 1989; Rak, 1986, 1992; Smith, 1976, 1984, 1989; Smith &

Paquette, 1989; Smith & Arensburg, 1977; Trinkaus, 1983a, 1984, 1986, 1988, 1989a). These hominids may be described as a relatively specialised group, possessing a series of autapomorphic features that allow confident identification of specimens. Most of the autapomorphic traits or combination of traits of the Neanderthals can be directly or indirectly related to the development and/or maintenance of:

(1) long and low crania, with occipital bunning;
(2) a large anterior dentition in the context of reductions in overall cranio-facial morphology (large and worn anterior teeth, associated with retreating zygomatic processes, retromolar spaces, posterior position of mental foramina, receding chins, mid-facial prognathism and a continuous supraorbital torus);
(3) overall post-cranial robusticity (thick bone cortices);
(4) environmental adaptations (wide and protruding noses, short distal limbs).

Some of the Neanderthal facial specialisations, but not the occipital morphology, are found among earlier Middle Pleistocene crania, suggesting that some of the behavioural (anterior tooth loading) and physiological (nasal form) adaptations generally associated with the Neanderthals of the last glaciation were under favourable selective pressures from an earlier date.

The Neanderthals share with *H. erectus* some aspects of post-cranial anatomy, like a medial ischial torsion, iliac pillars accompanied by increased cortical thickness on the gluteal surface of the ilium, and femoral cortical thickness, although the latter is more pronounced in *H. erectus* (Day *et al.*, 1991; Kennedy, 1973, 1985, 1992; Rak, 1990, 1992; Rak & Arensburg, 1987). Trinkaus (1989a) interprets the Neanderthal post-cranial morphology as related to strength and endurance locomotor patterns. Such interpretation is supported by the lack of planning depth and organised hunting inferred from the archaeological record, suggesting relatively continuous population movement (Trinkaus, 1989a). However, Stringer (1989b) contests this interpretation on the basis of the association of modern post-crania with Middle Stone Age technologies in Africa, similarly lacking modern subsistence patterns.

European early modern fossils are, in comparison to other regions, relatively recent. For purpose of analysis they have been generally divided into Western and Central European samples on the basis that the Central European group shows what have been considered Neanderthal reminiscent features. However, this division is most probably artificial, related to historical factors, theoretical tendencies and the fragmentary nature of some fossils. All the European modern specimens show what is generally considered a 'modern' morphology, i.e. rounded and high crania, short faces, relatively high frontal bones, reduced supraorbital and occipital tori, mental eminencies, reduced dentition, and gracile post-crania (Howells, 1970; Stringer, 1978). The earliest sample, associated with Aurignacian industries, is characterised by robust specimens, some presenting pronounced supraorbital ridges (Smith *et al.*, 1989) and cranial dimensions 10–30% larger than more recent crania (Kidder *et al.*, 1992). This robusticity may also be expressed in some specimens by the presence of a scapular axillary border, possibly reflecting a larger more powerful *M. teres minor* (Trinkaus, 1977).

As mentioned above, many researchers see evidence of morphological continuity from Neanderthals to the Central European early modern fossils. Neanderthal specimens, like Kulna, Sipka, Vindija G3 and possibly Sala, have been described as showing relative facial reduction, the presence of canine fossae, vertical mandibular symphyses and reduced brow ridges, interpreted by Smith (1984), Smith *et al.* (1989) and Wolpoff (1992a) as 'progressive' features transitional to the anatomically modern form (but see Bräuer, 1989). Smith *et al.* (1989) point to the fact that what is observed in the Vindija fossils is not only supraorbital torus reduction, argued by some to be caused by a large number of females and juveniles in the sample, but a particular pattern of reduction with a progressive mid-orbital diminution that links these fossils to later modern humans in the area. Accordingly, the robust aspect of some of the early modern specimens, like Velika Pecina, Mladeč 1 and 5, Pavlov 1 and the Hahnofersand frontal, is seen as reminiscent of a Neanderthal morphology (Frayer, 1984, 1986; Jelinek, 1976; Smith, 1984). Two arguments have been put forward against this interpretation: firstly, that the Neanderthal-related cranial features observed in the modern sample are plesiomorphic traits observed elsewhere (Stringer *et al.*, 1984); and secondly, that the fragmentary nature of all Central European fossils, together with the lack of associated post-crania, does not allow certainty in the inferences drawn from the sample. In Western Europe, the evidence for gradual evolution of the 'archaic' *H. sapiens* forms into the 'classic' Neanderthals and the replacement of these by modern forms is very strong (Howell, 1984), with no Neanderthal traits showing any change towards the modern direction (Stringer *et al.*, 1984; Smith, 1985). Frayer *et al.* (1991) and Wolpoff (1992a) stand alone in recognising transitional features in the occipito-mastoid region of the St. Cesaire cranium. Furthermore, Waddle (1994) has recently examined statistically the likelihood of a Neanderthal versus a fully African ancestry in the formation of recent European populations. She concluded that even when modelling situations of strong gene flow, the best morphological correlates of European Upper Palaeolithic crania were found not in European Neanderthals, but with the early African/Middle Eastern modern fossils. Waddle's results strongly refute the view of hybridisation as the source of early Upper Palaeolithic cranial diversity in Europe.

Middle East

The late Middle Pleistocene fossil record of the Middle East is represented by four fossils (Azykh, Tabun B, Zuttiyeh and Gesher Benot Ya'acov), of which the Tabun B mandible is comparatively modern in morphology. This early sample is followed by two robust modern human series, the Skhūl and Qafzeh remains, which in turn are followed by a Neanderthal sample variously dating to the last glaciation (Kebara, Amud, Bisitun, Kiik-Koba, Teshik-Tash, Zaskalnaya VI, Shanidar, Tabun C2). It is not clear if the Tabun C1 partial skeleton derives from layers C or D of the Tabun sequence (Bar-Yosef, 1992a).

Vandermeersch (1981, 1989) places the Middle Eastern archaic fossil of Zuttiyeh at the ancestry of the Skhūl and Qafzeh remains, mainly based on Zuttiyeh's facial flatness. However, Simmons *et al.* (1991), Smith & Paquette (1989) and Trinkaus (1984, 1987) have shown that the condition found in Zuttiyeh and modern orthognathism are not homologous. The Skhūl and Qafzeh fossils, like the early European Upper Palaeolithic sample, show a high degree of variation in levels of robusticity, with some specimens distinctly more robust than most recent humans. However, the overall morphological pattern, including the post-crania, allows their identification as anatomically modern. Corruccini (1992) has pointed out the presence of primitive features in some of the Skhūl material, suggesting that the less complete specimens have been overlooked and are not fully modern. However, the features he cites are mostly plesiomorphic traits, which rather than excluding the Skhūl fossils from a modern *Homo sapiens* hypodigm indicate that there have been important fluctuations in temporal and spatial levels of variability. Within the framework of a single origin of modern humans, the Skhūl and Qafzeh remains are interpreted as evidence of an early geographical expansion of a robust modern population from Africa, also supported on biogeographical grounds (Tchernov, 1992a,b,c).

The origin of Middle Eastern Neanderthals is a matter of controversy. Vandermeersch (1981) and Condemi (1985) interpret the similarities of the occipital and temporal regions of the skull between some of the Middle Eastern Neanderthals and the 'proto-Neanderthals' of Western Europe such as Saccopastore, as the result of migration of European forms into the Middle East during the early Upper Pleistocene. This early Upper

Pleistocene separation of the European and Middle Eastern Neanderthals would account for the more generalised features in the Middle Eastern fossils (Bar-Yosef, 1988, 1989a). However, Rak (1992) argues that the differences between European and Middle Eastern Neanderthals represent the range of variation of this group, reflecting possible respiratory adaptations to the different climates (Aiello, 1992; Franciscus, 1989; Laitman *et al.*, 1992). On the other hand, Bräuer (1989) considers the less specialised features of Middle Eastern Neanderthals as the result of admixture with early modern populations represented by Skhūl and Qafzeh. Another view sees the Middle Eastern Neanderthals as the result of local evolution from archaic forms such as Zuttiyeh, passing through an intermediate facial morphology as seen in Shanidar 2, 4 and Tabun 1, to the Neanderthal face as seen in Shanidar, Amud and possibly Tabun 2 (Trinkaus, 1983b, 1984, 1986, 1988, 1989a). It is evident that these different hypotheses, as well as the interpretation of the Central European material, are attributing different evolutionary significance to the appearance or change of different Neanderthal features.

East Asia

The East Asian Middle Pleistocene fossil record is probably the most complex, with the first 'archaic' *Homo sapiens* fossils broadly contemporaneous with late *H. erectus* populations (Pope, 1992). The dating of these samples is not absolute, and therefore, whether these two groups were fully contemporaneous is difficult to establish. However, Chinese *H. erectus* survived significantly longer in East Asia than in Europe and Africa.

Coon (1962), Aigner (1976, 1978) and Wolpoff (1980a, 1986b,c, 1992a,b) have stressed the regional character of East Asian fossils, suggesting there is evidence of morphological continuity throughout most of the Middle–Upper Pleistocene. This interpretation is generally favoured by Chinese palaeoanthropologists. However, since this hypothesis was first proposed by Weidenreich (1938, 1939, 1943a, 1946), who compared the Zhoukoudian *H. erectus* with modern East Asian populations, the Chinese fossil record has increased significantly. The new material relates mainly to 'archaic' *H. sapiens* fossils (Yinkou, Maba, Dali, Xujiayao, Yunxian), and also to *H. erectus* remains (Hexian). Wolpoff *et al.* (1984) discuss the evidence for continuity between these two groups, with the fossils of Dali and Yinkou playing a central role. However, other workers consider that although clear affinities exist among the various *H. erectus* fossils, the 'archaic' *H. sapiens* crania deviate from a continuity pattern in a number of features (Howell, 1984; Stringer, 1988; Stringer & Andrews, 1988). The

Chinese fossils of Dali, Jinniu Shan and Maba lack a number of distinct universal *H. erectus* apomorphies (Groves & Lahr, 1994), and share certain features with other 'archaic' *H. sapiens*, which not only differentiates them from typical *H. erectus*, but removes them from a continuous evolutionary lineage in East Asia. Furthermore, the contemporaneity of the late *H. erectus* and early 'archaic' *H. sapiens* makes a direct ancestor–descendant relationship between the two groups very unlikely, and a possible external (European) ancestry for the Chinese 'archaic' *H. sapiens* material has been suggested (Stringer, 1988; Groves, 1989b).

The Yunxian *H. erectus* crania described by Tianyuan & Etler (1992) have been interpreted by these authors as showing strong evidence of regional continuity. These Middle Pleistocene fossils show some typical *H. erectus* traits, such as long, low, sharply angled and tent shaped vaults, low position of maximum parietal breadth, hyper robust tympanic regions, long and narrow mandibular fossae without a developed articular tubercle and poor basicranial flexion. Although showing features typical of the Zhoukoudian (LC) remains, they lack some of the features that form the taxon's East Asian clade (which is considered as showing no progressive features in the *sapiens* direction by workers opposing a regional origin of *H. sapiens*). Furthermore, they are described as showing facial traits both of 'archaic' and modern *H. sapiens* not present in the Neanderthals, such as non-inflated maxillas, canine fossae, flat faces with moderate alveolar prognathism, high origin of the zygomatic root and incisura malaris (Tianyuan & Etler, 1992). However, before these descriptions are assessed, it should be noted that this material is seriously distorted and flattened, and any interpretations relating to shape have to be taken, at best, with caution. Instead of *H. erectus* with transitional features towards a *sapiens* morphology, the Yunxian crania may be considered as an 'archaic' *H. sapiens* on the basis of the derived features. Therefore, rather than de-characterising Asian *H. erectus*, these hominids may reflect a high level of variability within Chinese 'archaic' *H. sapiens*.

The early modern sample, represented by Zhoukoudian Upper Cave, Liujiang and Minatogawa fossils, are much later – all under 30 ka. This sample shows features such as alveolar prognathism, nasal form, receding frontals and length of palate, that are outside the range of recent Chinese populations and discontinuous with the archaic forms, but resembling characteristics of other early modern humans from outside East Asia (Kamminga & Wright, 1988).

The East Asian archaic fossil record is very variable, encompassing two partly contemporaneous hominid forms. This record is separated from the earliest modern fossils by some 100 ka, possibly more. The first moderns depart strongly from the recent East Asian morphology, and show similarities to early modern humans elsewhere in the world.

Southeast Asia

The Southeast Asian late Lower–Middle Pleistocene hominid fossil record is restricted to Java, while late Upper Pleistocene remains are known from a broader area. Several archaeological and hominid bearing sites of possible Middle Pleistocene age found in Vietnam, Cambodia and Laos are difficult to interpret because of uncertainties in relation to the dating techniques (mostly biostratigraphy, but also tool typology) and quality of archaeological assemblages. Olsen & Ciochon (1990) have reviewed this evidence and concluded that several of these early dates should be treated with caution, although some of the sites, like Tam Hang cave in Laos, where a hominid calotte was found, may have earlier Pleistocene deposits. However, too little information is available to include these sites in the present discussion.

Besides the problem of establishing the provenience of the Javanese hominid fossils (only Sambungmachan and possibly Perning 1 were found *in situ*; Pope, 1988), the dating of the Javanese hominids is very uncertain. The stratigraphy of the Sangiran Dome sediments is very complex and the entire sequence is not exposed at any single locality, allowing for serious disagreements in interpretation (Curtis, 1981; De Vos, 1985; Hooijer, 1983; Pope, 1985, 1988; Sondaar, 1984). Some of the early Javanese material has now been re-dated (Swisher *et al.*, 1994), obtaining results close to 2 Ma (millions of years). These new dates double the time span of hominid occupation of Asia, and raise a number of important questions in relation to affinities and continuity of occupation. However, in terms of the origins of modern humans, they only affect the length of time during which the perfect balance between gene flow and differentiation should have been maintained for multiregional evolution to occur. The Ngandong fauna, previously believed to be of Middle Pleistocene age, has been reinterpreted as probably Upper Pleistocene (Barstra *et al.*, 1988; Pope, 1983, 1984). Barstra *et al.* (1988) obtained a Th/U date of 100–75 ka.

Taxonomically, all Javanese archaic fossils are considered *H. erectus* (Pope, 1985, 1988). They form a relatively homogeneous sample, presenting varying degrees of robusticity. The group shows expanded 'tent' shaped cranial vaults, thick bones, basal cranial pneumatisation, thick horizontal supraorbital torus, separated from the frontal squama by a broad sulcus and expanded nuchal muscle attachments (Wolpoff, 1980a). The Ngandong fossils are considered by most authors as a relic *H. erectus* population (Santa-Luca, 1980; but see Pope, 1988), which presents an increase in cranial size (average cranial capacity of *c.*1250 cm^3) while maintaining a *H. erectus* cranial shape. The first modern human fossils in Southeast Asia are all younger than 40 ka, and fully modern in morphology.

Australia

While there are no claims to link the fossils of Niah and Tabon to Javanese *Homo erectus*, the opposite is certainly true in Australia. A strong and long-standing argument for a Javanese archaic origin of Australian aborigines exists. Even at maximum low sea-level stands, the colonisation of Sahul land necessarily implied an important sea-water crossing (Birdsell, 1977). However, there is evidence that both New Guinea and Australia were occupied relatively early (Allen *et al.*, 1988; Roberts *et al.*, 1990; Roberts & Spooner, 1992), some 20 ka before Europe or the earliest modern remains in East Asia.

Australian hominid fossils are all modern, but present great variability of shape, size and robusticity, and two groups of fossils, one gracile and one robust, have been identified (Habgood, 1985; Thorne, 1977). The variability observed in Australian crania has been interpreted in several ways, advocating one, two or three ancestral sources which would have inter-mixed within Australia to form recent aboriginal morphological patterns (Abbie, 1966, 1968, 1975; Birdsell, 1967, 1977, 1993; Freedman & Lofgren, 1979a,b; Habgood, 1985, 1987; Macintosh & Larnach, 1976; Thorne, 1977, 1980a,b). However the interpretation, most workers recognise at least some degree of continuity between the Javanese *H. erectus* fossils and the Australian crania, especially with the robust sample.

According to Weidenreich (1951), Jacob (1976), Thorne & Wolpoff (1981) and Wolpoff *et al.* (1984), the robust Australian fossils resemble Indonesian *H. erectus* in a number of specific features, especially facial traits that resemble those of Sangiran 17 (Wolpoff, 1989b). Even though Wolpoff *et al.* (1984) recognise that no fossil exactly resembling Ngandong has ever been found in Australia, they still find that some features reflect their close affinities and give support to a continuity argument. Furthermore, Thorne (1984) argues that the very robust WLH50 cranium, with a possible date similar to Lake Mungo (Flood, 1983), represents an intermediate form between the Javanese and Australian fossils. However, even disregarding the fact that Sangiran 17 is the only Southeast Asian archaic fossil that preserves the face, and thus immediately attributing to a sample of one a 100% incidence of the facial traits, a strong argument against the usefulness of these features to show a Southeast Asian *H. erectus* origin of Australians has been recently put forward: Groves (1989b) and Habgood

(1989, 1992) have shown that many of these traits are plesiomorphous and found among archaic populations elsewhere (although Habgood, 1992, recognises that the occurrence of a combination of some of them make up the continuity pattern). Lahr (1992, 1994) has also shown that these traits are not a Southeast Asian/Australian evolutionary clade, since they occur elsewhere among modern populations of the world. Furthermore, the incidence of some of the features related to robusticity may be functionally determined by the combined effect of the size of the masticatory apparatus and spatial arrangements of facial and vault size and shape (Lahr, 1992, 1994; Lahr & Foley, 1992). But these points will be discussed more fully in the following section of this book.

Archaeological evidence

The archaeological evidence from the African late Middle and Upper Pleistocene reveals that early modern humans in Africa were manufacturing Middle Stone Age (MSA) artefacts, contradicting the traditional assumption that the appearance of modern *Homo sapiens* was associated with the emergence of Upper Palaeolithic (Europe) or Late Stone Age (LSA; Africa) technologies. The African MSA covers the period from 200–40 ka (Clark *et al.*, 1984; Rightmire, 1978; Singer & Wymer, 1982; Vogel & Beaumont, 1972), and is found associated with both modern and archaic hominids (Clark *et al.*, 1984). The MSA is, as a techno-complex, very close to the Middle Palaeolithic of Europe (Allsworth-Jones, 1986, 1993). However, there is some evidence for the production of microlithic blade industries in the African MSA: blades at Howieson's Poort in South Africa, 90–60 ka (Clark, 1982), and pre-Aurignacian blades in Libya (Gowlett, 1987; McBurney, 1967). Some aspects of MSA adaptations seem to have been more developed than in the temporally equivalent European Middle Palaeolithic: there is a higher degree of regional and chronological variability, a more elaborate and intentional production of projectile points, increased use of foreign raw material, more specialised hunting techniques, blade technology and backing, and the presence of some carved and decorated objects (Brooks & Yellen, 1989). Behavioural interpretations of this evidence vary from those that consider that it reflects archaic behaviours (Binford, 1984) to those that consider it modern (Deacon, 1989). Deacon & Shuurman (1992) suggest that the small but significant stylistic changes observed in the Howieson's Poort industries at Klasies River Mouth, such as novel raw material and backed tools, are indicative of modern subsistence and behaviour. Furthermore, the African MSA shows a high level of diversity and specific environmental adaptations (Masao, 1992; Phillipson, 1985). In North Africa, Aterian sites may be very early, like the fossil lakes at Bir Tarfawi dated to 140–120 ka by various methods, like uranium series (Schwarcz, 1992), OSL (Aitken & Valladas, 1992), thermoluminesence and electron spin resonance (Wendorf *et al.*, 1990). The later Aterian industries at sites like Dar-es-Soltane show a Middle Palaeolithic technology with some advanced characteristics, such as retouched tools with stems and tangs (Close & Wendorf, 1990; Debenath *et al.*, 1986; Ferring, 1975). However, Klein (1992) suggests that interpretations of the behavioural significance of these industries should be conservative, since the only differences between the Howieson's Poort and Aterian from a typical Middle Palaeolithic assemblage are the backed pieces and tangs, with no real art or bone artefacts.

The Middle–Upper Palaeolithic transition in the Middle East occurred around 45 ka (Bar-Yosef & Cohen, 1988; Gilead, 1989; Marks, 1988; Solecki, 1963). The early modern humans from the sites of Skhūl and Qafzeh are, like the early African moderns, associated with an archaic industry (Mousterian) (McCown & Keith, 1939; Vandermeersch, 1981, 1989). However, contrary to the African archaeological record that can be interpreted as the behaviour of a single population, whether modern or archaic, the Middle Eastern evidence relates equally to both the early moderns and Neanderthals of the area. However, recent studies of the chronology of the sites in the area (Bar-Yosef, 1989a,b, 1992a,b) suggest that there are identifiable differences among the Israeli Mousterian series, so that Tabun C-type would be associated with the early moderns and Tabun B-type with the Neanderthal remains. An interpretation in terms of behavioural differences reflected in the hominid–tools association, depends on the stratigraphic position of the Tabun C cranium. If this Neanderthal fossil derives from Tabun B levels, then Neanderthals and modern humans were making somewhat different tools, possibly reflecting different behaviours and subsistence patterns, but if this fossil derives from Tabun C levels, the association falls apart.

The European Upper Pleistocene archaeological record shows a clear distinction between Middle and Upper Palaeolithic industries. This transition is characterised by differences in stone tool technologies, with the development of blades and use of bone as raw material, and in most non-artefactual aspects, like the appearance of modern subsistence patterns (Binford, 1981, 1982, 1984) and symbolic behaviour, reflected not only in art and personal ornamentation (Bahn & Vertut, 1988; White, 1989a, b), but in burial practices and in the design and shape of stone tools (Klein, 1992; Mellars, 1989, 1991; but see Simek, 1992 and Simek & Snyder, 1988). This sharp transition in the Upper Pleistocene European archaeological record is mainly interpreted as the result of population replacement (Harrold, 1989; Hublin & Tillier, 1992b; Klein, 1992; Mellars, 1989, 1992; Stringer *et al.*, 1984), although it has also been recognised that there are significant behavioural changes that occurred later, around 28–26 ka, within the Upper Palaeolithic (Chase, 1989; Clark, 1987). Some workers interpret the differences between Middle and Upper Palaeolithic archaeological assemblages as reflecting the development of language in modern human populations (Binford, 1989; Lieberman, 1989).

Two pieces of archaeological evidence support the replacement hypothesis of the European archaic populations. The first, is a temporal East to West cline of Aurignacian industries observed in Europe (Mellars, 1989, 1992), starting around 45 ka in Central European sites such as Istallösko, Hungary, with dates of 44.3 ± 1.9 ka and 39.7 ± 0.9 ka (Vogel & Waterbolk, 1972; but see Gaboni-Csank, 1970; Hahn, 1977), and from Samuilica Cave and Bacho Kiro Cave in Bulgaria, with dates of 42.8 ± 1.3 ka and >43 ka respectively (Strauss, 1994), but where complete recognition of the industry as Aurignacian is uncertain (Smith, 1984). It has to be borne in mind though, that if indeed real, an East to West temporal cline for the Aurignacian discloses a very rapid process, since the earliest Aurignacian dates in Western Europe are now in the order of 44–41 ka (Strauss, 1994). The second piece of evidence supporting a replacement hypothesis is the Neanderthal skeleton of St. Cesaire, dated to 36 ka (Mercier *et al.*, 1991), and associated with a Chatelperronian lithic industry (Leveque & Vandermeersch, 1981). This skeleton established the Neanderthal identity of the Chatelperronian tool-makers. The Chatelperronian industry is represented in at least 35 sites, possibly 66; it occurs in the Perigord and Charente between 35–32 ka, and up to *c.* 30 ka (Arcy–Denekamp Interstadial) in north-central France. These dates extend the Neanderthal presence in Western Europe to a period when this group had already possibly disappeared from more eastern areas. All over Europe, Mousterian industries are, if overlain by subsequent occupation, followed by Upper Palaeolithic industries (in the Franco-Cantabrian region by the Chatelperronian). There is never an Upper Palaeolithic level overlain by a Mousterian complex (Howell, 1984). Aurignacian 1 has always been found above Chatelperronian levels, but in at least two instances (Le Piage and El Pendo), there is an Aurignacian 0 level beneath the Chatelperronian, suggesting contemporaneity of Neanderthals and modern humans (Howell, 1984). In Cantabrian Spain, the Aurignacian seems to be very early, dated to 40–37 ka in El Castillo and 39–37 ka in L'Abreda (Bischoff *et al.*, 1989; Cabrera & Bischoff, 1989). Furthermore, the latter site presents evidence of very fast replacement of the local Mousterian industries (Bischoff *et al.*, 1989; Strauss, 1989), although now it seems probable that isolated Neanderthal groups manufacturing Mousterian tools, co-existed with these early Aurignacian peoples for a long time as evidenced by the site at Zafarraya, Spain (Hublin, 1994). Other European industries interpreted as terminal Neanderthal are the Ulluzzian in Italy and Szeletian in Central Europe, currently dated 43–33 ka (Valloch, 1976; Vogel & Waterbolk, 1972). The latter interstratifies with the Aurignacian (36–27 ka) during the last part of the Podhraden Interstadial.

Possible origins of the Upper Palaeolithic are to be found in the Middle East (Mellars, 1992) in sites like Boker Tachtit, in the Negev Desert, where local Mousterian is replaced by LSA between 47–38 ka (Marks, 1990), and Ksar Akil, Lebanon, where a juvenile skeleton dated to 37 ka shows a modern human associated with an early Upper Palaeolithic industry (Bergman & Stringer, 1989). However, archaeologists largely consider the Aurignacian a European event, the specific source of which has not been identified yet. In the Ukraine, a number of sites, with interstratified Mousterian and Aurignacian levels have been described (Gladilin, 1989; Sitliviy, 1989).

There are no Upper Palaeolithic (or equivalent) industries in China, a chopper/chopping tool technology surviving until terminal Pleistocene (Pope, 1988). This has been taken as implying that there was no major population movement into this area for most of the Pleistocene (Pope, 1988). However, Klein (1992) has suggested that the Chinese late archaeological evidence can be criticised on the basis of the small number of sites and lack of controlled excavations. Furthermore, Klein (1992) argues that the appearance of Upper Palaeolithic assemblages in Siberia, and possibly Northeast Asia, reflects the movement of modern people into the area. In the Indian subcontinent, the site of Batadomba Lena, Sri Lanka, has yielded modern human fossils associated with a sophisticated tool industry, dated to 28 ka (Kennedy & Deraniyagala, 1989).

The earlier Australian archaeological sequences do not show clear affinities with Indonesian assemblages (Hutterer, 1977), or indeed to anywhere else. However, few if any stone tools can be associated with *H. erectus* in Java (Barstra, 1982; Soejono, 1982), so that comparisons are difficult. The earliest secure date available in Southeast Asia is for the Lang Rongrien cave in Thailand, where two horizons have been found, one early Holocene with heavy Hoabinian tools, and one dated to 37 ka, with unifacially retouched flake tools (Anderson, 1987).

In summary, we may identify a number of points in the archaeological evidence that are very relevant to the origins of modern humans. Firstly, each of the Pleistocene hominid groups mentioned has been traditionally associated with a specific lithic technology, namely Acheulean with *H. erectus* and 'archaic' *H. sapiens*, Mousterian with Neanderthals, and Upper Palaeolithic industries with modern humans. It is now clear that this association is not always true, as seen in the absence of Acheulean hand-axes in Asian *H. erectus* sites, the association of early moderns with archaic industries both in Africa and the Middle East, and the lack of a modern lithic technology in Australia. These mismatches strongly indicate the need for searching for more complex explanations. The now seemingly earlier divergence date of Javanese *H. erectus* from a pre-Acheulean African source could explain the absence of hand-axes in Southeast and East Asia, although it also implies relative isolation of Java from western hominids for nearly two million years and that *H. erectus* left Africa towards Eurasia at least twice. A similar scenario of multiple events could possibly explain the prolonged absence of Upper Palaeolithic-like tools in the same areas at the end of the Pleistocene, i.e. an earlier divergence date of a group of modern humans ancestral to Australians from a MSA African context, prior to the expansion of the Upper Palaeolithic/LSA traditions. Such a model would be possible in the context of variability and relative subdivision within the African MSA populations of the late Middle to early Upper Pleistocene. That is what the archaeological data suggest, and is also consistent with recent genetic interpretations.

Genetic evidence

Genetic techniques have been applied, for a long time, to studies of population relationships, firstly in terms of variation in the frequencies of gene polymorphisms, such as proteins and blood groups, and more recently by nuclear and mitochondrial DNA (mtDNA) studies. The latter have had an enormous impact on studies of modern human origins by proposing a recent African ancestor for all modern peoples.

The pattern of evolution and inheritance of mtDNA (Avise, 1986; Brown, 1985; Brown *et al.*, 1979; Brown *et al.*, 1982; Giles *et al.*, 1980; Gyllensten *et al.*, 1991; Wilson *et al.* 1985) renders it a very useful tool for phylogenetic reconstructions (Cann *et al*, 1987; Johnson *et al.*, 1983; Stoneking & Cann, 1989). Cann *et al.* (1987), based on a rate of substitution of 2–4% per million years, suggested that modern human populations first began diverging from their common maternal ancestor 290–140 ka and that this ancestral mother lineage was most probably living in sub-Saharan Africa. The latter conclusion is based on the fact that living Africans present the greatest diversity in mtDNA lineages among modern populations, in the order of at least five times more diverse (Cann, 1988), and that resulting geneological trees have an African versus an African/non-African split at its first bifurcation (Cann *et al.*, 1987; Merriwether *et al.*, 1991; Vigilant *et al.*, 1991).

This theory of a single origin for all living humans on the basis of mtDNA diversity has been criticised on three grounds. Firstly, Avise *et al.* (1988) argue that even assuming constant mutation rates, variation in mtDNA lineages is also affected by the loss of unique lines in every generation without daughters. This fact has been used by Thorne & Wolpoff (1991) and Wolpoff & Thorne (1991) to suggest that in a situation of a population of few individuals with many daughters, the number of mtDNA lineages would be reduced without any association to genetic diversity, whether originating from archaic or modern gene pools. Secondly, studies by Gyllensten *et al.* (1991) have shown a small amount of paternal transmission of mtDNA lineages in mice, although the amount of paternal contribution was minimal and has not been shown to occur in other species. And lastly, Maddison (1991), Maddison *et al.*, (1992) and Templeton (1992), re-analysing the data originally used by Stoneking and Cann, found that the evolutionary trees these authors obtained were not the most parsimonious statistical solution possible, and that the nature of the data did not allow statistically significant tests of human origins from mtDNA data. It should be noticed, however, that their results still show the greater diversity of mtDNA lineages in Africa.

Although these criticisms have undermined the strength of the genetic argument for a recent origin in terms of its statistical significance, the theoretical basis still stands. There is no evidence for an accelerated rate of mutation in Africa (Cann, 1988), and therefore, the greater diversity in Africa implies either greater antiquity or a significantly larger size of African populations in relation to other modern humans. Since the early 1990s, Harpending, Rogers and co-workers have provided some new interpretations of the mtDNA data in terms of mismatch distributions, variability and palaeo-demography (Harpending, 1994; Harpending *et al.*, 1993; Rogers, unpub. data a,b; Rogers & Harpending, 1992; Rogers & Jorde, 1995; Sherry *et al.*, 1994). These authors obtain a 'signature' in each of the world's large populations which reflect the points in time when these populations expanded demographically. The African 'signature' leads those of other groups, suggesting an earlier expansion of this population. Furthermore, these interpretations suggest that modern humans underwent a severe demographic bottleneck in the late Pleistocene, followed by a period of population differentiation. According to their results, these newly diverged populations remained relatively isolated from each other until they subsequently expanded and gave rise to major world population stocks. Such a model would be consistent with both the morphological and archaeological variability in the late African Middle Pleistocene, and through the suggested period of comparative isolation, major populations could have acquired the different levels of biological and technological differences that characterised the Upper Pleistocene. These ideas will be pursued in later chapters.

The accuracy of studies of evolutionary relationships based on nuclear DNA and protein and blood group polymorphisms was reduced due to the problems of following the fate of a particular gene lineage through time (Cann, 1988) and determining the effects of selection on the gene products (Cavalli-Sforza, 1965; Lewontin & Krakauer, 1973). Most of the results from studies on blood group polymorphisms obtained genetic distance trees that showed an East–West division (Cavalli-Sforza & Bodmer, 1971; Cavalli-Sforza & Edwards, 1967; Cavalli-Sforza *et al.*, 1974). However, when electrophoretic techniques were later applied to blood proteins used as genetic markers, the results obtained showed a division between Africans and the rest of the world, and were therefore similar to those of mtDNA studies (Bowcock *et al.*, 1987, 1991a,b, 1994; Cavalli-Sforza *et al.*, 1988; Nei & Roychoudhury, 1982, 1993; Wainscoat *et al., 1986*). Cavalli-Sforza *et al.* (1988) found that a first split of populations separated Africans and non-Africans, and in all analyses four populations always clustered together as the African group (Pygmies, West Africans, Bantu and Nilo-Saharan). San and Ethiopians were variously joined with the sub-Saharan Africans above or with the 'Caucasoid' group in different analyses. The next bifurcation in their tree separates two 'superclusters': the first a North Eurasian group, subsequently divided into 'Caucasoids' and Northeast Asians plus Amerindians, and a second Southeast Asian group. The latter includes the Australians, that together with the New Guineans, separate from other Southeast Asian populations very early. These results differ from mtDNA data in that the latter shows an early separation by Australo-Melanesians, followed by the separation of Eurasian clusters. A very early separation of the !Kung population of southern Africa and the West African pygmies were obtained through direct sequencing and mapping of hyper variable parts of mtDNA (Vigilant *et al.*, 1989) and Y-chromosome DNA differences (Lucotte, 1992). Furthermore, when populations within Papua New Guinea were studied (Stoneking *et al.*, 1990), the results showed marked differences between coastal and highland groups, the former typically Southeast Asian while the highland populations were very distinct and homogeneous in terms of number of mtDNA lineages, but very different from one another. These results support other evidence of an early origin of highland New Guineans followed by significant isolation, arguing against continuous and extensive gene flow between past populations. Tishkoff and co-workers have examined patterns of linkage desequilibrium between alleles of two polymorphic sites in chromosome 12, a short tandem repeat (STRP) and an Alu deletion at the CD 4 locus, in 1600 individuals from around the world (Tishkoff *et al.*, 1995). They found that all 29 non-African populations exhibit only three haplotypes at significant frequencies and a specific linkage desequilibrium pattern, while in sub-Saharan African populations the pattern was totally different, with very pronounced variability in haplotypes and desequilibrium patterns. Like the other genetic studies mentioned above, the results obtained by Tishkoff and co-workers strongly support an African ancestry of all modern humans, with a relatively recent expansion between 140–90 ka. Dorit *et al.* (1995) have also obtained coalescence estimates in the late Middle Pleistocene (approximately 270 ka) for an ancestral Y-chromosome lineage.

The combination of the results from nuclear genetic studies with the greater mtDNA diversity in Africans, and the new analyses reported by Harpending *et al.* (1993) and Tishkoff *et al.* (unpub. data), form a very strong body of genetic evidence in support of a single African origin of modern humans.

Linguistic evidence

In recent years, a controversy has arisen among historic linguistics related to the notion of 'proto-languages' that may pre-date dispersal events. The prehistoric existence of 'proto-languages' has been suggested and supported by Greenberg (1987) and Ruhlen (1990), and before them by the joint long-standing work about the 'Nostratic' proto-language of V. Illich-Svitych in the former USSR and Dolgopolska in Israel. Renfrew (1991) reviewed various theories of historical linguists and suggested possible interpretations in terms of known archaeological events that could accommodate such ancestral languages before a dispersal/divergence event. He concluded that if a proto-Nostratic ancestral language existed, its dispersal and development could be related to the origins of agriculture, with further superfamily divisions with the introduction of farming in the various areas (e.g. Indo-European in Europe at 7000 BC). In terms of the origins of modern humans, the linguistic evidence relates to too recent events. However, its relevance is that linguistic monogenesis is generally interpreted as implying a single origin of all modern humans. This is supported by the work of Cavalli-Sforza *et al.* (1988), who correlated a human genetic tree with language superfamilies. They obtained a significant match which they suggest reflects parallel evolutionary processes.

Discussions of continuity versus discontinuity

The previous lack of a strict chronological framework for the late Pleistocene allowed some generalisations to be made about hominid evolution. These may be summarised in three main points. Firstly, that after *Homo erectus* populations had become established in the various regions of the world, there followed a 'grade' in human evolution generally called 'archaic' *H. sapiens* typical of the late Middle Pleistocene; secondly, that in Europe and the Middle East these 'archaic' *H. sapiens* were followed by the Neanderthals, who preceded modern populations; and thirdly, that all modern humans appeared at a similar time throughout the Old World, around 40 ka.

On the basis of the above sequence some workers developed and refined the Multiregional Model of modern human origins, first proposed by Weidenreich in 1938 and 1943 with a comparatively much smaller number of fossils. Weidenreich considered that modern regional differences in morphology were a reflection of the parallel evolution of different modern populations from regional archaic hominids. In East and Southeast Asia, the regional differences would be represented by a group of features present in Chinese or Javanese *H. erectus* and modern Chinese and Australian aborigines respectively, therefore showing an ancestor-descendant relationship in each of these areas. Current proponents of this model have emphasised the role played by gene flow in maintaining 'grade' similarities and preventing speciation, while allowing for regional differentiation (Frayer *et al.*, 1993; Wolpoff, 1989a, 1992a; Wolpoff & Thorne, 1991; Wolpoff *et al.*, 1984).

However, new data, both chronological and morphological, strongly refute these original generalisations, and consequently, some of the basis for a Multiregional Model. Firstly, by showing temporal differences among the 'archaic' *H. sapiens* populations (earlier in Africa and Europe than in China) and the possibility of contemporaneity of *H. erectus* and 'archaic' *Homo sapiens* in China, it is shown that the evolution of new hominid forms in the Middle Pleistocene was not a world-wide event characterising 'grade' stages in later human evolution, but rather regional processes with distinct evolutionary outcomes as evidenced in the resulting morphological patterns

(Ngandong, Neanderthals, AMH). Secondly, by establishing the presence of anatomically modern humans in the Middle East before the Neanderthals, it is shown that these groups were contemporaneous forms with separate regional origins. Thirdly, by recognising the earlier dates of modern humans in Africa and the Middle East some 60 ka before other regions of the world, indicating temporal patterns for the first appearance of modern humans world-wide. And lastly, by showing that the space of time separating the last Neanderthals in Western Europe from the first Upper Palaeolithic populations would be too short for an *in situ* evolution of modern Europeans. By integrating the recent chronological advances with the morphological evidence available, it is possible to interpret the evolution of the contemporaneous modern and Neanderthal morphologies as two parallel and independent processes arising from similar, and possibly phylogenetically related, 'archaic' *H. sapiens*. These two morphologies may be seen as the best physiological and behavioural solutions to the problems presented by the respectively cold/arid and tropical/arid European and African environments.

Therefore, a review of the literature relating to the origins of modern humans suggests that a single, African and recent origin is the model best supported by the available evidence, although there are a number of inconsistencies and many questions that remain to be answered. This book deals with two of these questions: if a single origin model of modern human origins is correct, how can the apparent morphological continuity from archaic to modern humans in certain regions of the world be explained?; and how could modern human diversity have developed from a single ancestral source? Most researchers would argue that the answer to both these questions lies in genetic admixture with regional archaic populations. Part II of this book examines the argument for morphological continuity between archaic and modern humans as the main mechanism of diversification, while Part III investigates the possible processes and mechanisms that could explain the evolution of cranial diversity from a single ancestral source without regional archaic genetic contribution, which is most compatible with the different lines of evidence available today.

II Multiregional evolution as the source of modern human cranial diversity

Proponents of the Multiregional Model of modern human origins argue that the diversity among regional populations of archaic hominids is an important source of modern cranial regional diversity. The validity of this model depends both on the gradual character of the regional fossil record and on the phylogenetic usefulness of the morphological traits that represent morphological continuity through time. A morphological feature will be of phylogenetic value only if a minimum of environmental (developmental, mechanical and climatic) factors affect its expression (i.e. has a strong genetic basis), and if the expression of that feature is either shared between evolutionary sister taxa (synapomorphies) or is unique to any one particular taxon (autapomorphy). The following four chapters deal with these points. Chapter 3 describes the morphological features considered to show regional morphological continuity by the proponents of the Multiregional Model. Chapter 4 examines the geographical distribution of each of these morphological traits among five modern human regional populations (Europeans, sub-Saharan Africans, Southeast Asians, East Asians and Australians), a regionally-restricted sample (Fueguian-Patagonians) and three sub-fossil modern groups of *Homo sapiens* (Afalou, Taforalt and Natufians), statistically testing the significance of the temporal and spatial distributions disclosed. Chapter 5 examines the incidence and distribution of these traits in a number of fossil specimens from the late Middle Pleistocene and Upper Pleistocene, testing the ancestral or derived character of these features. Chapter 6 examines the relationship of these traits with other metrical and non-metrical characters of the skull, investigating the levels of inter-trait correlation and possible functional influences on their degree of expression. Finally, Chapter 7 discusses why multiregional evolution is an inadequate explanation for the origin of modern cranial diversity.

3 *The morphological basis of the Multiregional Model*

The group of skeletal features that forms the morphological basis of the Multiregional Model was first proposed as such by Weidenreich in 1939 and 1943, who like other anthropologists of the time, was studying and identifying a number of skeletal traits to characterise human populations and races. Therefore, most of the morphological traits proposed by Weidenreich as regional characters of East Asian and Australian aboriginal populations had been in use as such in most anthropological studies of racial relationships of the time. Weidenreich's work not only summarised this knowledge for these two regions, but through his personal study of the Zhoukoudian and Ngandong *Homo erectus* fossils, concluded that the modern inhabitants of East Asia and Australia shared their regional morphology with these two groups of hominids, providing a link through time and evidence of independent evolution of the two lineages. In later years, workers like Coon (1962), Aigner (1976,1978) and Wolpoff (1980a,1987) developed this model further and added other features of continuity to the list originally compiled by Weidenreich. This hypothesis has remained the main interpretation of East Asian and Southeast Asian hominid evolution in the past decades.

The Northeast Asian morphological pattern

Table 3.1 presents the list of morphological characteristics proposed by Weidenreich and others as providing evidence of regional continuity in the evolution of Chinese *Homo erectus* into modern East Asian populations. Wolpoff *et al.* (1984) attribute features 12, 13 and 14 (Table 3.1) to Aigner (1976), however these traits were recognised as showing continuity by Weidenreich himself (Weidenreich, 1943a:p.253). Likewise, the traits suggested in Wolpoff *et al.* (1984) are listed by Weidenreich (1943a) as part of 121 features that characterise the *Sinanthropus* skull, even though these particular ones are not listed in his section of 'Special Affinities' (Weidenreich, 1943a:pp. 246–256). Wolpoff *et al.* (1984) conclude that the traits that most consistently show continuity in East Asia are shovel-shaping of the upper

Table 3.1. *Morphological features proposed by Weidenreich and others as evidence for regional continuity in East Asia*

1	Mid-sagittal crest and parasagittal depression (Weidenreich, 1939)
2	Metopic suture (Weidenreich, 1939)
3	Inca bones (Weidenreich, 1939)
4	'Mongoloid' features of the cheek region (Weidenreich, 1939; facial flatness of Wolpoff *et al.*, 1984)
5	Maxillary exostoses (Weidenreich, 1943a)
6	Ear exostoses (Weidenreich, 1943a)
7	Mandibular exostoses, forming a mandibular torus (Weidenreich, 1939)
8	High degree of platymerism in the femur (Weidenreich, 1939)*
9	Strong deltoid tuberosity in the humerus (Weidenreich, 1939)*
10	Shovel-shaped upper lateral incisors (Weidenreich, 1939)
11	Horizontal course of naso-frontal and frontomaxillary sutures (Weidenreich, 1943a)
12	Profile contour of the nasal saddle and nasal roof (Weidenreich, 1943a; Aigner, 1976; low nasal profile of Wolpoff *et al.*, 1984)
13	Pronounced frontal orientation of the malar facies and the frontosphenoidal process of the maxilla (Weidenreich, 1943a; Aigner, 1976)
14	Rounded infraorbital margins (Weidenreich, 1943a; Aigner, 1976)
15	Facial flatness (Wolpoff *et al.*, 1984)
16	Forehead profile – with development of a frontal boss (Wolpoff *et al.*, 1984)
17	Associated distinct angulation in the zygomatic process of the maxilla and anterior orientation of the frontal process of the zygomatic (Wolpoff *et al.*, 1984)
18	Lack of anterior facial projection (Wolpoff *et al.*, 1984)
19	Low nasal profile (Wolpoff *et al.*, 1984)
20	Shape of the orbits (Wolpoff *et al.*, 1984)
21	High frequency of M3 agenesis (Weidenreich, 1939)
22	Small frontal sinuses (Weidenreich, 1939)
23	Reduced posterior dentition (Weidenreich, 1943a)

* As this study concentrates only on the cranium, East Asian traits 8 and 9 will not be considered any further.

incisors, followed by a trend in facial height reduction, posterior dental reduction, M3 agenesis, flatness of the mid-face and nasal saddle, and more forward direction of the frontosphenoidal process of the zygomatic. Other traits, like sagittal keeling, horizontal position of the frontonasal and frontomaxillary sutures, frequency of inca bones and mandibular exostoses, decrease through time, but are still found to a higher degree in Mongoloid populations today than in any other modern group (Wolpoff *et al.*, 1984).

The Southeast Asian and Australian morphological pattern

Following his observations on the *Sinanthropus* and *Pithecanthropus* fossils, and his studies of the Ngandong sample, Weidenreich (1951)

Table 3.2. *Morphological features proposed by Weidenreich and others as evidence for regional continuity in Australasia*

1	Well-developed superciliary ridges
2	Pre-lambdoid depression
3	Short sphenoparietal (5–8 mm) articulation at pterion
4	Deep and narrow infraglabellar notch
5	Frontal flatness
6	Posterior position of minimum frontal breadth
7	Relatively horizontal orientation of the inferior supraorbital border
8	Presence of a distinct pre-bregmatic eminence
9	Low position of maximum parietal breadth
10	Marked facial prognathism
11	Presence of malar (zygomaxillary) tuberosity
12	Eversion of the lower border of the malar
13	Rounding of the inferolateral border of the orbit
14	The lower border of the nasal aperture lacks a distinct line dividing the nasal floor from the subnasal face of the maxilla
15	Marked expression of curvature of the posterior plane of the maxilla
16	Very large, rounded zygomatic trigone
17	Absence of a distinct supratoral sulcus
18	Presence of a suprameatal tegmen
19	Transverse course of squamo-tympanic fissure
20	Angling of the petrous to the tympanic in the petro-tympanic axis
21	Marked, ridge-shaped occipital torus
22	External occipital crest emerging from the occipital torus
23	Marked occipital supratoral sulcus

proposed a similar hypothesis of a continuous line of evolution in Southeast Asia from *Homo erectus*, through Ngandong, to modern Australians. This hypothesis was further developed by Jacob (1976), Thorne & Wolpoff (1981) and Wolpoff *et al.* (1984), and a list of traits reflecting continuity has also been proposed (Table 3.2). A large part of the evidence for regional continuity in facial morphology in Southeast Asia and Australia depends on a single archaic fossil, Sangiran 17, the only *H. erectus* specimen in the region with a preserved face. This is a major drawback for the claims being made, because of the obvious problems of using a single specimen as representing an archaic population covering close to two million years.

Descriptions of the features of continuity

One obvious aspect of the majority of the features used by the Multiregional Model – the 'regional continuity traits' – is their vague definition, usually

Table 3.3. *Final list of regional continuity traits being studied – 10 East Asian and 20 Southeast Asian (five of which are shared by both populations, although with the exception of rounding of the infero-lateral margin of the orbit, the other four features in common for both East and Southeast Asian lines of continuity represent opposite expressions in either region)*

*EA*1	Sagittal keeling
*EA*2	Course of the nasofrontal and frontomaxillary sutures
*EA*3	Profile of the nasal saddle and nasal roof
*EA*4	M3 agenesis
*EA*5	Reduced posterior dentition
*AUS*1	Development of the supraorbital ridges
*AUS*2	Lambdoid depression
*AUS*3	Size and type of pterion
*AUS*4	Profile of the infraglabellar notch
*AUS*5	Position of minimum frontal breadth
*AUS*6	Position of maximum parietal breadth
*AUS*7	Development of a zygomaxillary tuberosity
*AUS*8	Eversion of the lower border of the malars
*AUS*9	Type of narial margins
*AUS*10	Degree of curvature of maxilla
*AUS*11	Development of the zygomatic trigone
*AUS*12	Presence of a suprameatal tegmen
*AUS*13	Angle between the petrous and the tympanic
*AUS*14	Development of an occipital crest
*AUS*15	Development of an occipital torus
*EA*6/*AUS*16	Degree of facial projection
*EA*7/*AUS*17	Lateral facial orientation
*EA*8/*AUS*18	Profile of the frontal bone
*EA*9/*AUS*19	Rounding of the inferolateral margin of the orbit
*EA*10/*AUS*20	Orbital shape

EA: East Asia; AUS: Australia (which also implies archaic Southeast Asia).

implying an empirical and subjective observation with no quantification. Therefore, an objective study of each trait is necessary if these features are to be used as the basis of any evolutionary theory, including the description of its variability of expression. Such a critical examination of the features of continuity results in a reduction in the number of traits considered useful (Lahr, 1992, 1994). This is either because of redundancies, or because the geographical distribution of a specific feature has already been studied as non-metrical cranial variants, and confirmed as neither East Asian nor Australian. A description and detailed explanation of those features excluded or merged is presented below, while the regional continuity traits to be considered in the present study are listed in Table 3.3.

Traits excluded

Extensive research, largely reported in the literature, of discontinuous trait variability, shows that seven features either fail to present highest incidence among Mongoloid populations as claimed by Weidenreich in his original study, or show no relevant geographical patterning. These traits are metopism, inca bones, maxillary exostoses, ear exostoses, size of frontal sinuses, absence of a distinct supratoral sulcus and course of the squamo-tympanic suture. Therefore, these traits have no bearing upon the problem of morphological continuity and were excluded from the study. An eighth trait – shovel-shaped incisors – was excluded because insufficient samples of anterior teeth were available in the cranial collections studied.

Metopism

This is by definition the abnormal persistence of the medio-frontal suture into adulthood. Work by Welcker (1862), Papillaut (1896–1902) and Comas (1942) suggested that metopism was the result of fronto-cranial increase, while Bolk (1920) and Maslovsky (1927) would rather see it as resulting from mechanical forces from the masticatory muscles. Sjovold (1984) has shown that metopism is a hereditary character, though with low heritability estimates, especially when compared with mice (Self & Lamy, 1978) and macaques (Cheverud & Buikstra, 1981). In primates, metopism is common among Strepsirrhines and Platyrrhines, rare in Cercopithecoids (with the exception of Colobines – 20%), and rare or absent in Hominoids (Olivier, 1965). Metopism can occur in all modern races, and sex differences are either inconsistent or absent (Cesnys & Konduktorova, 1982; Corruccini, 1974; Czarnetzki, 1975; Dodo, 1974; Finnegan, 1972; Pietrusewsky, 1985; Vecchi, 1968). The highest incidences of metopism are found among Hindus (10–16%) and Europeans (10–15%) (Hauser & de Stefano, 1989, and references therein), and the single incidence reported by Weidenreich in *Sinanthropus* skull XI bears no importance to the relationships between *Homo erectus* and modern people (Weidenreich, 1943a).

Inca bones

These are formed by a transverse suture in the occipital squama at the level of the superior nuchal line. They result from the partial or complete lack of fusion of the two parts of the occipital squama along the mendosal suture during embryonic growth, while the presence of additional ossification centres leads to further subdivisions, both longitudinal and/or transverse (Hauser & De Stefano, 1989). Pedigree studies by Torgersen (1951) suggest that inca bones are inherited as a dominant trait, with approximately 50% penetrance. Strong genetic control is supported by studies in mice (Deol & Truslove, 1957).

Weidenreich (1943a) found inca bones in four out of five *Sinanthropus* skulls (X, XI, XII, II?), which he considered 'as a peculiarity found only in certain races of modern man; among them it occurs in high percentage only in certain Mongolian groups'. (Weidenreich, 1943a:p.253). A survey of the literature on inter-population variation is difficult because of the different scoring methods employed, with some authors including ossicle at lambda as equivalent. However, a review of the reported frequencies presented by Hauser & De Stefano (1989) shows the highest incidence to be among Europeans (Bohemians: 18.0%, N=657; Czarnetzki, 1975), followed by Nigerians and other European groups (Alamannes and French), while the highest incidence among any Mongoloid population, in protohistoric Japanese, was much lower (7.1%, N=103, Yamaguchi, 1985). The frequencies mentioned by Groves (1989b) are in the order of 45–60%, and must therefore reflect the incidence of other sutural ossicles in the occipital bone, besides inca bones. The ossicles at lambda and lambdoid ossicles have a very different embryological origin, representing fontanelle ossicles characterised by a later onset developmentally (Davida, 1914) and the premature fusion of secondary additional centres of ossification along the lambdoid suture (Ranke, 1899). Both may occur postnatally (Kadanoff & Mutafov, 1968) and may be thus subjected to mechanical influences. Accordingly, a relationship between incidence and/or distribution of sutural bones and artificial cranial deformation has been suggested (El-Najjar & Dawson, 1977; Lahr & Bowman, 1989; Ossenberg, 1970).

Maxillary and mandibular exostoses/tori

These are bony increments on the lingual margin of the molar portion of the alveolar process, only rarely extending to the buccal side (Kajeva, 1912; Schreiner, 1935) or to the area of premolars and canines (Woo, 1950). They are usually found bilaterally (Cosseddu *et al.*, 1979; Milne *et al.*, 1983), and sex differences are either inconsistent or absent (Berry, 1975; Corruccini, 1974; De Villiers, 1968; Milne *et al.*, 1983; Scarsini *et al.*, 1980; Van der Broek, 1945). Variation among populations shows the highest incidence to be in Lapps (16.3%, N=308, Van der Broek, 1945), followed by Africans and Europeans. Milne *et al.* (1983) report very high side incidence among Australians.

There are both genetic and functional explanations for the formation of mandibular and palatal tori (Furst & Hansen, 1915; Hrdlička, 1910). The functional hypothesis (Hrdlička, 1910) argues that these structures are the result of heavy chewing, supported by studies on hunter–gatherer groups like Eskimos and South African bushmen (Drennan, 1937; Mayhall, 1970; Waugh, 1930). The parallel high incidence of tori in northern European populations in Greenland and Iceland (Broste *et al.*, 1944; Fischer-Moller, 1942; Hooton, 1918) and Eskimo (Furst & Hansen, 1915; Hrdlička, 1910, 1940; Oschinsky, 1964), while lacking in most Europeans (Hrdlička, 1940), also support functional arguments. However, a certain genetic component of these traits also exists, as shown by the studies of Moorrees *et al.* (1957) and Suzuki & Sakai (1960).

Ear exostoses

These are bony increments within the external auditory meatus, and were extensively studied by Hrdlička (1914, 1935), who considered these features to be caused by a genetic disorder of the neuro-vascular control of the meatal area. Auditory exostoses are absent in small infants (Ascenzi & Balistreri, 1975), and their highest incidence is among middle-aged individuals (Gregg & Bass, 1970; Hrdlička, 1935; Meyer, 1949; Roche, 1964; Ruttin, 1933). It is considered that prolonged irritation with consequent inflammation resulting from cold water swimming can cause auditory exostoses (Belgraver, 1938; Fowler & Osmum, 1942; Harrison, 1962). There has been a suggestion that this type of mechanical force only causes deep meatal exostoses (Mann, 1984), while other, more superficial types could be under stronger genetic control. Research by Kennedy (1986) on the latitudinal distribution on ear exostoses, strongly supports the notion of an environmentally (cold water swimming) determined incidence, which renders this feature useless for populational studies.

Frontal sinuses

Weidenreich (1943a) noticed that the frontal sinuses of four *Sinanthropus* skulls (II, X, XI and XII) were small and strictly confined to the nasal portion, never extending laterally to the roof of the orbits, while in the only *Pithecanthropus* skull (1) in which the feature could be observed and in the Ngandong crania, the sinuses were all large, extending upwards to the glabella and laterally over the medial part of the orbital roof. In modern human crania, Weidenreich (1943a:p.31) states that Australians have very small sinuses or they are missing altogether; Turner (1901) reports the absence of frontal sinuses in 30.4% of Australians, 37% of Maori skulls and 7.5% of Europeans; Troitzky (1928) found sinuses in only 3 out of 22 Buriat skulls and lacking in 12% of Russians; while Sitsen (1931) reports small and large sinuses in all races. Among Neanderthals, large sinuses are present in all specimens, some of which are very large in all dimensions, like

La Chapelle, Le Moustier and Saccopastore (Boule, 1911; Hofman, 1933; Keith, 1927; Martin, 1928; Schwalbe, 1899; Sergi, 1930–32; Weinert, 1936; Weidenreich, 1943b). Among the apes the size varies from large (gorillas and chimps), to small (orangs) or absent (pigmy chimp) (Weidenreich, 1943a:p.166). Recent work among modern humans by Szilvassy (1982, 1986) shows that frontal sinuses have a strong genetic determination of size and shape. Among 92 Mongolian crania, 52 African, 79 Khoisan, 109 Oceanic and 21 Fueguian, the majority (36–87%) of cases had frontal sinuses ranging between 0–3 cm^2 in area, while 7–50% had sinuses with an area between 3 and 6 cm^2 (Szilvassy, 1986). Only the European sample (N = 366) showed a majority (39–44%) of cases with sinuses 3–6 cm^2, while 24–42% had sinuses 0–3 cm^2. The Europeans were the only group consistently to have cases of frontal sinuses larger than 12 cm^2 (Szilvassy, 1986). These results clearly show that small sinuses generally characterise all modern groups, but that there is a large variation in size. Furthermore, considering the fossil evidence, the small frontal sinuses of *Sinanthropus* are possibly a specialisation of that population. The evidence does not support a special relationship between *Sinanthropus* and modern Mongoloids on the basis of this feature.

Supratoral sulcus

A supratoral sulcus on the frontal bone is a *Homo erectus* character, not present in all later Pleistocene hominids. Accordingly, Australian aborigines, like all other modern humans, do not show this trait. The fact that the Ngandong fossils also lack this feature does not reflect a special relationship with modern Australians (or indeed with any other modern group), but suggests instead a possible effect of cranial enlargement on a basic *H. erectus* cranial structure.

About the *course of the squamotympanic fissure* Weidenreich wrote (1943a:p.53):

> Klaatsch (1902) recorded for the two Spy skulls that the Glasserian fissure runs at right angles to the mid-sagittal line of the skull, whereas in modern man it forms an acute angle, a statement verified by Loritz (1916) on one hundred Bavarian skulls. An acute angle varying from 56° to 80° was found by this author in 82.5%, and an angle from 80° to 92° in 17%. It follows from this investigation that a transverse course of the Glasserian fissure also occurs in modern man, though the acute angle predominates. In the four *Sinanthropus* cases in which this angle can be measured it varies between 87° and 90° . . . Since the angle, however, is about the same in anthropoids, the direction of the Glasserian fissure fails to furnish a reliable criterion for the actually existing differences in the orientation of the tympanic plate.

This trait was not included in the present study.

Size and type of pterion articulation

In the case of the size and type of pterion articulation, only the former was included in the study, because the regional variation in the type of pterion has already been well-established. The articulation at pterion may take one of three forms (when sutural ossicles are not involved): the parietals articulate with the great wings of the sphenoid ('H'), the frontal articulates with the temporals ('I'), or the four bones meet at a single point ('X'). Among non-human primates, all types of pterion articulation can be found, with 100% 'H' in *Lemur*, *Propithecus* and *Nycticebus*, and 100% 'I' in Sumatran orang-utans (Borneo orangs show only 27.9% 'I' type). Other genera show various types with predominance of one or other (Anoutchine, 1878; Collins, 1925; Ranke, 1899). Among human populations, the spheno-parietal type of articulation ('H') is the most common, but an articulation between the frontal and temporal bones may occur in up to 30% in some groups (Collins, 1926; Murphy, 1956). However, the incidence of the 'I' type is generally less than 10% (Table 3.4). Sjovold (1984) found a very high heritability estimate for this trait.

Traits merged

When examining the definition and anatomical interpretation of the features of regional continuity it is possible to recognise redundancies, either because two or more traits represent different aspects of a complex feature (like facial flatness), or because two or more traits represent different expressions of the same feature in East Asia and Southeast Asia, such as development of a frontal boss and frontal flatness.

Facial flatness

Among the East Asian features listed in Table 3.1, traits 4, 13, 15, 17 and 18 are considered to represent one single characteristic, i.e. facial flatness associated with frontal orientation of the lateral portion of the face. The correlation among these features was recognised by Weidenreich (1943a:pp.85–6), who stated that

> . . . in skulls of the Mongolian racial types . . . the malar surface faces forwards rather than laterally, and the outline takes a more vertical direction. Both peculiarities are among other factors mainly responsible for what has been called broad and prominent cheeks and a 'flat face' . . . In addition, in Mongolian types the part of the zygomatic bone which

Table 3.4. *Incidence of temporo-frontal articulation in modern populations*

	Collins (1926)		Ranke (1899)		Anoutchine (1978)	
Population	%	N	%	N	%	N
American whites	1.5	265				
Europeans	0.4	236	1.5	11000	1.6	9867
Egyptians (12th dyn.)	1.7	600				
Siberians	1.3	78				
Mongolians, Urga	1.1	371	2.0	1200	3.7	596
Japanese	2.4	42				
Chinese	0.0	26				
Asiatic			2.0	1200	1.9	1194
Malay	3.8	105				
Eskimo	1.8	734				
American Indians:			1.7	2520	1.6	2335
Aleutians	0.0	106				
NW Coast and Alaska	0.8	244				
California	0.3	746				
Sioux	0.0	339				
Dakota Mounds	2.1	239				
Plains	0.0	241				
Mounds, Mid. W.	2.2	367				
NE U.S. and Canada	0.0	244				
Southern Mounds	1.0	834				
Pueblo	0.1	1252				
Peru, Pachacamac	1.6	3946				
Peru, Chicama	1.1	1614				
Peru, misc.	1.6	1550				
Bolivia	20.0	20				
Patagonia	0.0	65				
Hawaiians	3.1	291				
New Zealand	0.0	54				
Australians	9.4	116	9.0	422	15.7	210
Tasmanians	26.2	42				
New Britain	28.9	13				
Fiji and Solomon Islands	4.5	22				
New Guinea	7.1	28	9.3	787	8.6	697
Africans	5.2	248	11.9	1231	12.4	884
Veddas, Tamils and Singalese	11.0	81				

> joins the maxilla first keeps to a frontal direction and then bends backwards in a moderate curve . . . One of the most striking features is the frontal orientation of the (frontosphenoidal) process . . . This peculiarity evidently is in accordance with the frontal orientation of the malar facies described above.

In order to study the degree of facial flatness in modern human populations, this feature was treated as two different, not necessarily correlated components, sagittal projection (prognathism) and lateral orientation, which gives strength to the notion that it is the associated occurrence of these two features that characterises the Mongoloid regional evolutionary line (and in its opposite expression, the Neanderthal facial form). This is supported by studies of developmental processes which show that the face presents two centres of growth, one alveolar-dental and one lateral-facial, related to the masticatory muscles (Enlow, 1982; Moore, 1982). These two centres of growth undergo different developmental processes (Enlow, 1966, 1982). Therefore, two traits are used here – lateral facial orientation and facial projection.

Profile of the frontal bone

Forehead profile with development of a frontal boss and frontal flatness are here treated together as profile of the frontal bone.

Rounding of the inferolateral orbital margin

From Weidenreich's (1943a:p.84) description of East Asian trait 14 (Table 3.1) – 'The margo infraorbitalis . . . (has) certain peculiarities (that) can be defined: (1) there is no sharply edged margin but a rounded one, (2) the facies orbitalis does not slope to a deep depression beyond the margin but keeps to the same even level . . .', and that of Larnach & Macintosh's (1966) description of Australian trait 13 (Table 3.2) – 'The border is described as rounded when the facial surface of the malar bone approaches the orbital margin and gradually curves inwards until it meets the orbital surface. When the margin is flattened there are three distinct surfaces at the margin: (1) the facial surface of the malar; (2) the wide flattened area along the top of the margin; and (3) the orbital surface of the malar', it is possible to see that these traits represent the same morphological feature. The fact that it is considered to typify each of the regional evolutionary lines is confusing. Although both Weidenreich (1943a) and Larnach & Macintosh (1966) recognised the occurrence of this trait elsewhere, they still considered it to be most common in North Asia and Australia respectively:

> I failed to find any data recording the frequency of this feature in different races but, as much as I was able to see, an even floor and a rounded inferior lateral angle, as they occur in *Sinanthropus*, are rather common features in Mongolian skulls. In Australian skulls rounded borders seem to be very frequent too. (Weidenreich, 1943:84.)

> However, it must be admitted that while this feature is frequent among Australian aborigines, it is by no means uncommon in Melanesian and New Guinea skulls. (Larnach & Macintosh, 1966.)

Other authors considered the trait specifically Australian – e.g. Klaatsch, 1908: 'When the border of the malar is found to be rounded, without a well-defined boundary for the eye-cavity, its aboriginal nature can be accepted.' Heretofore, a single feature, 'rounding of the inferolateral orbital margin', is used.

Orbital shape

Australasian trait 7 (Table 3.2), 'horizontal orientation of the supraorbital border', is considered as an aspect of the variation observed within the shape of the orbits, East Asian trait 20 (Table 3.1).

Occipital torus

Following Hublin's description, an occipital torus develops between the superior and supreme nuchal lines, marking a change between the nuchal and occipital planes (one concave, the other convex), whereby the occipital or supra-iniac zone is always thicker than the nuchal zone and therefore more pronounced in profile (Hublin, 1978a). However, these characteristics are not sufficient to form an occipital torus. The existence of the torus is related to a sulcus on the superior border ('sulcus supratoralis' of Weidenreich, 1940), which corresponds to the supreme nuchal lines of Merkel (1871). It follows that, by definition, Australasian traits 21 and 23 (Table 3.2) are part of the same morphological feature, i.e. development of an occipital torus. Such a definition has the further implication that a fully developed torus with a regular, continuous supratoral sulcus (or depression along the relief of the supreme nuchal lines), will not show an external occipital protuberance (EOP), as is the case in archaic hominids. However, an EOP may occur superimposed on an occipital torus. In this case, the EOP may be positioned well apart from the *tuberculum linearum* (the position of *inion* at the junction of the superior nuchal lines), causing either a medial depression along the superior toral margin, or a supra-iniac fossa if the EOP is pronounced and oriented towards the superior nuchal lines (Hublin, 1978a).

Scoring methods

In order to assess the regional incidence and phylogenetic value of the 'regional continuity traits' (Figure 3.1) an objective measurement of these traits in modern crania is necessary. It should be stressed that although these features have been used in evolutionary studies as evidence of regional continuity for some 50 years, their scoring and measuring has not

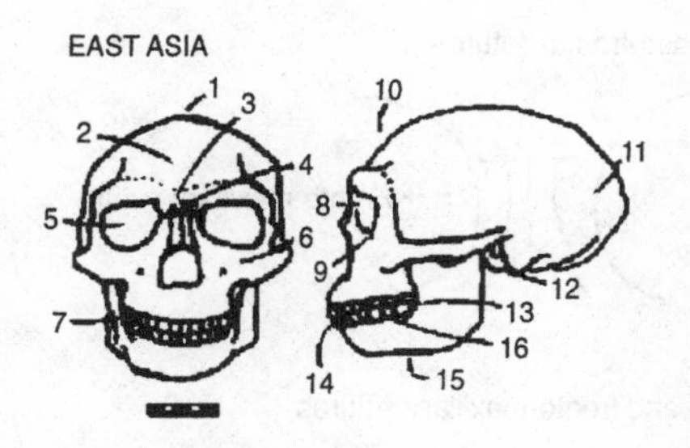

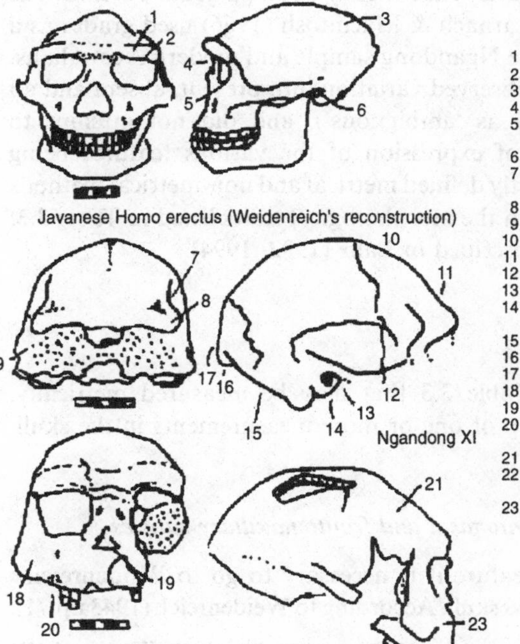

Figure 3.1 East Asian and Southeast Asian cranial features of morphological continuity marked on an East Asian *Homo erectus* (Weidenreichs reconstruction of Zhoukoudian *Homo erectus*), and three Javanese *Homo erectus* specimens – a reconstruction by Weidenreich, Ngandong XI and Sangiran 17.

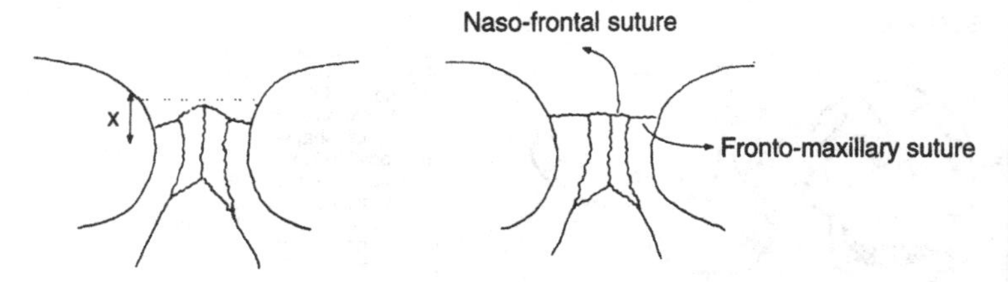

Figure 3.2 Measurement of the course of the naso-frontal and fronto-maxillary sutures. 'x' represents the vertical distance that reflects inclination of the sutures.

been consistent or standardised, or even performed in some studies. Weidenreich (1943a) described the means by which he recorded these features in *Sinanthropus*, but he relied mostly upon personal observations with no statistical testing. Larnach & Macintosh (1966) used grades as a means of scoring traits in the Ngandong sample and modern Australians. These grades classified the observed variation into present, absent and an intermediate score (defined as 'ambiguous'), and did not attempt to establish different degrees of expression of the various features being studied. This study uses strictly defined metrical and non-metrical methods for recording the variation in the morphological traits listed in Table 3.3. These methods were first described in Lahr (1992, 1994).

Metrical traits

There are 12 features in Table 3.3 that may be measured metrically, generally as the combination of one or more measurements in the skull. These are described below.

EA2: Course of frontonasal and frontomaxillary sutures

For the definition of this feature it is necessary to go to Weidenreich's description of the *Sinanthropus* skull. According to Weidenreich (1943a:p.71):

> The nasofrontal suture is completely preserved in skulls III, X and XII and partly so in Skulls II and XI. In all these skulls, this suture together with the frontomaxillary suture takes a continuous, horizontal course: The same course is evident in the Rhodesian skull, while in the European Neanderthalians . . . the suture more or less ascends toward the midline or seems interrupted because the nasal bones extend further upward than the frontal process of the maxilla.

For the purpose of this study, this trait was measured by drawing a straight

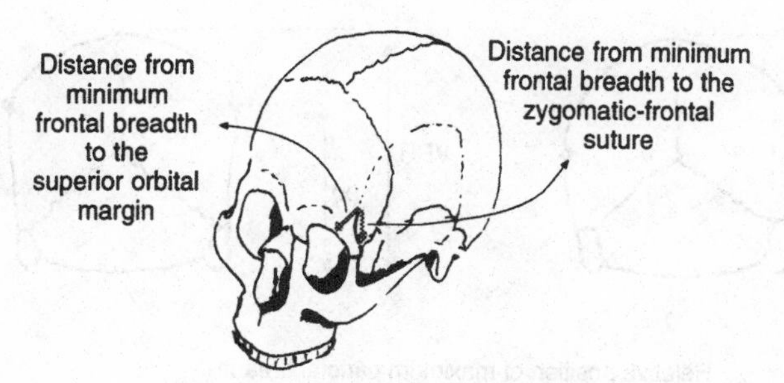

Figure 3.3 Measurement of the relative position of minimum frontal breadth through the distance between the point of minimum frontal breadth along the temporal line and both the superior orbital margin and the zygomatico-frontal suture.

line from nasion to the orbital rim, then measuring the distance (x) in mm from the point of intersection of this line with the orbital margin to the orbital end of the fronto-maxillary suture (Figure 3.2). In cases where the naso-frontal and fronto-maxillary sutures follow a horizontal course, the distance (x) should be nil. It is clear that, although measured in a horizontal dimension, this feature also reflects the flatness of the nasal bones at the naso-frontal suture.

EA5: Reduced posterior dentition

The relative size of the posterior dentition was recorded through standard buccolingual and mesiodistal measurements of molar and premolar teeth. The difference between area of M1 and M3 was then used as a measure of *posterior dental reduction*. The difference between buccolingual M1–M3 and mesiodistal M1–M3 values was also tested.

AUS3: Size and type of pterion

The size of the pterion articulation was recorded.

AUS5: Position of minimum frontal breadth

This was taken by marking the point of minimum frontal breadth along the temporal line with a pencil, and then taking two measurements – one corresponding to the distance from minimum frontal breadth to the zygomatico-frontal suture, and the other to the distance from minimum frontal breadth to the superior orbital margin (Figure 3.3).

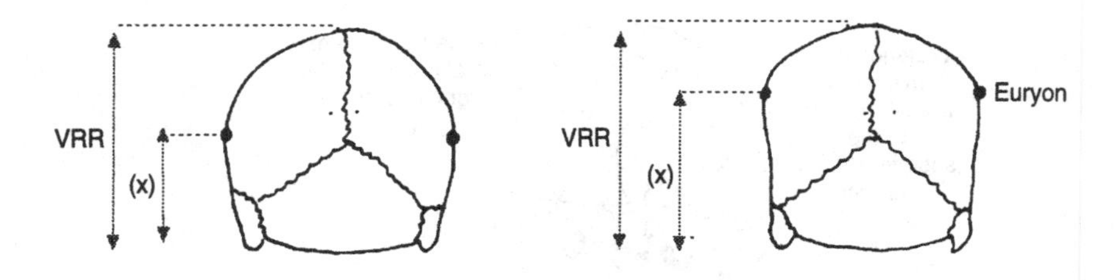

Figure 3.4 Measurement of the relative position of maximum parietal breadth by calculating an index of the relative position of euryon to cranial height. Position of maximum parietal breadth measured from euryon to porion; cranial height taken as vertex radius (VRR).

AUS6: Position of maximum parietal breadth

This was taken by marking euryon with a pencil while taking maximum parietal breadth, and then measuring the distance from euryon to porion (x) (Figure 3.4). To standardise this measure in terms of cranial height, the value (x) was transformed into a proportion of the vertex radius ($x^1 = x * 100/VRR$).

AUS8: Eversion of the lower border of the malars

This was measured similarly to Larnach & Macintosh (1966), as the difference between (a) the distance between the orbital edge of one zygomatico-maxillary suture to the same point on the other side (x1), and (b) the distance from a point approximately halfway along the lower free edge of the malar bone on its outer surface to the corresponding point on its fellow on the opposite side (x2), giving the total flare of the two malar bones (Figure 3.5). Larnach & Macintosh (1966) use as landmark for x1 the orbital edge of the fronto-malar suture. However, it was considered that a purely facial measurement (not involving the variation in frontal bone breadth) was preferred.

AUS10: Degree of curvature of maxilla

The orientation of the alveolar plane of the maxilla is here defined in relation to the Frankfort Plane as approximately parallel to the latter or rising gradually towards the basicranium, and thus defining a slight curvature. This feature was measured as the difference between the distance prosthion to the Frankfort Plane (x1) and the posterior end of the maxillary margin to the Frankfort Plane (x2), taken to reflect the degree of curvature of the posterior margin of the maxilla (x) (Figure 3.6).

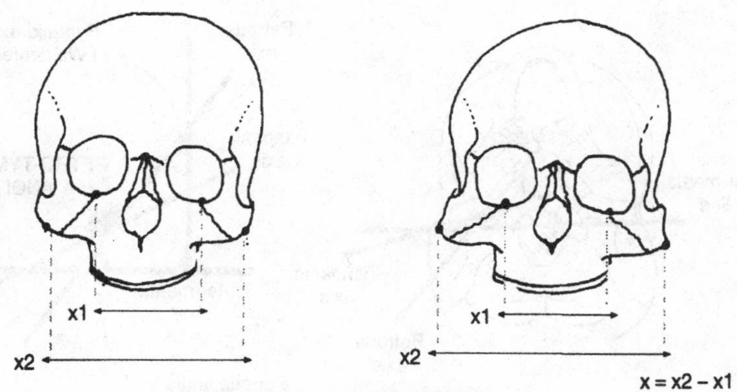

Figure 3.5 Measurement of the eversion of the lower border of the malars represented by the difference between the distance from zygo-orbitale to zygo-orbitale and from approximately the mid-point on either side of the lower free border of the malars.

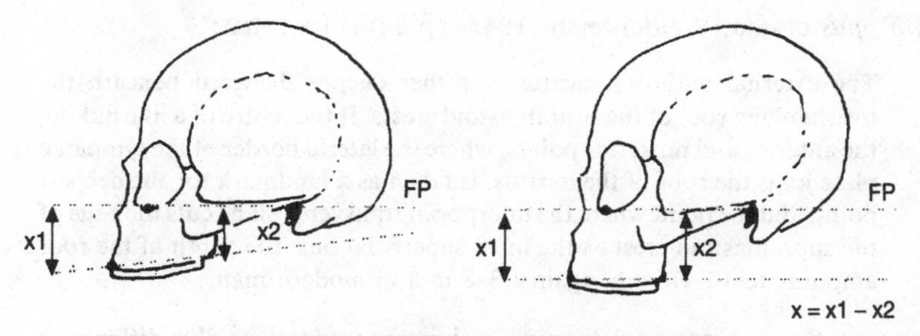

Figure 3.6 Measurement of the curvature of the posterior alveolar plane of the maxilla, taken as the difference (in lateral view) between the alveolar plane and the Frankfort plane (FP) at the posterior-most and anterior-most (prosthion) points of the maxilla.

AUS12: Presence of a suprameatal tegmen

The *tegmen tympani* can be described as

> . . .a small part of the petrous portion of the temporal bone that forms the roof over the middle ear that becomes trapped on the external aspect of the base of the cranium during development in modern humans. This small lip of bone projects between the tympanic plate and the glenoid fossa into the squamotympanic fissure. (Aiello & Dean, 1990:p.45)

The *suprameatal tegmen* is taken to be the continuation of the portion of petrous bone towards the lateral wall of the skull. In describing the

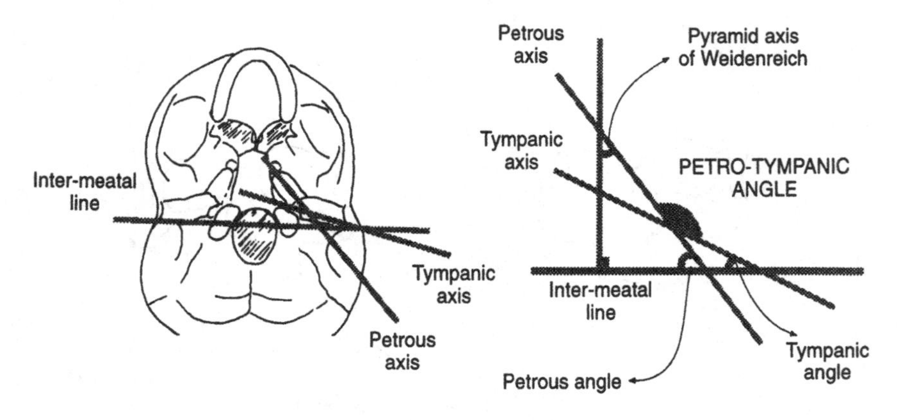

Angle of the petrous to the tympanic in the petro-tympanic axis

Figure 3.7 Measurement and definition of the angle between the petrous and the tympanic in the petro-tympanic axis using an inter-meatal line as reference for the measurement of the petrous and tympanic angles.

Sinanthropus crania, Weidenreich (1943a: p.22) stated that:

> The external auditory meatus is rather deeply sheltered beneath the overhanging root of the supramastoid crests. If the centre of a line linking the anterior and posterior points, where the lateral border of the tympanic plate joins the roof of the meatus, is taken as a landmark for the deepest point, and the point where the interporial transverse plane cuts the edge of the supramastoid crest as the most superficial one, the depth of the roof amounts to 10–18 mm, against 3–8 mm in modern man.

The size of the suprameatal tegmen is here measured as the difference between biauricular and intermeatal breadths, divided by two. The latter measurement was taken from the most lateral margin on the tympanic plate on each side.

AUS13: Angle between the petrous and the tympanic

The angle between the tympanic plate and the petrous bone was recorded by measuring the direction of the tympanic plate and the petrous bone in relation to an intermeatal line (Figure 3.7). The various angles formed were then calculated.

EA6/AUS16: Degree of facial projection

Facial projection (or prognathism) was measured as the Gnathic Index (basion-prosthion × 100/basion-nasion) and as the difference between prosthion, subspinale and nasion radii (Figure 3.8). The different measurements are used because while the Gnathic Index reflects total facial

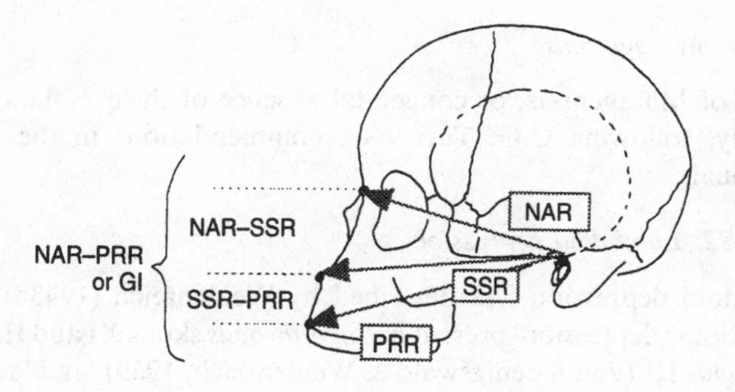

Figure 3.8 Four measurements of facial prognathism. Three represent differences between facial radii: (1) total facial prognathism: nasion radius (NAR) – prosthion radius (PRR); (2) mid-facial prognathism: nasion radius (NAR) – subspinale radius (SSR); (3) alveolar prognathism: prosthion radius – subspinale radius. The fourth measurement, Gnathic Index (GI), also represents total facial prognathism (BPL * 100/BNL).

projection, the differences among the radii reflect the relative contribution of the nasal and maxillary segments of the face to the total prognathism observed.

EA7/AUS17: Lateral facial orientation

Lateral facial orientation (or coronal facial flatness) is measured by the naso-frontal subtense and the difference between nasion and ectochonchion radii.

EA8/AUS18: Profile of the frontal bone

Forehead profile was measured as frontal angle, following the description in Howells (1973), i.e. computed from nasion-bregma chord, nasion-bregma subtense and nasion-subtense fraction.

Non-metrical traits

Discontinuous traits are by definition all those features that can be measured as categories according to either presence/absence or degree of development in a grading scale. Among the 'regional continuity traits', we find both, 16 in total.

Traits scored as present or absent

Two features, M3 agenesis (EA4) and lambdoid depression (AUS2), were measured as present/absent.

EA4: M3 agenesis

The scoring of M3 agenesis, or congenital absence of third molars, was done visually, following C.G. Turner's recommendations in the ASU Dental Manual.

AUS2: Lambdoid depression

A pre-lambdoid depression was described by Weidenreich (1943a) as a 'distinct obelionic depression' present in *Sinanthropus* skulls XI and III and *Pithecanthropus* III (von Koenigswald & Weidenreich, 1939), and lacking in skulls II, X and XI. However, this trait is apparently only clear on skull III, since in skull XI Weidenreich describes it '. . . more like a general flatness which also involves the entire apical region of the occipital squama' (Weidenreich, 1943a:p.27). Such obelionic depressions in modern man have been observed, and variously called 'depression praelambdoidea', 'fovea verticalis' or 'trigonum supralambdoideum'. According to Weidenreich (1943a:p.35) none of these is exactly as found in *Sinanthropus*. Le Double (1903) regards them as the result of disturbances in the ossification of the parietal bone, but no further studies have confirmed this. A pre-lambdoid or obelionic depression was scored visually as present or absent, although this feature varies markedly in how distinct and clear it may be.

Traits scored as grades

A scoring system of grades was developed for nine of the morphological features that represent specific cranial shapes and cranial superstructures (Lahr, 1992, 1994; Appendix I). This scoring system aims at obtaining a quantified measure of the degree of expression of these features. The features were scored as categories reflecting minimum, maximum and intermediate degrees of development. The minimum number of scores recorded for any feature is three and the maximum seven. The grades were defined on a sample of one hundred skulls from various regions of the world (Table 3.5). The best examples of each category were chosen and reproduced in plaster casts, so as to be used as reference plaques during the process of scoring. The nine features scored as grades are listed in Table 3.6. In the case of the occipital torus, the scoring was not done only on the basis of degree of expression, i.e. size. In an attempt to include the variability in morphological pattern, the classification was based on both the form and extent of the superior and supreme nuchal lines and whether there was a torus/sulcus formation. The best examples of each category were also reproduced in plaster casts to be used as reference. A detailed description of these features, scores and number of grades are in Appendix I.

Table 3.5. *Cranial sample used for developing the scoring system of grades and categories of cranial morphological features*

DC Eu.1.5.109 – Lanhill, Scotland; Neolithic	DC Inuit C – Angmacsalik
DC Eu.1.5.103/JT 278 – Gatcombe; Neol., UK	DC Inuit II
DC Eu.1.4.96 – Scotland; Bronze Age	DC Am.1.0.3 – Greenland
DC Eu.1.4.97 – Aldwincle; Bronze Age, UK	DC Am.0.0.1 – Greenland
DC Eu.1.3.379 – Wandlebury; Iron Age, UK	DC 1832 – Greenland
DC Eu.1.3.178 – Cambridge; Rom-British, UK	DC 1842 – Vancouver
DC Eu.1.3.112 – Maiden Castle; Rom-Br, UK	DC xx 5928 – Iroquois, Ontario
DC Eu.1.0.1043/W 7049 – Late Medieval, UK	DC 4498 – Tierra del Fuego, Patagonia
DC TEAN V – Cornwall, UK	DC 4211 – Chinese from Surinam
DC 1209/1177 – Finland	DC 4213 – Chinese from Surinam
DC 1171 – Lapp, from Paddeby, old mounds	DC Oc.11.0.28 – Papua New Guinea
DC Eu.31.00.1/4612-3 – Russia	DC Oc.11.0.45 – Papua New Guinea
DC Eu.31.00.2/5498-4 – Russia	DC Oc.11.0.56 – Papua New Guinea
DC 1205/1178 – Sweden or Denmark	DC Oc.12.0.1 – New Britain
DC Eu.26.00.2/JT 220 – Germany	DC Oc.13.0.1/54 – Solomon Is.
DC Eu.42.00.5/JT 244 – Ostia, Rome, Italy	DC Oc.14.0.1 – Sta. Cruz, Ovaova Is.
DC 1119 – Sardinia	DC 3 – Australia
DC Af.11.5.235 – Naqada, Egypt	DC 4 – Australia
DC Af.11.5.238 – Naqada, Egypt	DC 5 – 'Berida', Australia
DC Af.11.5.7 – Badari; Pre-Dynastic, Egypt	DC 33 – Australia
DC SUD 62 – Kerma, Sudan; 12–13th Dynasty	DC 42 – West Australia
DC E 864 – Gizeh; 19–26th Dynasty, Egypt	DC 44 – Australia ?/ Torres Strait ?
DC 4161 – Egypt	DC 51 – Mackay, Queensland, Australia
DC 695 NU GR 85 112/119x – Nubia, ASN	DC 'A' – Australia
DC Af.23.0.23/ K 9 – Haya Tribe, Tanzania	DC 2164 – Australia
DC 1774 – Yao Tribe, Uganda	DC 2165 – Australia
DC 1777 – M'Zadi, Congo	DC 2133/1168 – NSW, Australia
DC 6096 – Yola, North Nigeria	DC 6081 – NSW, Australia
DC 5060 – Osheba, French Congo	DC Oc.4.0.4 – Queensland, Australia
DC 1758 – Bushman, Cape Town	DC Oc.0.0.1 – Australia
DC 1738 – Bushman	DC 2112 – Sandhills, Australia
DC 6110 – South Africa, ?Botswana	DC 1215 – Australia
DC 6109 – Amaxosa, South Africa	DC 2100 – Tasmania
DC 4205 – black mulatto from Surinam	DC 2096 – Tasmania
DC 4208 – mulatto from Surinam	DC 2099/1170 – Tasmania
DC 4201 – negress from Surinam	DC 1212 – Punjab, India
DC 5323 – Africa	DC 1225 – Bihar, India
DC xxx1 – Africa	DC xxx2 – India
DC BU 41 – Moulmein, South Burma	DC 2131/272 – Polynesian
DC BU 53 – Moulmein, South Burma	DC 47 – North Is., New Zealand
DC 5w – Andaman Islands	DC 4496 – North Is., New Zealand
DC As.21.0.4 – China Kowloon, Urn Burials	DC 6099 – Maori, New Zealand
DC As.21.0.7 – China Kowloon, Urn Burials	DC Oc.31.0.4 – Moriori
DC 1761 – China	DC Fol. 35, Acc. Book – Easter Island
DC As.17.0.2 – Tibet	DC 1763 – Kamchatka
DC 5334 – Dokihol, Tibet	DC Inuit I – Pend Is.
DC 2377 – Japan	DC Inuit IV – Cape Simpson

DC: Duckworth Collection.

Table 3.6. *Features measured as grades of development*

Trait	Features	Grades*
*EA*1	sagittal keeling	SK1 – SK2 – SK3
*EA*3	the profile of the nasal saddle and nasal roof	NS1 – NS2 – NS3 – NS4
*AUS*1	supraorbital ridges	ST1 – ST2 – ST3 – ST4 – ST5
*AUS*4	infraglabellar notch	IN1 – IN2 – IN3 – IN4
*AUS*8	zygomaxillary tuberosity	ZT1 – ZT2 – ZT3 – ZT4
*AUS*12	zygomatic trigone	TR1 – TR2 – TR3 – TR4
*AUS*15	occipital crest	OCR1 – OCR2 – OCR3
*AUS*16	occipital torus	OT1 – OT2 – OT3 – OT4 – OT5 – OT6 – OT7
*EA*9/*AUS*19	rounding of the inferolateral margin of the orbits	RO1 – RO2 – RO3

*See Appendix I for trait and grade description.

Traits scored as categories

Two features were scored as categories, lower narial margins and orbital shape.

AUS9: The type of narial margins

That is whether there is a sharp line or fossa dividing the nasal floor from the subnasal portion of the maxilla, was found to be extraordinarily variable. Several classificatory systems have been proposed (Burkitt & Lightoller, 1923; Gower, 1923; Johnson, 1937; Macalister, 1898). However, none of these classifications was found to represent accurately the variability observed. In order to record this variation, drawings of the narial margins were made, trying to identify line origins (whether from the lateral margin or from the central nasal spine, interiorly or exteriorly), and the relative position of the pre-nasal fossa to the other elements of the nose and maxilla. These were organised into 10 categories, described in Figure 3.9.

EA10/AUS20: Orbital shape

The shape of the orbits was not treated as a single character in terms of variation ('square', 'rounded', 'rectangular'), but as a complex trait composed of three elements: (1) a frontal element or superior orbital margin; (2) a zygomatico-frontal element or superior-lateral corner; and (3) a zygomaxillary element or infero-lateral margin. The variation in each

(1) Single margin

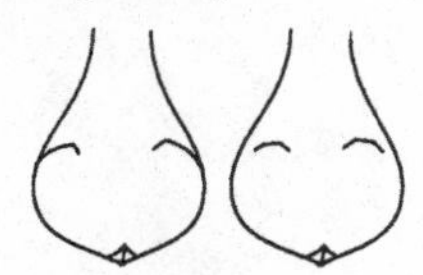

As recognised by Gower (1923), the single ridge or Anthropine/oxycraspedote of Macalister (1898) can be the result of two combinations: first, complete fusion of all three crests; second, the fusion of lateral and spinal crests, forming an anterior ridge but with the inconspicuous turbinal crest within the nasal cavity

(2) Double margin, joined medially by the nasal spine, with or without formation of a lateral fossa

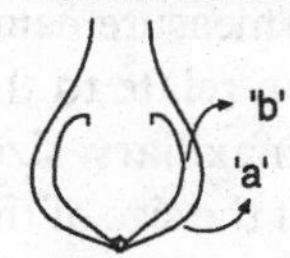

'a': crista maxillaris of Holl (1882)
crista nasalis anterior of Bonin (1912)
crista lateralis of Gower (1923)

'b': crista intermaxillaris of Holl (1882)
crista nasalis posterior of Bonin (1912)
dvided into crista spinalis and crista turbinalis by Gower (1923)

(3) 'b' forms the margin, with an ill-defined fossa

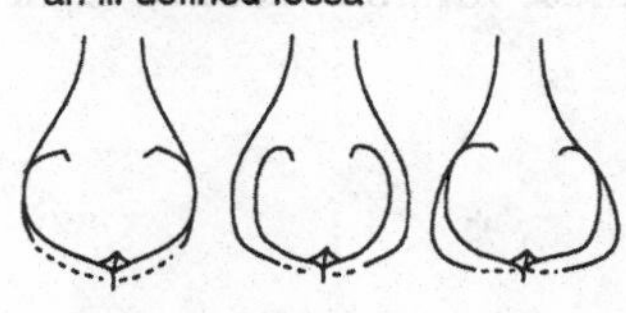

(4) 'b' forms the margin, with a well-defined fossa

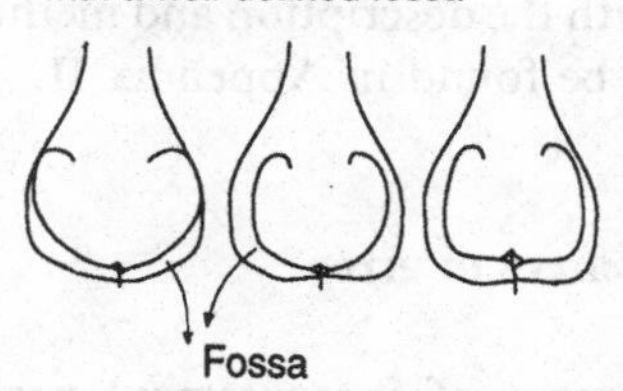

(5) 'a' and 'b' form a double margin

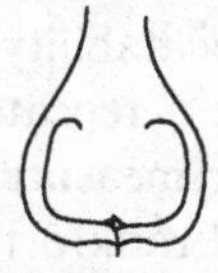

(6) 'a' and 'b' form a double margin, with a well-defined internal fossa

'a' and 'b' at the same level

Fossa

(7) 'a' forms the margin and determines a maxillary fossa

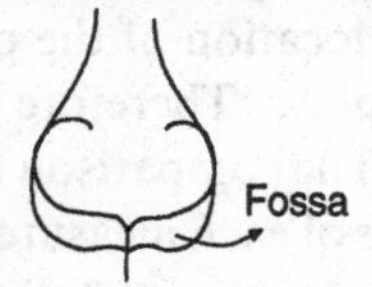

(8) 'a' forms the margin, 'b' separated internally, ill-defined external fossa

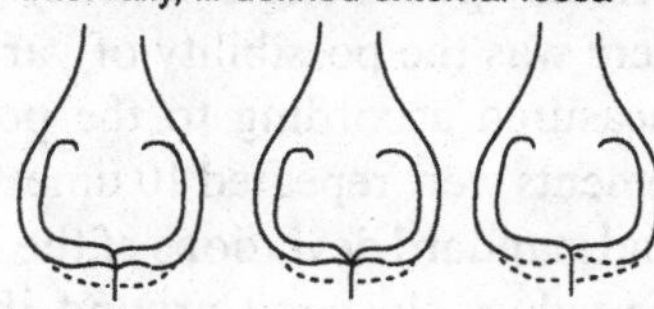

(9) Ill-defined external margin

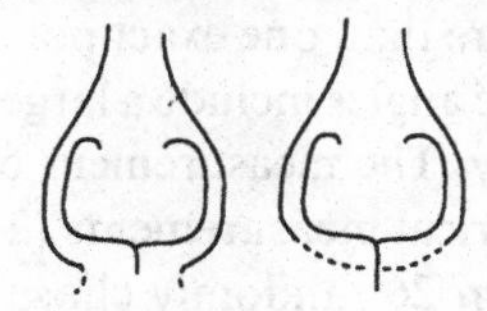

(10) No external margin

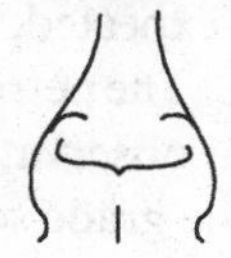

Figure 3.9 Categories of lower narial margins.

of these elements (shape/angle) was combined to obtain orbital shape (Figure 3.10).

Other measurements taken

In order to examine the relationship between the 'regional continuity traits' and general shape and size parameters, a series of standard cranial measurements supplemented by other measures considered relevant were taken on each skull. The choice and description of the standard measurements were taken from Howells (1973). Other measurements taken relate to the position of inion in relation to lambda and opisthion; maxillary size; relative size and angle of basion-hormion-palate; measures of the size of the temporal fossa and temporal muscle lines; dental measurements; and dimensions of the parietal bones. A complete list of measurements taken, together with the description and methods of those variables defined in this study, can be found in Appendix II.

Sources of error

The assessment of measurement precision was done as a measure of intra-observer reliability (observer's consistency or repeatability) and inter-observer reliability. The greatest concern was the degree of repeatability of the scoring system of the grade features. For the four measurements taken by drawing on an acrylic plate (distance between the Frankfort Plane and prosthion, distance between Frankfort Plane and the posterior-most lateral alveolar point, angle of the tympanic axis, and angle of the petrous axis), there was the possibility of varying the exact location of the points being measured according to the position of the plate. Therefore these measurements were repeated 10 times on each skull. The comparison of the means and standard deviations of the 10 observations of each measurement show the values clustered around the mean in the case of the alveolar measures, while a larger error was involved in the petro-tympanic angles. These latter results were expected, since there is more than one exact point for tracing the angle lines. The petrous and tympanic angles include a larger error than other metrical observations in this study. The measurement of intra-observer error of the grade scores and categorical measurements (21 in total) was done by scoring each feature twice on 26 randomly chosen skulls. Of the 546 scores, 45 (8.2%) of observations were different between the first and second scoring session. The differences, however, were mostly concentrated in a few variables, namely the zygomatic trigone, superior

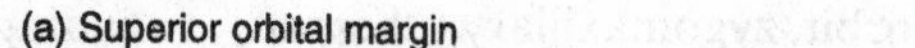

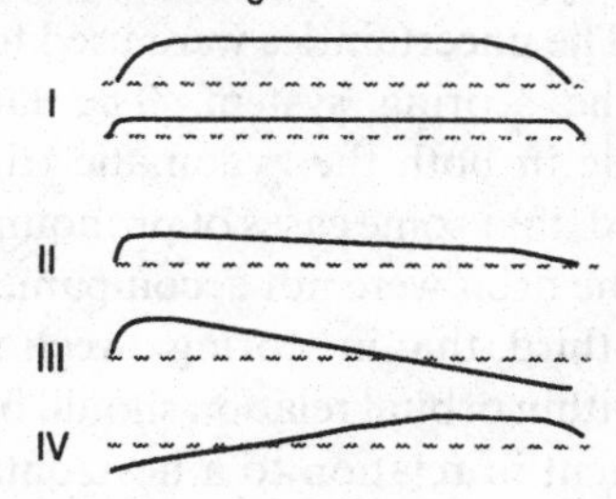

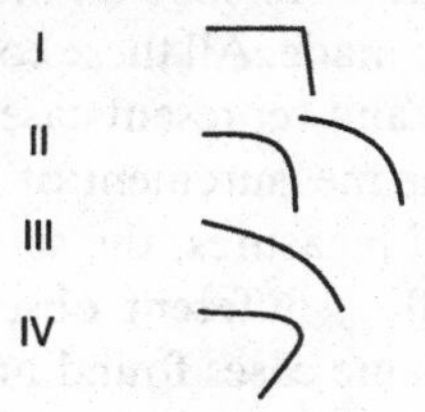

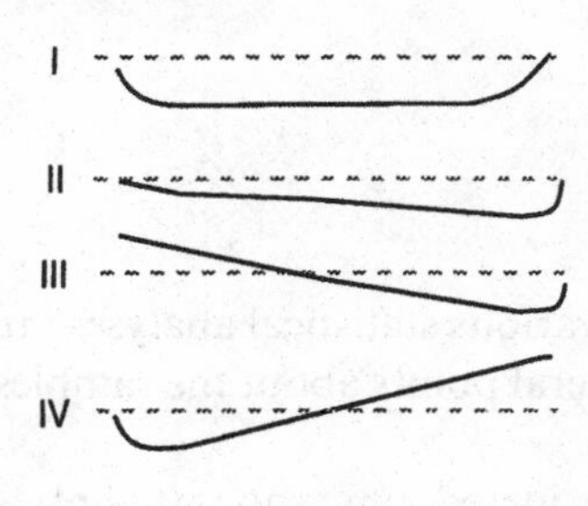

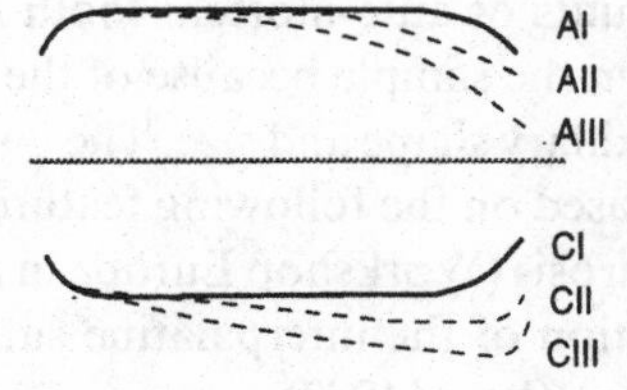

Figure 3.10 Scoring of orbital shape through the relative inclination of the superior, lateral and inferior orbital elements.

and inferior margin of the orbit, zygomaxillary tuberosity, rounding of the orbit and lambdoid flattening. The uncertainties were used to improve the definition of the grades in the scoring system. The latter involved recognising first, a 'slight' grade in both the zygomatic trigone and the zygomaxillary tuberosity; second, that some cases of pronounced rounding of the infero-lateral margin of the orbit were not accompanied by levelling with the floor of the orbit; and third, that in scoring direction of superior and inferior orbital margins, a within orbital relation should be maintained, i.e. the inclination of each element in relation to a horizontal line and to each other defines the overall orbital shape observed. With these corrections in the grade descriptions, a third score was performed on the 26 skulls and 3.3% (18/546) different observations were made. All these corresponded to cases that were scored differently before, and represent cases in which the scoring is not completely certain. For the measurement of inter-observer error of the grade scores and categorical measures, the same skulls were scored by a second person, and 22 (4.0%) different observations were made, 20 of which corresponded to the same cases found uncertain in the intra-observer error tests, confirming that the scoring method and grade description of these traits are a reliable measure of the variation in almost 100% of cases.

Cranial samples

The cranial samples used in the various statistical analyses are described in each chapter. However, two general points about the samples used should be made here.

Only adult crania were considered, and no attempt was made at subdividing the sample into further age categories. Older individuals or any skull that showed large amounts of ante-mortem tooth loss and alveolar resorption were excluded from the sample because of the known effects of this process on facial and maxillary shape and size. The distinction between adults and adolescents was based on the following features: (1) the closure of the basisphenoid synchondrosis (Workshop European Anthropologists, 1980); and (2) some obliteration of the interpalatine suture and/or more than half of the incisive suture (Bass, 1987).

In order to use sex as an analytical variable and study patterns of sexual dimorphism among populations, a reliable and precise sex diagnosis has to be obtained. However, associated post-cranial material was mostly unavailable, especially for odd skulls that were not part of larger skeletal series. If the more reliable methods of sexing skeletal material through morphology of the pelvis (Phenice, 1969; Washburn, 1948) are not possible, sex diagnosis has to be obtained through the skull. While the diagnostic features of the pelvis are functional in origin, and therefore reflect reproductive and locomotor adaptations of modern humans in general, the characters for sex determination in the skull are related to levels of robusticity, such as, for example, the size and direction of mastoid processes, temporal lines, supra-orbital ridges, size of mandibular ramus and presence of supramastoid crests (Workshop European Anthropologists, 1980). There are two reasons why these characters were not used. Firstly, the level of overall robusticity varies markedly between different populations, and standards derived from one cannot be applied to another. This problem is not so serious as long as there are enough individuals from each single population to determine what should be considered robust and gracile, but in the case of the sample used in this study, a large number of skulls are the single or one of a few representatives of a particular group. This issue is exemplified by the number of studies that show the problem of sexing the very robust and variable Australian aboriginal material through standards basically drawn from European samples (Brown, 1989; Fenner, 1939; Hrdlička, 1928; Klaatsch, 1908; Larnach & Freedman, 1964). Secondly, many of the features that reflect cranial robusticity (and possible sexual differences) are the cranial superstructures under study.

On the basis of these considerations, the material was not sexed. Although sexual dimorphism is a potential source of variability in the distribution of the features being examined in this study, it was decided not to consider it further, for the following reasons:

(1) As already mentioned, the choice of individual skulls was done randomly, taking availability and state of preservation into account. At no point did degree of robusticity or expression of the various traits influence this choice. The only possible sources of sexual bias relate to the higher likelihood of preservation of robust skulls, and the preference of early collectors for large or 'typical' crania. However, as a whole the sample is considered to be sexually random.
(2) In his extensive studies of population relationships and distance, Howells (1969, 1973, 1989) found that the two sexes agree to a considerable extent, especially in terms of population discrimination.
(3) The use of such a geographically varied sample, which precludes reliable sexing, was important for representing broad modern human variation. Furthermore, since the results of this work are to be interpreted within an evolutionary perspective, the sample used to represent modern *Homo sapiens* should be comparable with other

samples of the hominid fossil record. Although attempts have been made at sexing hominid fossils, and indeed sexual dimorphism has been used to interpret observed variability, the problems of sexing individual fossil crania are even more pronounced than those between modern populations.

Therefore, the variation due to sexual differences is not identified in either recent or fossil populations, but treated as part of the range of variation observed in temporal and spatial hominid groups.

Testing the morphological basis of the Multiregional Model

This chapter reviews the definition of those morphological features that have been used as evidence of an archaic source of regional differences among modern humans. An examination of the anatomical meaning of some traits reveals redundancies, so that when these are taken into consideration, the number of traits applicable to the problem of regional continuity is reduced. Another group of traits was dismissed for not representing a specific North Asian or Australian incidence on the basis of recent studies of non-metrical variation in human skulls. A final list of regional continuity traits is established as 10 East Asian and 20 Southeast Asian/Australian (Table 3.3), and methods of quantifying these cranial features defined.

The Multiregional Model of modern human origins argues for continuity in morphology from archaic to present regional populations, and therefore explains the origin of modern cranial diversity as the consequence of differential phylogenetic relationships with archaic hominid groups. This model is based on the 30 features described above and how these features are distributed in past and present hominid populations. The following chapters test this morphological evidence by examining three aspects of these features: (1) the regional distribution in recent populations; (2) the temporal and spatial distribution in the Upper Pleistocene; and (3) the independence of expression from each other and from other cranial dimensions.

4 *The regional expression of the East Asian and Australian 'continuity traits'*

As we have seen, proponents of the Multiregional Model claim that the 46 traits listed in Tables 3.1 and 3.2, reduced to 30 in Table 3.3, show a specific regional patterning in distribution and expression. They propose that these patterns are evidence of a genetic link between regional archaic populations and recent inhabitants of those same regions. Multiregional evolution does not allege that these traits occur exclusively in those populations in which they show regional continuity, but that they occur in those populations with a higher frequency or with a significantly different mean value than in other groups.

The review of the regional continuity traits presented in Chapter 3 shows that there has been relatively little research on the comparative incidence and distribution of these traits across populations. There have been a number of studies that compare the incidence of these features between a given recent population and the archaic hominids found in the same region. However, these studies rarely include other recent groups or even archaic hominids from outside the region under study in the comparisons. The inclusion of other recent populations is important in order to investigate the 'uniqueness' of the regional distribution/incidence being claimed, while the comparison to archaic hominids outside the region under study is necessary in order to investigate the derived character of any specific archaic–modern regional pattern observed.

In this chapter, the basic assumption of the Multiregional Model, i.e. the regionality of the continuity traits listed in Table 3.3 is tested statistically. An examination of the frequency and distribution of the 'regional continuity traits' in nine recent cranial samples was carried out. The samples represent five broad geographical populations (Europe, sub-Saharan Africa, South-east Asia, East Asia and Australia) and four narrowly-defined populations, either geographically (Fueguian and Patagonian) or temporally (the Epi-Palaeolithic Afalou, Taforalt and Natufian), described in more detail below. (See Figure 4.1.)

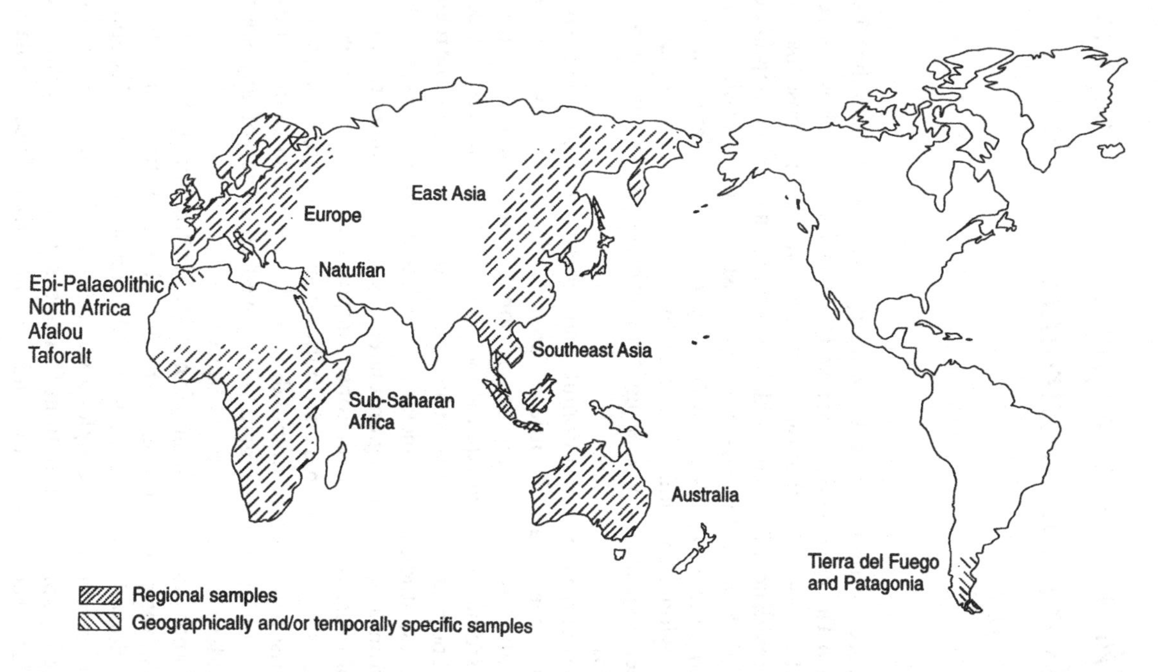

Figure 4.1 Geographical location (shaded areas) of the nine recent and sub-recent cranial samples used in the analyses.

Geographical distribution of the regional continuity traits in nine recent cranial samples

The incidence of the Asian and Australian regional continuity traits in modern populations was assessed by analysing statistically the variation within and between five regional samples. The samples used represent Europeans (N=25), sub-Saharan Africans (N=26), Southeast Asians (N=26), East Asians (N=43) and Australians (N=39). These samples differ from previous works (Lahr, 1992, 1994) in that the African sample is composed solely of populations south of the Sahara, the East Asian sample does not include Amerindian and Inuit populations and the Australian sample does not include Melanesians. It is intended that by using samples broadly defined on a regional scale, the problem of sampling a single spatial and temporal group as representing the entire modern descendants of an evolutionary lineage would be avoided. The composition of the five recent samples is described in Table 4.1.

A further four complementary samples are used for comparison with the five regional groups. One group consists of a sample of 29 skulls from the aboriginal populations from the Argentinian side of Tierra del Fuego and southern Patagonia (Onas/Selknam, Tehuelches). These were terrestrial hunter–gatherers and very different biologically, culturally and linguistically from the eastern coastal groups who inhabited the channel regions of present southern Chile, and from the agriculturalist populations to the north. Two groups are formed by samples from the Epi-Palaeolithic sites of Afalou-bou-Rhummel, Algeria (N=37) and Taforalt, Morocco (N=22). These populations of robust hunter–gatherers date to the end of the last glacial period (^{14}C 14–8.5 ka, Chamla, 1978; Vallois, 1969). The last group is composed by crania from several Natufian sites in Israel, dating from the Pleistocene–Holocene period (N=30). The Natufian population of the Middle East already exhibits many trends towards the early agriculturalists (Bar-Yosef & Valla, 1991). These four specific groups were not included in either the analyses of variance or the chi-square tests, as they represent narrowly defined spatial and temporal populations that do not represent modern samples on a regional scale. However, they provide a measure of the degree of differentiation of closely defined modern human groups. In the case of the *circum*-Mediterranean fossil populations, the sample sizes vary markedly depending on the feature being examined because of the fragmentary nature of some of the specimens.

Table 4.1. *Regional (Europe, sub-Saharan Africa, Southeast Asia, East Asia and Australia) and specific (Tierra del Fuego-Patagonia, Afalou, Taforalt, Natufian) cranial samples*

Europe (N=25)	
DC Eu.1.5.109, Lanhill, Scotland, Neolithic	DC 1205/1178, Sweden or Denmark
DC Eu.1.5.103/JT 278, Gatcombe, Neolithic	DC Eu.31.00.1/4612-3, Russia
DC Eu.1.4.96, Northton, Is. of Harris, Scotland, Bronze Age	DC Eu.31.00.2/5498-4, Russia
DC Eu.1.4.97, Aldwincle, Bronze Age	DC Eu.26.00.2/JT 220, Germany
DC Eu.1.3.379, Wandlebury, Iron Age	DC Eu.26.00.1/JT 219-1133, Germany
DC Eu.1.3.178, Arbury Road, Cambridge, Romano-British	DC 1170, Austria
DC Eu.1.3.112, Maiden Castle, Romano-British	DC 1150, SERB/AUSTRIAN, Balkans
DC NNI, Sedgeford, Saxon 4-7th C AD	DC Eu.42.00.2, Italy
DC Eu.1.0.1043/W 7049, Whitechapel, Late Medieval	DC 1114, Paestum, Italy
DC Eu.1.00.75, Stumps Cross 2	DC Eu.42.00.5/JT 244, Ostia, Rome, Italy
DC TEAN V, Cornwall	DC 1119, Sardinia
DC 1209/1177, Finland	DC 2332, France
DC 1171, LAPP, Paddeby old grave mounds	
Sub-Saharan Africa (N=26)	
DC 1774, YAO, Uganda	DC 4201, Surinam
DC Af.21.0.4, TURKANA, Kenya	DC 6109, AMAXOSA, South Africa
DC Af.23.0.23/K 9, HAYA, Bakava Cave, Musira Is., Tanzania	DC 1758, found in sand nr. seashore nr. Wynberg, Cape Town
DC Af.23.0.22/K 8, HAYA, Bakava Cave, Musira Is., Tanzania	DC 6110, South Africa/Botswana
DC Af.23.0.201, HAYA, E. Africa	DC 1733, AMAPONDA, bushman
DC Af.23.0.219, HAYA, E. Africa	DC 1743, bushman, from Strand Louper, Port Elizabeth
DC Af.21.0.26/KY 16, TEITA	DC 5102, bushman
DC 6096, YOLA, North Nigeria	DC no #xxx7, bushman
DC 5419, KAGORO, North Nigeria from Modakia	DC 1738, bushman
DC 1730/1165, Guinea	DC 5323, African
DC 5039, Senegal	DC no # xxx1, African
DC 1777, M'ZADI, Upper Congo River, Congo	DC 4208, mulatto, Surinam
DC 5060, OSHEBA, French Congo	DC 4205, mulatto, Surinam

Southeast Asia (N=26)

DC As.55.0.1, Andaman Islands
DC 5w, Andaman islands
DC 9b, Andaman Islands
DC 11, Andaman Islands
DC As.33.6.11, Gua Cha, Kelantan, Mesolithic
DC As.44.0.43, Mulu Cave, Sarawak
DC As.44.0.9, Sarawak
DC 4546, ORANG BUKIT, Sarawak
DC 5936, MABUIAG, Sarawak
DC 4556, MURUT, Sarawak
DC 4516, LEPUASING, Sarawak
DC no # xxx5, Pangai Sakei, Skeat Expedition
DC 1787, WRAPER, North Borneo
DC 4419, Salweenand Mekong, Thailand
DC BU 14, Burma
DC BU 19, Burma
DC BU 23, Burma
DC BU 26/As.57.0.26, Burma
DC BU 38, Burma
DC BU 41, Moulmein, South Burma
DC BU 53, Moulmein, South Burma
DC BU 71, Burma
DC BU 76/As.57.0.76, Burma
DC BU 97, Burma
DC BU 139, Moulmein, South Burma
DC 1756/1173c, Malay mulatto, Diepo Negoro

Eastern Asia (N=43)

DC 1759, China
DC 1761, China
BMNH – As.60/468, China
BMNH – As.60.759/758a, China
BMNH – As.60.470, China
BMNH – As.60.753, China
BMNH – As.60.750, China
BMNH – 84.9.11.1, China
BMNH – 85.12.3.1, China
BMNH – As.60.749, China
BMNH – As.60.469, China
BMNH – 7.6932, China
DC As.21.0.6, China Kowloon, Urn Burials
DC As.21.0.7, China Kowloon, Urn Burials
DC As.21.0.4, China Kowloon, Urn Burials
DC As.21.0.3, China Kowloon, Urn Burials
DC As.21.0.5, China Kowloon, Urn Burials
DC 1760, Macassar, China
DC 1762, Macassar, China
DC 4213, Chinese from Surinam
DC 4211, Chinese from Surinam
BMNH – 79.11.21.77.5, Macao, China
DC 5932, ATAYAL, Formosa, Taiwan
BMNH – As.62.7.5.7.1,ft, Hong Kong, China
BMNH – 7.5305, Tibet
BMNH – 7.5311, Tibet
DC As.17.0.2, Tibet
DC 5334, DOKIHOL, Tibet
BMNH – 7.5112, Kham Province, Tibet
BMNH – 7.5113, Kham Province, Tibet
BMNH – 7.5120, Kham province, Tibet
BMNH – As.80/530, Japan
DC 5430, Japan
DC 2377, Japan
BMNH – 66.2.7.3, Japan
BMNH – 1882.5.31.2, Japan
BMNH – As.80.760, Japan
BMNH – 1866.2.7.4a, Japan
BMNH – As.80.761, Japan
BMNH – 66.2.7.2, AINU, Japan
BMNH – 1866.2.7.1, AINU, Japan
BMNH – 82.5.31.1, AINU ? Japan
DC 6, AINU, Japan

Table 4.1 (cont.)

Australia (N=39)	
DC 44, ? Australia / ? Torres Strait	DC 5, Berida, Australia
DC 'A', Australia	DC 42, West Australia
DC 3, Australia	DC Oc.4.0.4, Queensland, Australia
DC 4, Australia	DC Oc.3.0.2, Queensland, Australia
DC 33, Australia	DC Oc.3.0.3, Queensland, Australia
DC 2164, Australia	DC Oc.3.0.4, Queensland, Australia
DC 2165, Australia	DC 51, Mackay, Queensland, Australia
DC Oc.0.0.1, Australia	DC 2154, North Queensland. Australia
DC 1215, Australia	DC 2133/1168, New South Wales, Australia
DC 2161, Australia	DC 6081, New South Wales, Australia
DC 2160, Australia	DC 2123/1084, New South Wales, Australia
DC no # xxx6, Australia	DC 2117/1073, South Australia
DC Oc.4.0.3, Australia	DC 2115/1080, South Australia
DC 5849, Australia	DC 2128/1082, Souh Australia
DC 2162, Australia	DC 2113, South Australia
DC 2163, Australia	DC 2128, Northwestern Australia
DC 2137/1294, Australia	DC 12/1390, Rose Bay, Port Jackson, Australia
DC Oc.0.0.3/4421, Australia	DC 2102/1086, lake nr. Murray River, Australia
DC 2159, Australia	DC 2112, Henley Beach, Sandhills, Australia
BMNH – 84.9.5.6/Ia20, Australia	

ADDITIONAL SAMPLES:

Tierra del Fuego-Patagonia (N=29)

DC 4498, Tierra del Fuego, Patagonia
BMNH – FC 1025/12.90, Tierra del Fuego, Patagonia
BMNH – FC 1025.1/12.90, Tierra del Fuego, Patagonia
BMNH – FC 1025.2/12.90, Tierra del Fuego, Patagonia
BMNH – FC 1025.3/12.90, Tierra del Fuego, Patagonia
BMNH – FC 1025.4/12.90, Tierra del Fuego, Patagonia
BMNH – FC 1025.5/12.90, Tierra del Fuego, Patagonia
BMNH – FC 1025.6/12.90, Tierra del Fuego, Patagonia
BMNH – FC 1025.7/12.90, Tierra del Fuego, Patagonia
BMNH – FC 1027/12.90, Tierra del Fuego, Patagonia
BMNH – 2.9.18.1, Tierra del Fuego, Patagonia
BMNH – 99.4.27.1/i.t.34, Tierra del Fuego, Patagonia
BMNH – 1033.6.15.1, ONA, Tierra del Fuego, Patagonia
BMNH – 1938.8.10.1, ONA, Viamonte, Tierra del Fuego, Patagonia
BMNH – Am.80-4, Navarino Is., Tierra del Fuego, Patagonia
BMNH – FC 1023/12.70, Sandy Point, Magellan Strait, Patagonia
BMNH – FC 1023.4/12.70, Magellan Strait, Patagonia
BMNH – FC 1021.1/12.70, Patagonia
BMNH – 1915.5.5.2, Patagonia
BMNH – 1915.5.5.3, Patagonia
BMNH – 1931.3.5.1, Patagonia
BMNH – Am.80-965, Patagonia
BMNH – 89.1.27.1/Id 30, Patagonia
BMNH – Am.80-964, Patagonia
BMNH – FC 1024.1/12.70, PUELCHE, Patagonia
BMNH – FC 1024.2/12.70, PUELCHE, Patagonia
BMNH – FC 1024.34/12.70, TEHUELCHE, Chubut, Patagonia
BMNH – 99.4.27.5, Chubut, Patagonia
BMNH – FC 1024.12/12.70, Pto.Madrin, Chubut river, Patagonia

Afalou (N=36)

IPH 2	IPH 15	IPH 23	IPH 31	IPH 40
IPH 3	IPH 14	IPH 24	IPH 32	IPH 43
IPH 51	IPH 15	IPH 25	IPH 33	IPH 44
IPH 71	IPH 17	IPH 27	IPH 34	IPH 46
IPH 91	IPH 18	IPH 28	IPH 36	IPH 47
IPH 10	IPH 20	IPH 29	IPH 37	IPH 48
IPH 11	IPH 22	IPH 30	IPH 38	IPH 49
IPH 12				

Taforalt (N=22)

IPH Tf I	IPH Tf IX	IPH Tf XV C2	IPH Tf XVII C1	IPH Tf XXIV C1
IPH Tf Ib	IPH Tf XI C1	IPH Tf XV C4	IPH Tf XVII C2	IPH Tf XXV C4
IPH R VIII	IPH Tf XII C1	IPH Tf XV C5	IPH Tf XX C1	IPH Tf XXVII C1
IPH Tf VIII C	IPH Tf XII C4	IPH Tf XVI C3	IPH Tf XX C2	IPH Tf XXV C3
IPH Tf X C4I	IPH Tf XIV			

Table 4.1 (cont.)

Natufian (N=54)

TASSM Ra H2, TA, Rakefet, Israel (frags)
TASSM Fa H8, TA, Nahal Oren, Israel
TASSM Fa H10, TA, Nahal Oren, Israel (vault)
TASSM Fa H13, TA, Nahal Oren, Israel (vault)
TASSM Fa H14, TA, Nahal Oren, Israel
TASSM Fa H16, TA, Nahal Oren, Israel
TASSM Fa H20, TA, Nahal Oren, Israel
TASSM Fa H23, TA, Nahal Oren, Israel
TASSM Fa H24, TA, Nahal Oren, Israel
TASSM Fa H34, TA, Nahal Oren, Israel (frags)
TASSM Fa H42, TA, Nahal Oren, Israel
TASSM Fa H43, TA, Nahal Oren, Israel (frags)
TASSM Fa H48, TA, Nahal Oren, Israel
TASSM EY H1, TA, Eynan, Israel (frags)
TASSM EY H6, TA, Eynan, Israel (frags)
TASSM EY H10, TA, Eynan, Israel (frags)
TASSM EY H16, TA, Eynan, Israel (frags)
TASSM EY H18, TA, Eynan, Israel (frags)
TASSM EY H22, TA, Eynan, Israel (frags)
TASSM EY H23, TA, Eynan, Israel (frags)
TASSM EY H32, TA, Eynan, Israel (frags)
TASSM EY H33, TA, Eynan, Israel (frags)
TASSM EY H34, TA, Eynan, Israel (vault, frags)
TASSM EY H34, TA, Eynan, Israel (vault, frags)
TASSM EY H37, TA, Eynan, Israel
TASSM EY H52, TA, Eynan, Israel
TASSM EY H60, TA, Eynan, Israel (vault)
TASSM EY H61, TA, Eynan, Israel (frags)
TASSM EY H66, TA, Eynan, Israel (vault)
TASSM EY H67, TA, Eynan, Israel (frags)
TASSM EY H69, TA, Eynan, Israel (vault, frags)
TASSM EY H70, TA, Eynan, Israel (frags)
TASSM EY H71, TA, Eynan, Israel (frags)
TASSM EY H80, TA, Eynan, Israel (vault, frags)
TASSM EY H87, TA, Eynan, Israel (frags)
TASSM EY H91, TA, Eynan, Israel (frags)
TASSM EY H93, TA, Eynan, Israel (frags)
TASSM EY H98, TA, Eynan, Israel (vault)
TASSM EY H101, TA, Eynan, Israel (frags)
TASSM EY H105, TA, Eynan, Israel (frags)
TASSM HA H2, TA, Hayonim, Israel (frags)
TASSM HA H4, TA, Hayonim, Israel
TASSM HA H8, TA, Hayonim, Israel (frags)
TASSM HA H9, TA, Hayonim, Israel (vault, frags)
TASSM HA H11, TA, Hayonim, Israel (frags)
TASSM HA H17, TA, Hayonim, Israel (frags)
TASSM HA H19, TA, Hayonim, Israel (frags)
TASSM HA H25, TA, Hayonim, Israel (frags)
TASSM HA H26, TA, Hayonim, Israel (frags)
TASSM HA H27, TA, Hayonim, Israel
TASSM HA H32, TA, Hayonim, Israel (frags)
TASSM HA H39, TA, Hayonim, Israel (frags)
TASSM EEA H2, TA, Erq el Ahmar, Israel
TASSM EEA H3, TA, Erq el Ahmar, Israel (frags)

DC: Duckworth Collection, Cambridge; BMNH: Natural History Museum, London; IPH: Institut de Paléontologie Humain, Paris; TASSM: Tel Aviv Sackler School of Medicine.

Statistical analysis

The regional distribution of each of the traits measured metrically was tested through analyses of variance, using the Cochrans and Bartlett's Box tests for homogeneity of variance, and the Scheffe Ranges test to find which groups were significantly different from others. All variables were tested for normality using visual observations of distribution plots and the Kolmogorov-Smirnov non-parametric test. The regional distribution of each of the discontinuous traits (present/absent, grade scores and categories) was analysed through Chi-square tests.

Throughout the analyses, a result was considered significant if its associated probability was below 5%. Although many researchers prefer to use probability levels below 1%, it was considered that disregarding results of regionality that showed $p<0.05$ could be criticised on the grounds of intentionally attempting to reduce the number of significant regional features, and therefore, biasing the results against the Multiregional Model. Three significant levels are used throughout the work: $p<0.05$, $p<0.01$, and $p<0.001$. It should be noted that, unlike the results of the analyses of variance in which the associated Ranges tests identify individually significant groups, the Chi-square statistic only expresses overall distribution significance. Therefore, considering whether or not a non-metrical feature followed the pattern proposed by the Multiregional Model, i.e. was most frequent in the region proposed, was done on the basis of relative incidence and not statistical significance.

East Asian regional continuity traits

EA1: Sagittal keeling

The degree of sagittal keeling (SK) was measured as three scores (App. I, Figure AI.1), representing absent (SK1), moderate (SK2) and pronounced keeling of the vault (SK3). The distribution of the three scores of sagittal keeling among the five recent regional samples was statistically significant ($x^2=70.6$, $df=8$, $p<0.001$). Figure 4.2a shows that keeling of the vault is not a rare condition and was present in approximately 40% of 274 crania examined. However, pronounced keeling is much less common (12%) and regionally localised. No cases of pronounced sagittal keeling were observed in the European, sub-Saharan African or Southeast Asian samples. Among the East Asian crania, there was a single case of pronounced sagittal keeling. Among the Australian crania examined, not only are there cases of pronounced sagittal keeling, but this sample is unique in that the most

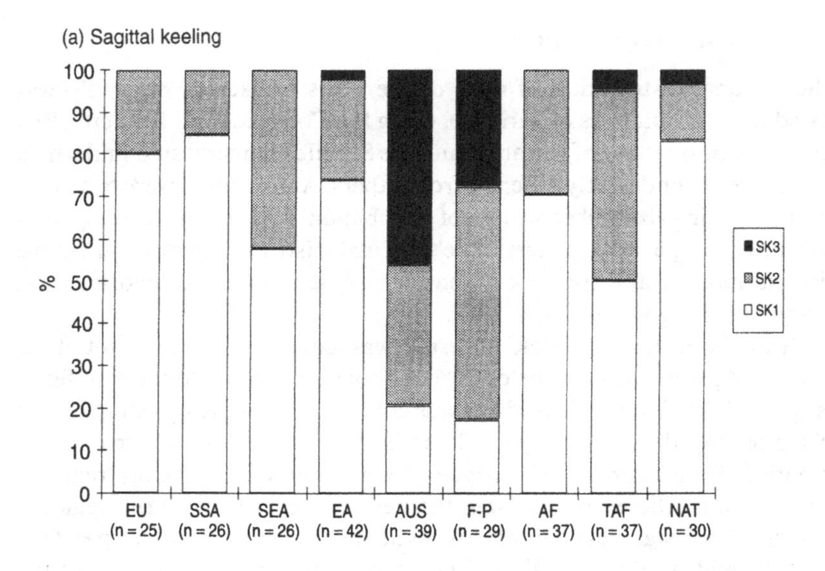

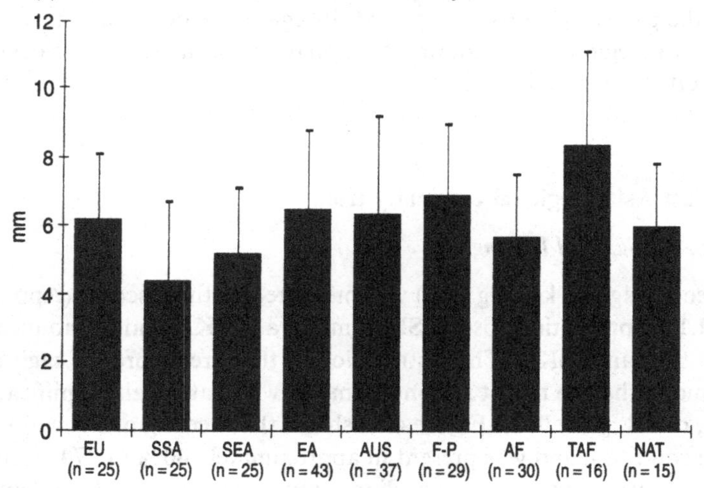

Figure 4.2 Incidence of the East Asian 'regional continuity traits' in the nine recent cranial samples.
(a) Sagittal keeling, with high incidence in Australia and the Fueguian-Patagonian sample.
(b) Course of the naso-frontal and fronto-maxillary sutures, showing that the most 'horizontal' pattern is observed in sub-Saharan Africa.

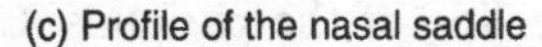

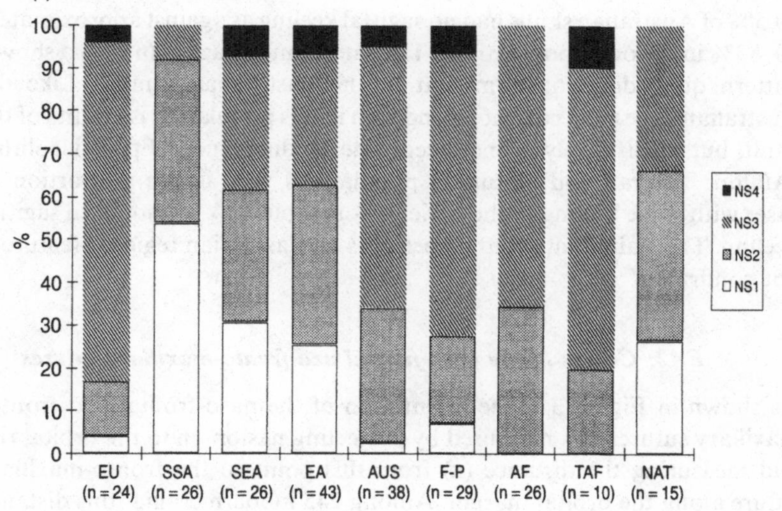

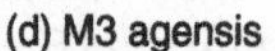

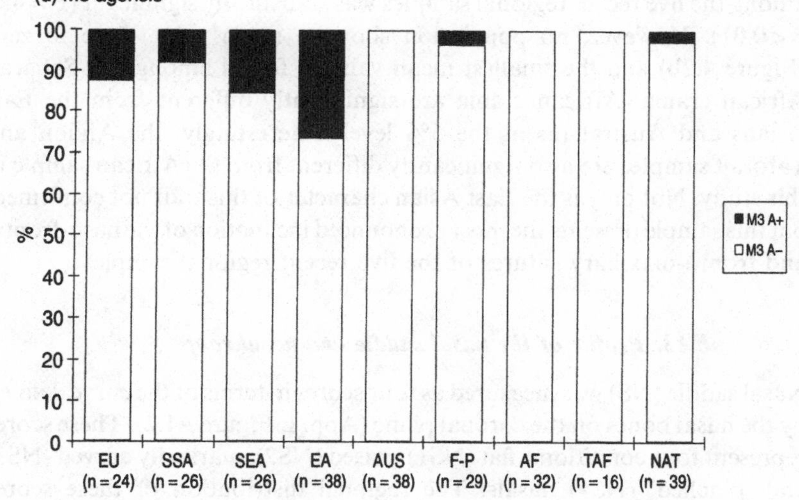

Figure 4.2 (*cont.*)
(c) Profile of the nasal saddle, showing that the largest number of flat nasal bones is found in sub-Saharan Africa.
(d) M3 agenesis.
EU: Europeans; SSA: sub-Saharan Africans; SEA: Southeast Asians; EA: Eastern Asians; AUS: Australians; F-P: Fueguian-Patagonians; AF: Afalou; TAF: Taforalt: NAT: Natufian

common (46.2%) condition is pronounced keeling of the parietals. Only 20.5% of Australian skulls had no sagittal keeling as against approximately 60–85% in the other populations. The Fueguian-Patagonian crania show a pattern quite different from that of the East Asian sample. Like the Australians, the most common condition in this population is keeling of the vault, but this is rarely pronounced. The Mediterranean Epi-Palaeolithic (Afalou, Taforalt and Natufian) populations vary in the proportion of cases with some keeling of the parietal bones, but lack pronounced sagittal keeling. The claim that sagittal keeling is an East Asian regional feature is not confirmed.

EA2: Course of the naso-frontal and fronto-maxillary sutures

As shown in Figure 3.2, the orientation of the naso-frontal and fronto-maxillary sutures was measured by projecting nasion on to the orbital rim and measuring the distance (x) from this point to the fronto-maxillary suture along the orbital margin. Among 245 modern crania, this distance (x) varied between 12 mm and nil, the latter representing a horizontal course of both the fronto-maxillary and naso-frontal sutures. The variation among the five recent regional samples was statistically significant ($F=4.6$, $p<0.01$). However, no population shows a mean value close to zero (Figure 4.2b) and the smallest mean value is found among sub-Saharan African crania. African crania are significantly different from the East Asians and Australians at the 5% level. Interestingly, the Afalou and Taforalt samples are also significantly different from the African sample in this study. Not only is the East Asian character of this trait not confirmed, but this sample presents the most pronounced inclination of the naso-frontal and fronto-maxillary sutures of the five recent regional samples.

EA3: Profile of the nasal saddle and nasal roof

Nasal saddle (NS) was measured as four scores in terms of the curve defined by the nasal bones on the coronal plane (App. I, Figure AI.2). These scores represent four conditions: flat (NS1), raised (NS2), markedly curved (NS3) and 'pinched' (NS4) nasals. The regional distribution of these scores among the five recent populations was significant at the 0.1% level ($x^2=49.2$, $df=12$, $p<0.001$). Pronounced curvature of the nasal saddle (NS3) is the most common condition and is observed in approximately 48% of the 235 modern crania examined (Figure 4.2c). The least common is 'pinched' nasals, found in only 3.8% of skulls. Flat nasals are relatively rare, observed only in 17.4% of the skulls examined. Of these, 34% were sub-Saharan African crania, 26.8% East Asian, 19.5% Southeast Asian, 9.7% Natufian, 4.9% Fueguian, 2.4% European and 2.4% Australian. The East Asian character of a 'flat nasal saddle and nasal roof' is not confirmed, since flat nasals are neither the most frequent East Asian condition nor is the incidence in the East Asian sample the highest among modern populations.

EA4: M3 agenesis

The regional incidence of congenital absence of the maxillary third molar among the five recent populations was marginally significant ($x^2=12.8$, $df=4$, $p<0.05$). This feature is most common in East Asian crania as proposed by the Multiregional Model (Figure 4.2d). The condition is however very rare, observed in only 7.8% of 268 crania examined and was completely absent in the Australian, Afalou and Taforalt samples. The East Asian character of this trait is slimly supported by these results. However, it should be noted that as a Mongoloid derived group, the Fueguian-Patagonian sample with only one case of M3 agenesis, departs from the East Asian condition. This highlights the variability in this feature, which according to the Multiregional Model remained stable for nearly a million years.

EA5: Posterior dental reduction

This was calculated as the difference in size between the first and third molars, measured through crown area, buccolingual and mesiodistal values. None of the three measures shows a significantly different regional distribution (area M1–M3: $F=1.3$, $p>0.05$; blM1–blM3: $F=1.8$, $p>0.05$; mdM1–mdM3: $F=0.95$, $p>0.05$). Of all modern samples examined, the Natufians, Europeans and Southeast Asians show the smallest third molars in relation to M1. The populations that consistently show the least difference in size between M1 and M3 are the sub-Saharan Africans, followed by the Taforalt. In buccolingual dimensions, the sub-Saharan African, Afalou and Taforalt samples have, on average, a larger M3 than M1. The reduction in M2 and M3 size observed in the Natufian sample is one of the distinguishing features between the North African and Middle Eastern Epi-Palaeolithic populations. The East Asian character of this feature is not confirmed.

East Asian traits EA6 to EA10 are considered below, since these are the same features as Australian traits AUS16 to AUS20, although in some cases representing the opposite expression.

Australian regional continuity traits

AUS1: Pronounced development of the supraorbital ridges

The development of supraorbital ridges was measured as five grades (App. I, Figure AI.4), ranging from totally flat glabellar and supraorbital areas (ST1) to complete torus formation (ST5). Three intermediate grades are recognised: presence of supraorbital ridges separated by a flat glabellar region (ST2), supraorbital ridges joined medially at glabella (ST3), and very pronounced supraorbital ridge and glabellar development without the formation of a torus (ST4), i.e. separated laterally from the zygomatic trigone. The distribution of supraorbital ridge development in the five recent regional samples was statistically significant ($x^2 = 89.1$, $df = 16$, $p < 0.001$). The Australian sample shows the highest incidence of ST4 (43.6%) among the five regional samples and the only four cases of supraorbital tori observed (Figure 4.3a). However, an even higher frequency of pronounced ridges (ST4: 58.6%) is found in the Fueguian-Patagonian group, while this feature is also very common (ST4: 37.8%) in Afalou crania. Therefore, although the most common condition in Australian crania is indeed very pronounced supraorbital ridges as proposed by the Multiregional Model, it occurs equally or more frequently in other modern groups. The uniquely Australian high frequency of this trait is not confirmed.

AUS2: Lambdoid depression

A flattening at around obelion is relatively common – observed in 22% of 270 modern skulls. Although the frequency of a lambdoid depression among the different populations varies (Figure 4.3b), the variation between the five regional samples is not statistically significant ($x^2 = 6.7$, $df = 4$, $p > 0.05$). The sub-Saharan Africans are the population with the largest number of cases of lambdoid depression (34.6%), followed by the Australians (33.3%) and the East Asians (31.7%). The Australian character of this trait is not confirmed.

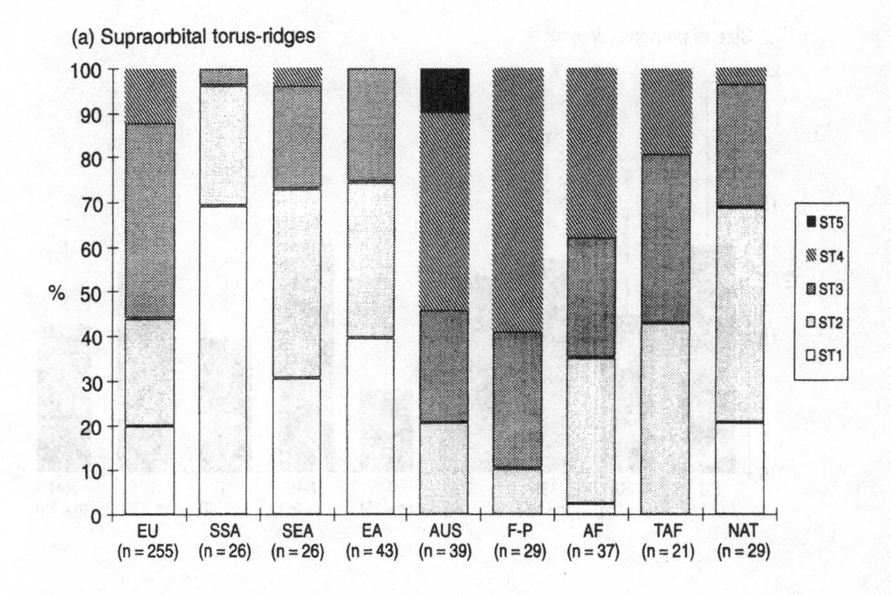

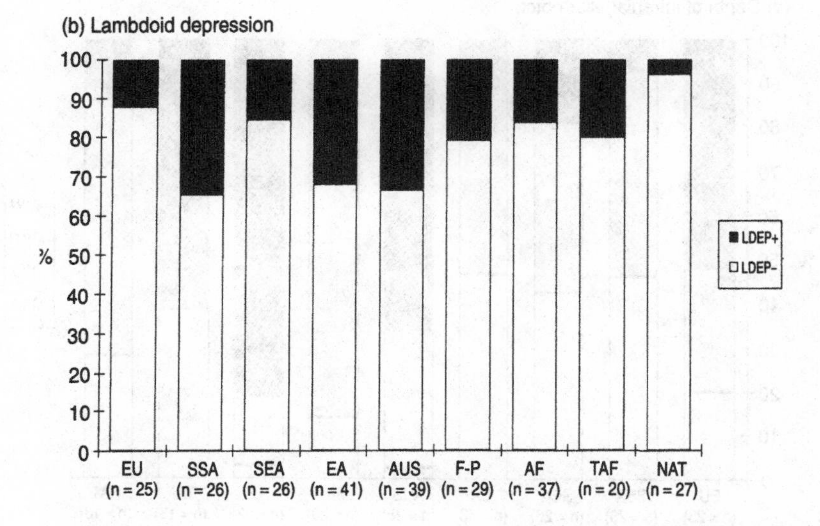

Figure 4.3 Incidence of the Australian 'regional continuity traits' in the nine recent samples.
(a) Supraorbital ridges-torus, showing that the population in which pronounced supraorbital ridges are most frequent is the Fueguian-Patagonian, followed by the Australian; also shows that supraorbital tori are extremely rare in all modern populations.
(b) Lambdoid depression, with similar small incidence in all populations.

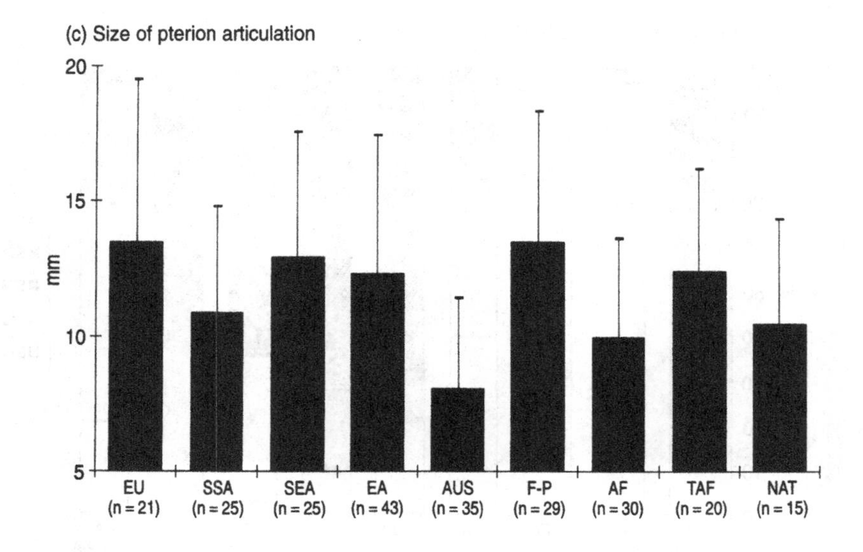

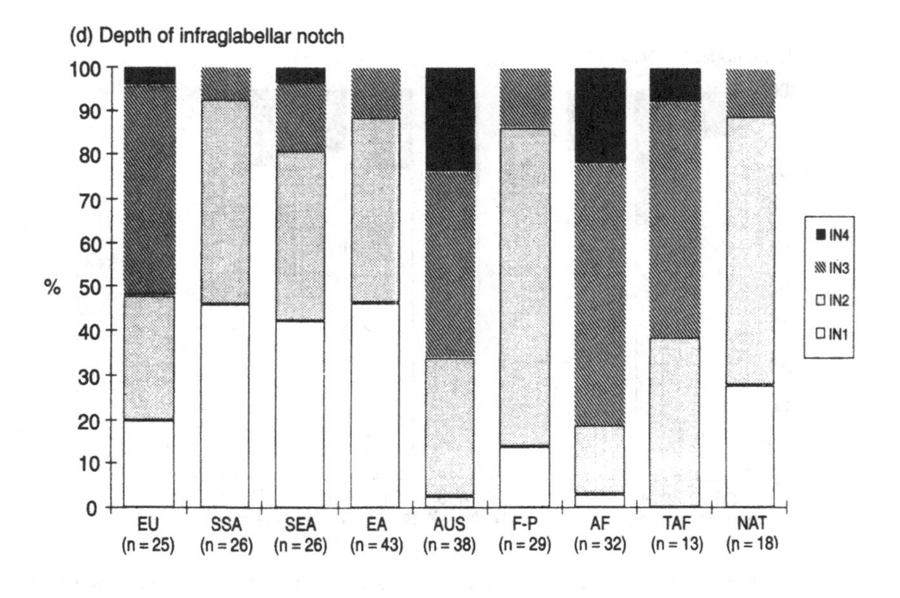

Fig. 4.3 (*cont.*)
(c) Size of the pterion articulation, confirming that Australians have a comparatively small articulation at pterion.
(d) Depth of the infraglabellar notch, showing that both the Australians and the crania from Afalou have a comparatively deep profile at nasion.

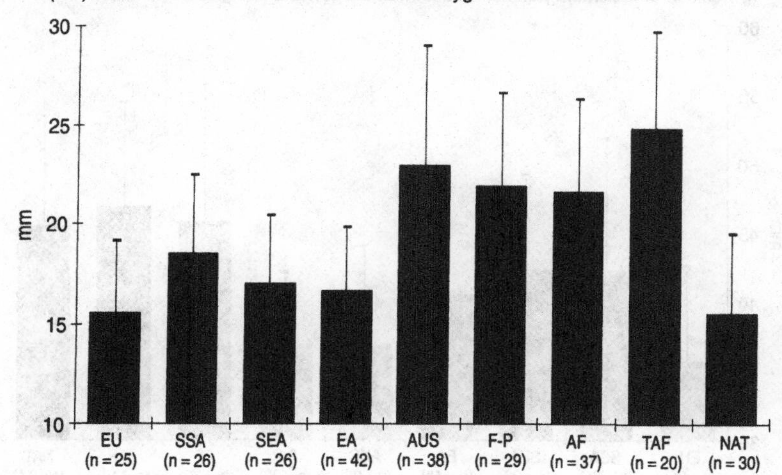

(e-2) Distance minimum frontal breadth to the superior orbital margin

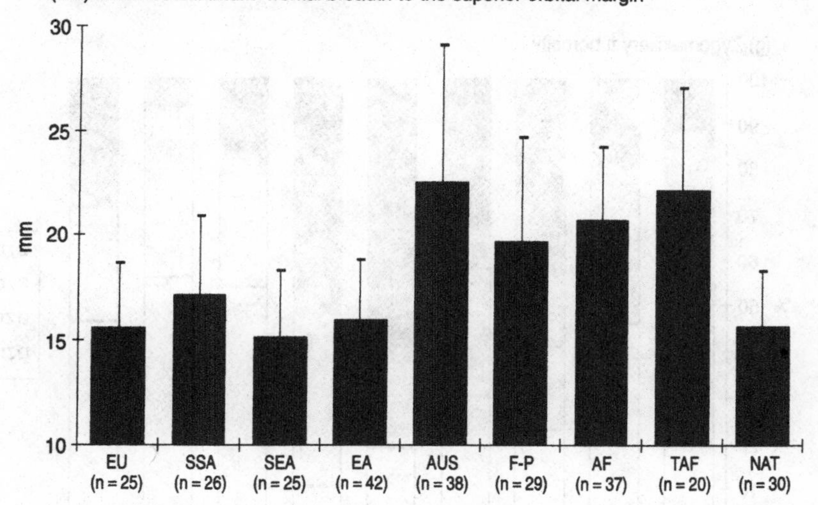

Fig. 4.3 (*cont.*)
(e) Relative position of minimum frontal breadth:
e - 1: distance between minimum frontal breadth and the zygomatico-frontal suture, showing that the greatest values are found in the Taforalt crania, although the Australian, Fueguian-Patagonian and Afalou crania also have a relative backward position of minimum frontal breadth;
e - 2: distance between minimum frontal breadth and the superior orbital margin, showing that Australian and Taforalt crania, and to a lesser degree the Afalou and Fueguian-Patagonian remains, have on average greater values than other recent populations.

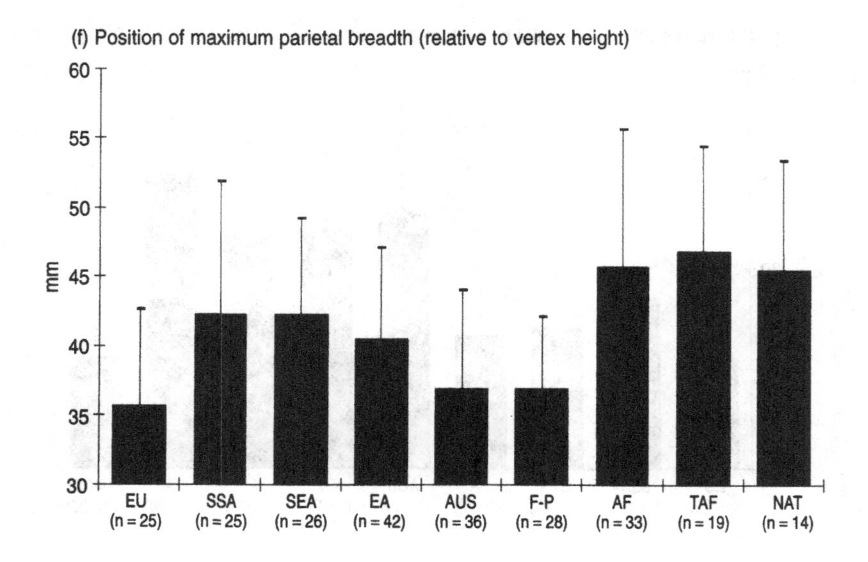

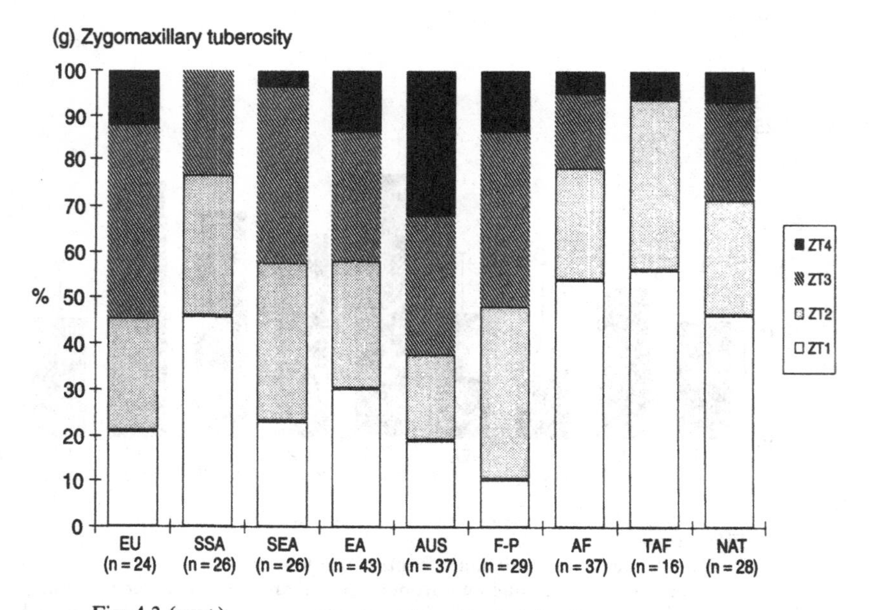

Fig. 4.3 (*cont.*)
(f) Relative position of maximum parietal breadth (in terms of vertex height), showing that the lowest mean values are observed in European crania, followed by the Australian and Fueguian-Patagonian remains.
(g) Development of a zygomaxillary tuberosity, showing that the Australians have the largest number of cases of pronounced tuberosities.

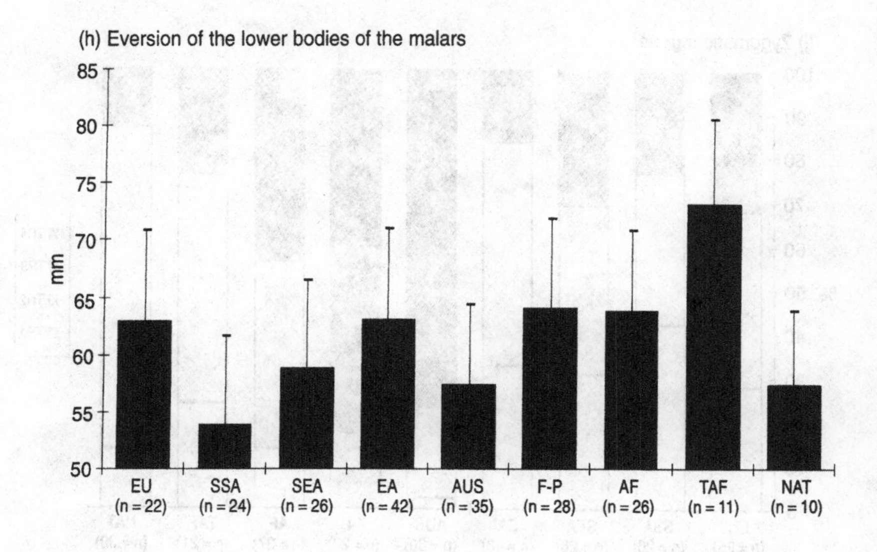

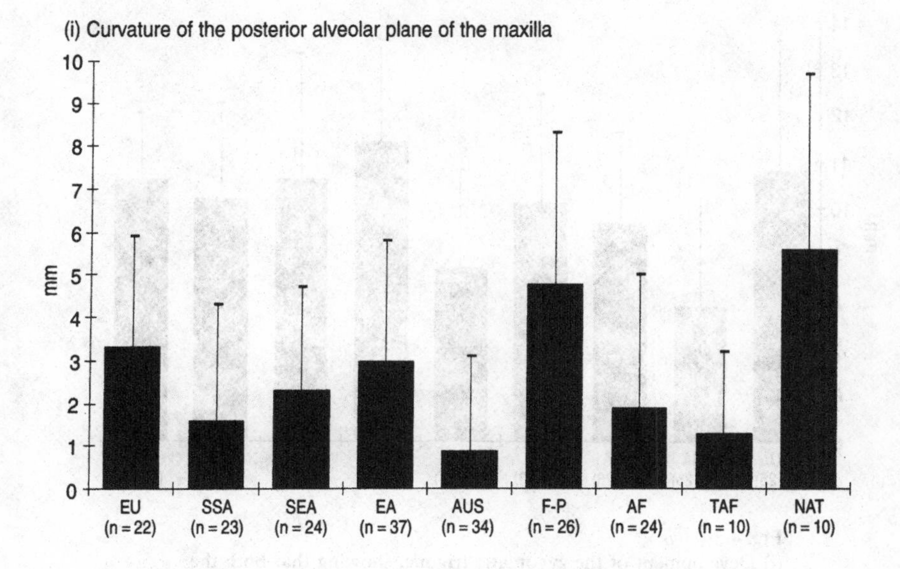

Fig. 4.3 (*cont.*)
(h) Eversion of the lower border of the malars, showing that the Taforalt crania have the most everted zygomatic bones.
(i) Curvature of the posterior alveolar plane of the maxilla, showing that the greatest values are observed in Natufian and Fueguian-Patagonian crania, while the Australians have the most levelled maxillary plane of all populations studied.

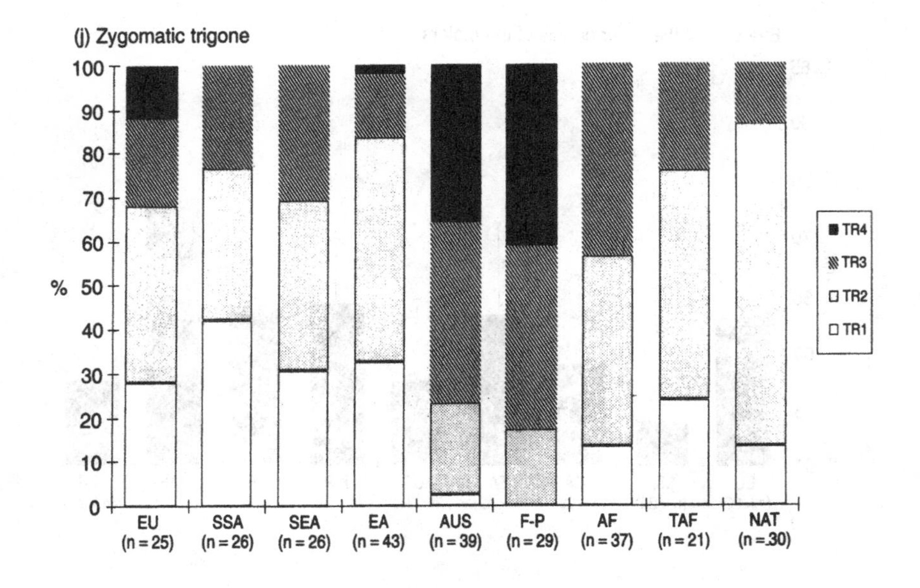

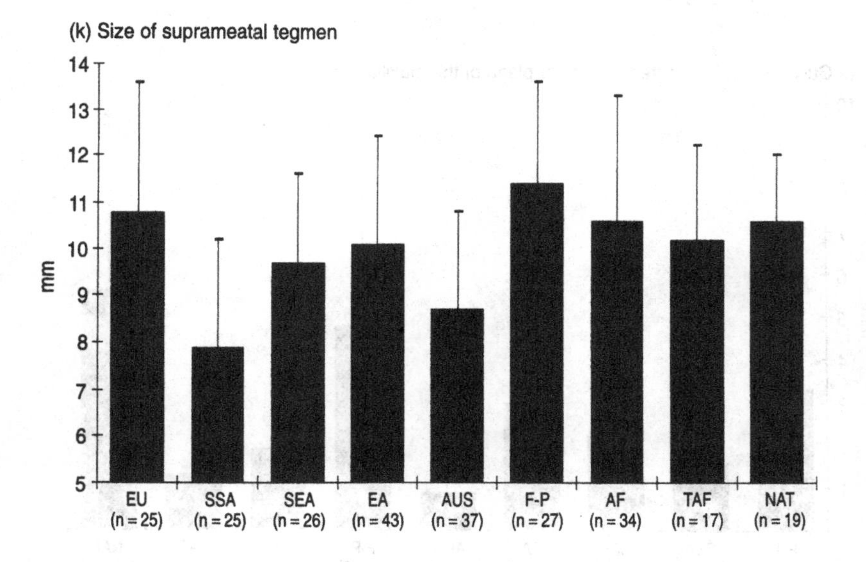

Fig. 4.3 (*cont.*)
(j) Development of the zygomatic trigone, showing that both the Fueguian-Patagonian and Australian remains have large numbers of cases of very pronounced development of the trigones.
(k) Size of the suprameatal tegmen, showing that the smallest values are observed in sub-Saharan and Australian crania, while a large suprameatal tegmen is most commonly observed in the Fueguian-Patagonian, European and Mediterranean Epi-Palaeolithic crania.

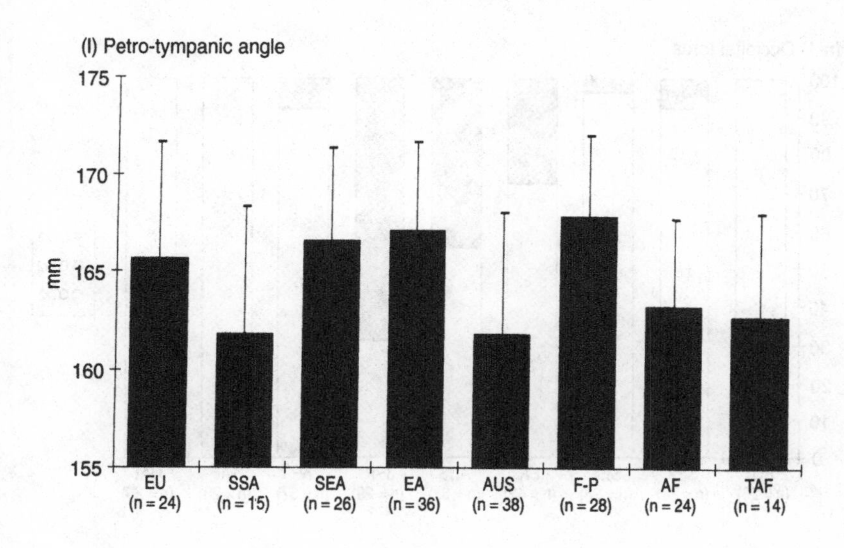

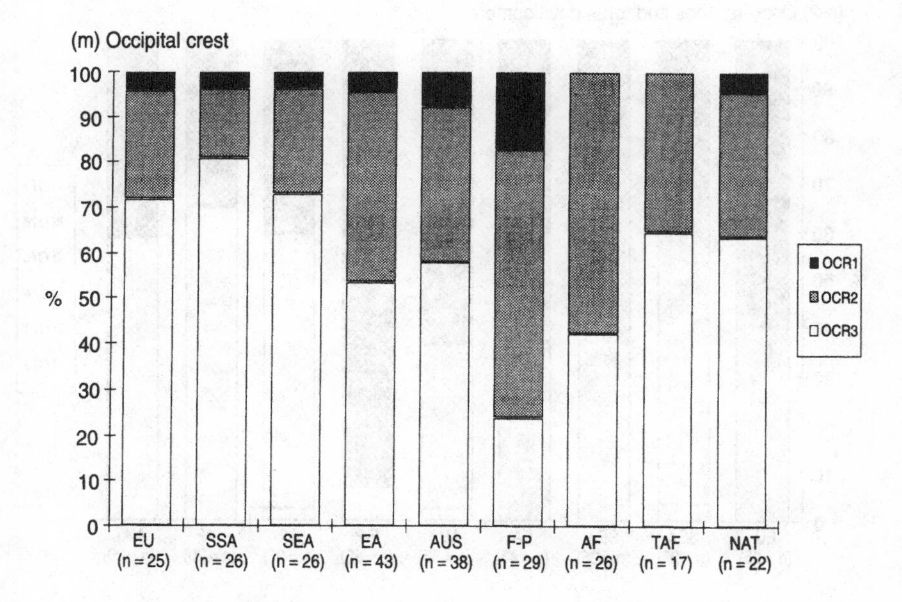

Fig. 4.3 (*cont.*)
(l) Petro-tympanic angle, showing that the lowest values are observed in sub-Saharan African and Australian crania.
(m) Development of an occipital crest, showing that a complete occipital crest is very rare in all populations studied, being observed at highest (but still absolutely low) frequency in Fueguian-Patagonian crania.

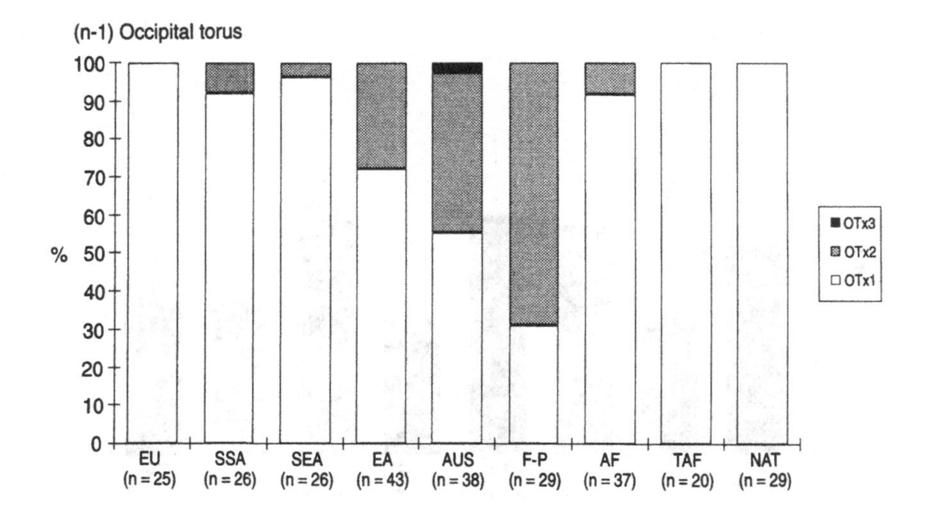

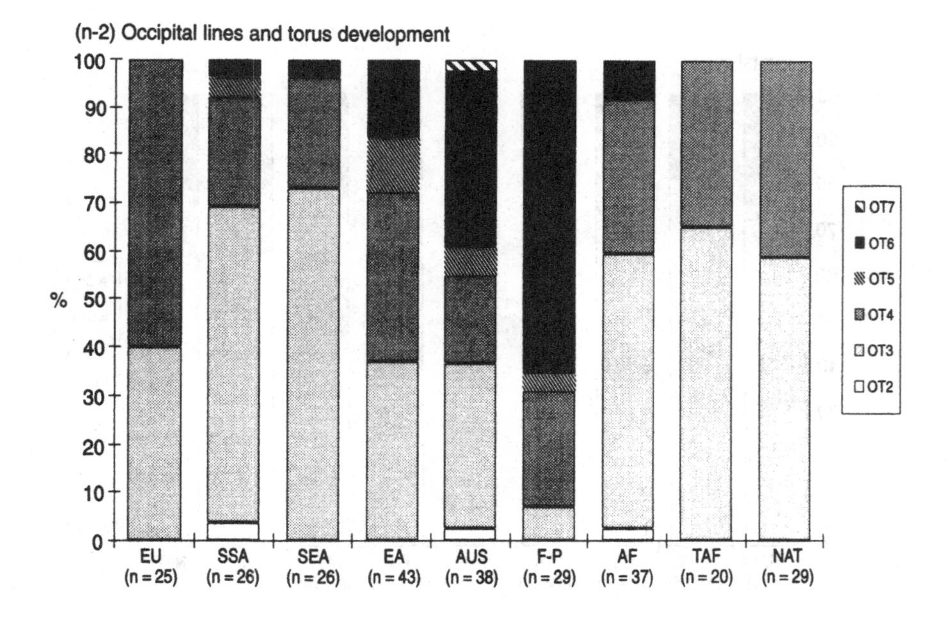

Fig. 4.3 (*cont.*)
(n) Development of an occipital torus:
n - 1: occipital torus, showing overall low incidence of this feature, mainly observed in Fueguian-Patagonian and Australian crania;
n - 2: nuchal lines and torus development.

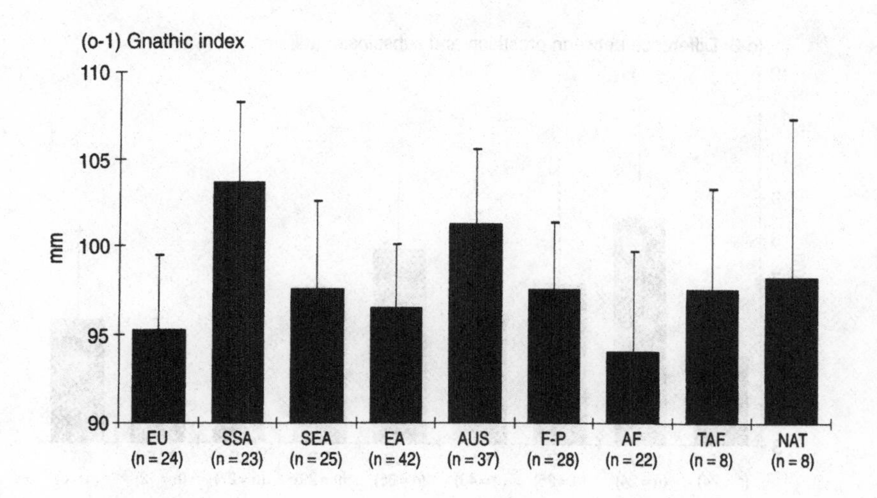

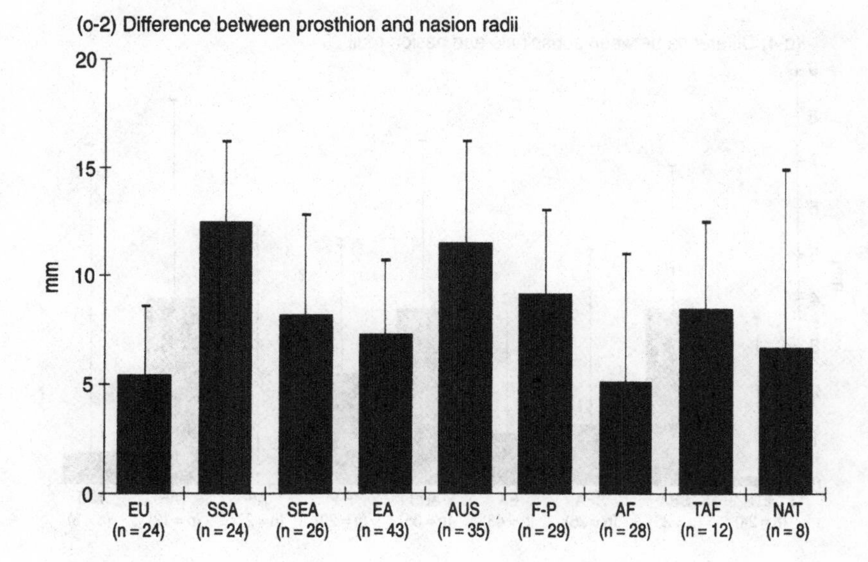

Fig. 4.3 (*cont.*)
(o) Facial prognathism:
o - 1: gnathic index, showing maximum total facial prognathism in sub-Saharan African crania;
o - 2: difference between prosthion and nasion radii, showing similar result as the Gnathic Index:

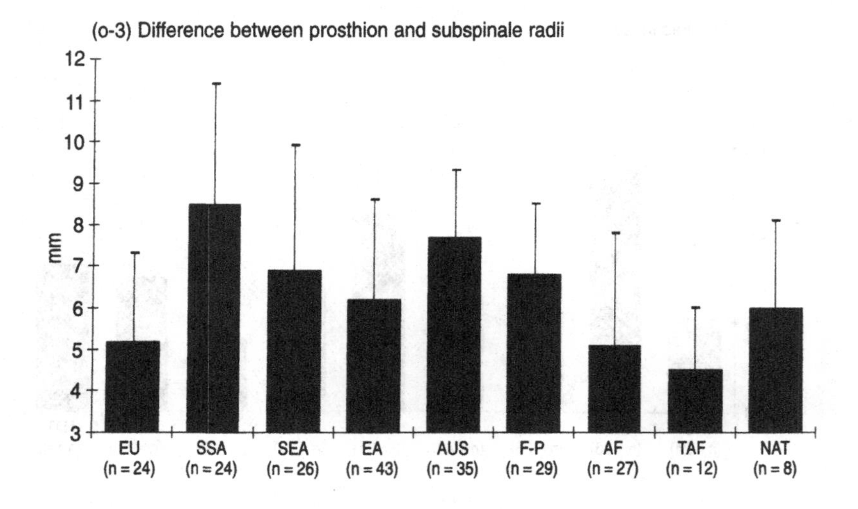

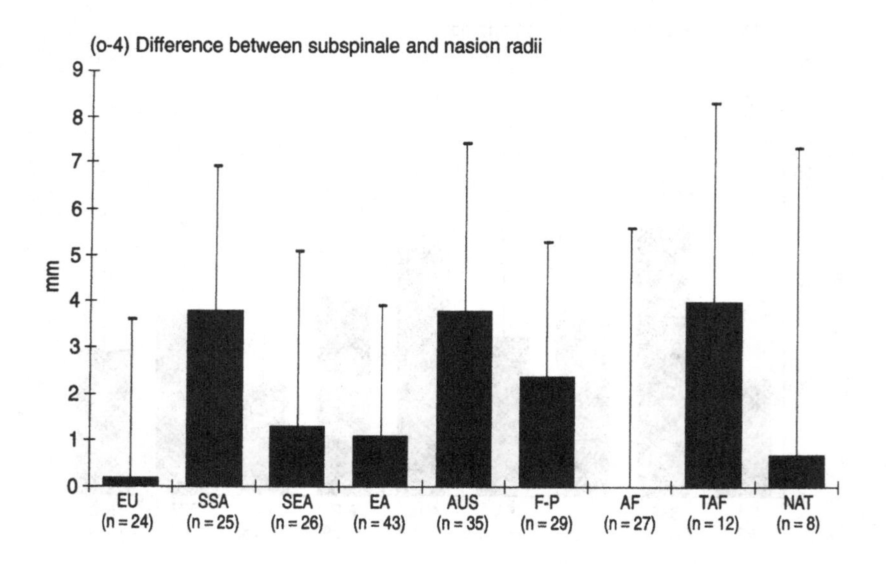

Fig. 4.3 (*cont.*)
o - 3: difference between prosthion and subspinale radii, showing that the sub-Saharan African crania have the greatest alveolar prognathism;
o - 4: difference between subspinale and nasion radii, showing that the Taforalt, Australian and sub-Saharan African crania have the greatest values of mid-facial prognathism, while the Afalou and European crania have a very flat mid-facial morphology.

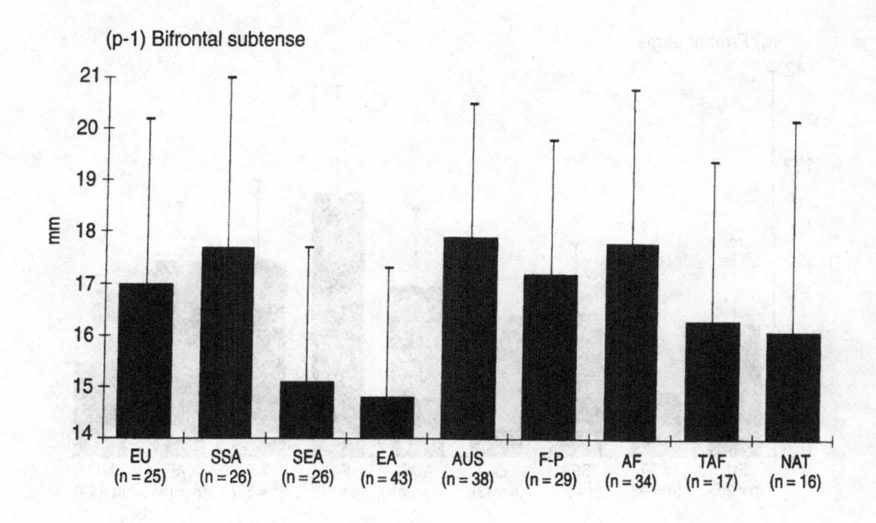

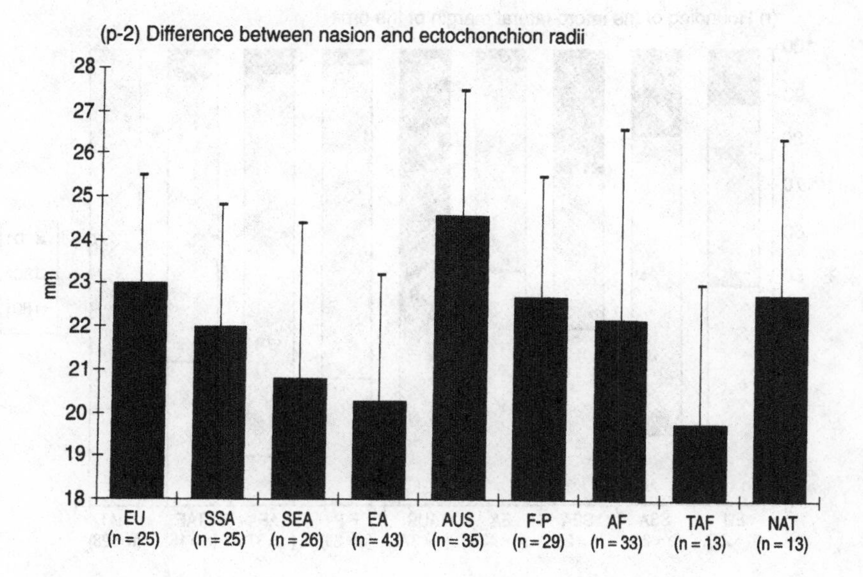

Fig. 4.3 (*cont.*)
(p) Lateral facial flatness:
p - 1: bifrontal subtense, showing that the least difference between nasion and frontomalare in the coronal plane, i.e. greatest flatness, is observed in the east Asian and Southeast Asian crania;
p - 2: difference between nasion and ectochonchion radii, showing that in terms of upper facial flatness (rather than sub-orbital), the lowest values are observed in the Taforalt remains.

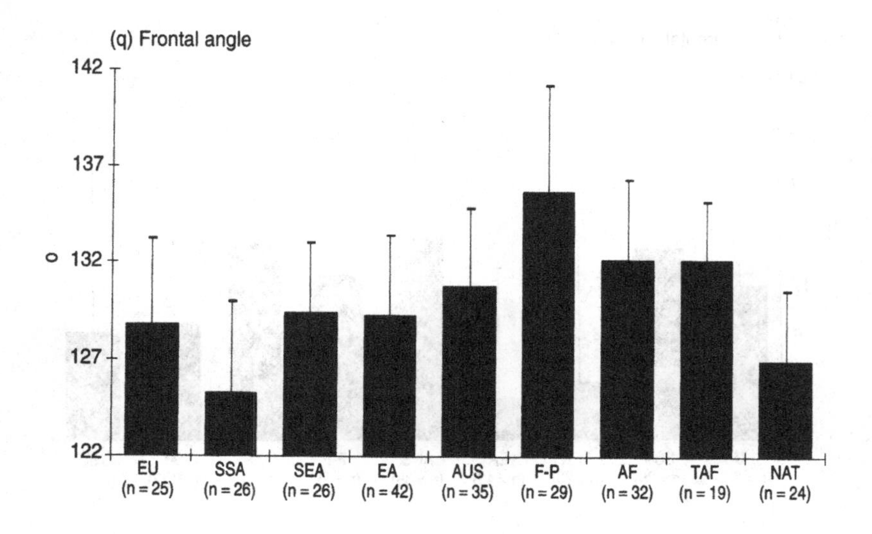

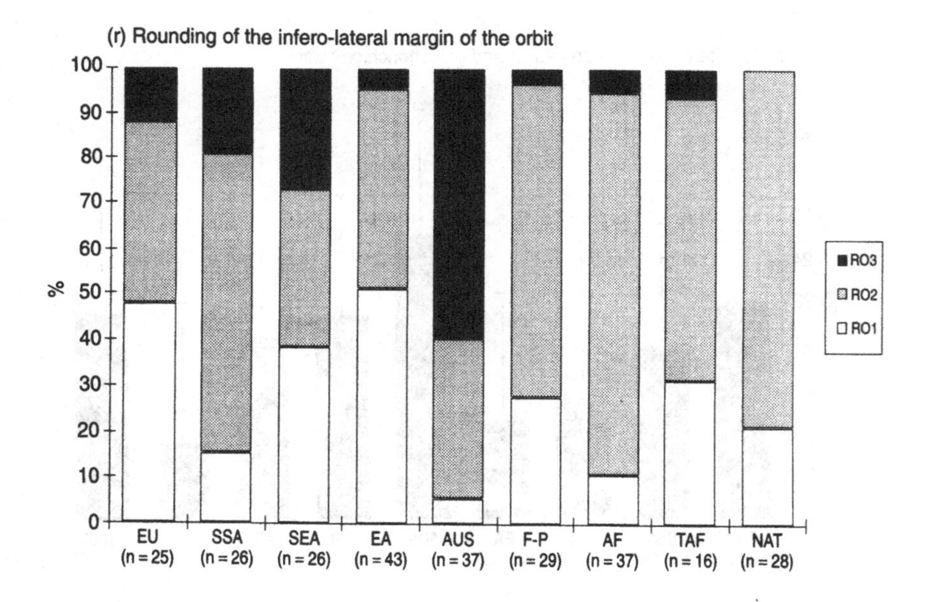

Fig. 4.3 (*cont.*)
(q) Frontal angle, showing that the flattest frontals are found in the Fueguian-Patagonian crania, and the most rounded ones in the sub-Saharan African sample.
(r) Rounding of the infero-lateral margin of the orbits, showing highest incidence in Australia.
(See Figure 4.2 for explanation of abbreviations.)

AUS3: Size of pterion articulation

Table 3.4 shows that, contrary to that claimed by the Multiregional Model, a fronto-temporal articulation at pterion does not occur most frequently in Australian crania. However, an examination of the size of the pterion articulation among the five regional samples shows that the Australians do have the shortest pterion articulation (Figure 4.3c). The regional variation observed is significant ($F = 6.7$, $p < 0.001$), and the Australians are significantly different from the Europeans, the Southeast Asians and the East Asians at the 1% level. The Fueguian-Patagonian and Epi-Palaeolithic samples show mean values similar to those of the other recent populations. The Australian character of this trait is confirmed.

AUS4: Deep and narrow infraglabellar notch

The profile of the infraglabellar notch (IN), i.e. the curve formed between glabella and the nasal saddle, was measured as four grades (App. I, Figure AI.3), ranging from completely flat (IN1) to very deep and narrow (IN4), with two intermediate grades. The distribution of this feature among the five regional samples was statistically significant ($x^2 = 56.7$, $df = 12$, $p < 0.001$). Cases of a deep and narrow infraglabellar notch are very rare, observed only in 7.6% of 250 crania (Figure 4.3d). They occur almost exclusively in Australian (23.7%) and Afalou (21.9%) crania. However, it should be noted that a deep and narrow infraglabellar notch is not the most common Australian condition. Like the European, Afalou and Taforalt samples, the most common profile of the infraglabellar notch observed in Australian crania is a pronounced broad curve (IN3). The Australian character of a 'deep and narrow infraglabellar notch' is not confirmed, for this feature is neither the most frequent Australian condition nor does the Australian sample show a uniquely high incidence of this trait among modern populations.

AUS5: Position of minimum frontal breadth

The position of minimum frontal breadth (WFB) relative to the face was measured as the distance between the point of minimum frontal breadth on the left temporal line and both the posterior end of the zygomatico-frontal suture (WFBZFS) and the orbital rim (WFBORB) (Figure 3.3). The range of values in all populations examined shows that there is considerable variation in both of these measures in modern humans. WFBZFS varies from 6 to 42 mm and WFBORB from 7 to 45 mm. There is also marked variation among the five regional samples (WFBZFS: $F = 16.3$, $p < 0.001$;

WFBORB: $F = 17.9$, $p < 0.001$), with the Australians showing the overall largest mean distances (Figures 4.3e-1, e-2). In the case of the distance between minimum frontal breadth and the zygomatico-frontal suture, the Australian sample is significantly different from the European, East and Southeast Asian populations at the 0.1% level, and from the African group at the 1% level. In the case of the distance between minimum frontal breadth and the orbital margin, the Australian sample is significantly different from the other four regional samples at the 0.1% level. These results would seem to corroborate the Australian character of this feature. However, an inclusion of the other modern samples in the comparison changes this picture. The relative position of minimum frontal breadth in the Fueguian-Patagonian, Afalou and Taforalt samples is very similar to the Australian group. Furthermore, the crania from Taforalt show a larger mean value of WFBZFS than the Australians. The uniquely Australian character of this trait is not confirmed.

AUS6: Position of maximum parietal breadth

The position of maximum parietal breadth (XPB) was measured as the distance from the point of maximum parietal breadth to porion, and expressed as a percentage of cranial height (vertex radius). The variation between the five regional samples is significant ($F \leq 4.8$, $p < 0.01$). In these five groups, the relative height of the broadest point of the vault along the parietals ranged from 16.2% to 60% of total height. The lowest absolute value recorded is for Africans, but this group actually has, on average, the highest position of maximum parietal breadth of the five regional groups (42.4%). The lowest mean value is not found in the Australian, but in the European sample (Figure 4.3f), both of which are significantly different from Africans and Southeast Asians at the 5% level. The Mediterranean Epi-Palaeolithic populations of Afalou, Taforalt and Natufian have the greatest mean values for position of maximum parietal breadth, and resemble current sub-Saharan Africans. The Australian character of low position of maximum parietal breadth is not confirmed.

An interesting aspect of the above results is that generally Australian crania give the visual impression of having a lower maximum breadth than other recent populations. However, if the relationship between maximum parietal breadth and biauricular breadth (AUB) is examined, we find that the Australians have the least mean difference, significantly different from Southeast Asians (0.1%), Europeans (1%) and sub-Saharan Africans (1%). This indicates that, although the Europeans show the lowest position of maximum parietal breadth, the larger difference between the latter and biauricular breadth results in the skull acquiring its modern-looking shape of enlarged parietals in relation to the base. In the case of the Australians, the small difference between XPB and AUB may give the overall impression of a low maximum cranial breadth.

It is of interest to note that all five regions presented at least one case in which maximum cranial breadth was not placed on the parietals. In the European sample, two skulls show maximum breadth on the temporals, two on the supramastoid crests, and two in which maximum parietal breadth equals that between the supramastoid crests; one African and one Southeast Asian skull show maximum breadth on the supramastoid crests; among East Asians, three skulls show maximum breadth on the temporals, three on the supramastoid crests, and one in which maximum parietal breadth and breadth of the supramastoid crests were the same; and lastly, among Australians, four skulls show maximum breadth on the temporal bone, of which three were even wider at the supramastoid crests, and nine skulls in which maximum cranial breadth was placed on the supramastoid crests. However, the Australian condition does not relate to a cranial base wider than the vault, but to a combination of very developed supramastoid crests superimposed on very narrow skulls. This is evident when the mean value of breadth at the supramastoid crests of the 12 Australian crania in which it corresponds to maximum cranial breadth ($\bar{x} = 132.4$ mm) is compared with the mean maximum parietal breadth of the other four regions (European: $\bar{x} = 141.7$ mm; sub-Saharan Africa: $\bar{x} = 133.2$ mm; Southeast Asia: $\bar{x} = 138.5$ mm; East Asia: $\bar{x} = 139.0$ mm). Therefore, if those cases in which maximum cranial breadth is positioned at the supramastoid crests are not considered, we find that two Europeans, three East Asians and four Australians present maximum cranial breadth on the temporal bone, rather than on the parietals.

AUS7: Pronounced zygomaxillary tuberosity

The development of a zygomaxillary tuberosity (ZT), which in its most pronounced expression takes the form of a ridge running parallel to the inferior border of the malar, was scored as four grades (App. I, Figure AI.5). The scores range from a smooth malar surface (ZT1), to the presence of a smaller or greater tubercle (ZT2, ZT3), to the development of a ridge (ZT4). The variation in incidence of this feature in the five regional samples was only slightly significant ($x^2 = 23.1$, $df = 12$, $p < 0.05$). All groups except the Africans show cases of all four grades (Figure 4.3g), but they vary as to which grade is the most frequent. In the European and Southeast Asian samples, a pronounced tubercle is the most common expression (41.7%

Table 4.2. *Incidence of the ten categories of lower narial margins in the nine modern cranial samples studied. (A) Distribution of cases by region; (B) Distribution of categories by region in order of frequency*

(A)	EU (N=24)%	SSA (N=26)%	SEA (N=25)%	EA (N=42)%	AUS (N=37)%	F/P (N=29)%	AF (N=30)%	TAF (N=17)%	NAT (N=15)%
N1 – Sharp	25.0	11.5	16.0	28.6	–	3.4	26.7	29.4	13.3
N2 – Double, joined by the nasal spine	12.5	11.5	8.0	7.1	5.4	10.3	33.3	11.8	26.7
N3 – Internal, ill-def. fossa	37.5	19.2	36.0	38.1	10.8	48.3	40.0	47.1	33.3
N4 – Internal, well-def. fossa	8.3	–	12.0	2.4	–	–	–	–	–
N5 – Double	4.2	15.4	8.0	2.4	2.7	–	–	5.9	–
N6 – Double, int. fossa	–	11.5	4.0	–	18.9	3.4	–	5.9	–
N7 – External, max. fossa	–	–	–	2.4	8.1	3.4	–	–	6.7
N8 – External, ill–def. fossa	12.5	23.1	16.0	16.7	37.8	24.1	–	–	20.0
N9 – Ill–defined	–	3.8	–	2.4	16.2	6.9	–	–	–
N10 – No margin	–	3.8	–	–	–	–	–	–	–

(B)	Categories of narial margins							
EU	N3	N1	N2=	N8	N4	N5		
SSA	N8	N3	N5	N1=	N2=	N6	N9	N10
SEA	N3	N1=	N8	N4	N2=	N5	N6	
EA	N3	N1	N8	N2	N4=	N5=	N7=	N9
AUS	N8	N6	N9	N3	N7	N2	N5	
F/P	N3	N8	N2	N9	N1=	N6=	N7	
AF	N3	N2	N1					
TAF	N3	N1	N2	N5=	N6			
NAT	N3	N2	N8	N1	N7			

EU: Europe; SSA: sub-Saharan Africa; SEA: Southeast Asia; EA: East Asia; AUS: Australia; F/P: Fueguian-Patagonian; AF: Afalou; TAF: Taforalt; NAT: Natufian.

and 38.5% respectively), while for both the African and East Asian samples, a smooth malar is the most common condition (46.2% and 30.2% respectively). Like the sub-Saharan Africans, the Mediterranean Epi-Palaeolithic groups (i.e. Afalou, Taforalt and Natufian) show a very high percentage of cases of completely smooth malars. The Australians differ from all other groups in that the most frequent expression is a complete zygomaxillary ridge (32.4%). The Australian character of a comparatively high frequency of this feature is confirmed.

AUS8: Eversion of the lower border of the malars

The eversion of the lower border of the malars was measured as the relative flare of the malar bones (Figure 3.5). This measure varied markedly among the five regional samples ($F=7.4$, $p<0.001$). Maximum eversion was found in the East Asian sample ($\bar{x}=63.1$ mm), a value very close to that of Europeans (Figure 4.3h). The group with least eversion of the malars were the sub-Saharan Africans ($\bar{x}=53.9$mm), followed by the Australians. The East Asian sample is significantly different from the Africans at the 0.1% level and to the Australian sample at the 5% level. The Europeans are significantly different from the Africans at the 1% level. Interestingly, the Fueguian-Patagonian, Afalou and Taforalt samples all show even greater mean values of malar eversion than the East Asians (64.2mm, 63.9mm, 73.2mm respectively). The North African Epi-Palaeolithic groups (i.e. Afalou and Taforalt) have the greatest eversion of the lower border of the malars among the nine modern samples studied. The Australian character of this feature is not confirmed.

AUS9: Absence of distinct narial margins

The conformation of the lower narial margins in modern human skulls is a very variable feature, several classificatory systems have been proposed (Burkitt & Lightoller, 1923; Gower, 1923; Johnson, 1937; Macalister, 1898). Here, the features identified in these various works are taken into account for defining 10 categories of lower narial margins (Figure 3.9). The distribution of incidence of these 10 categories in the nine modern samples studied is listed in Table 4.2. The variation between the five regional samples is statistically significant ($x^2= 76.4$, $df=36$, $p<0.001$).

Two forms of narial margins, N3 and N8, account for more than half of the cases in the nine modern samples examined (51.5%). These types differ in relation to which 'line' forms the actual edge of the nasal cavity, whether it is the *crista intermaxillaris* (Holl, 1882) in N3, or the *crista maxillaris* (Holl,

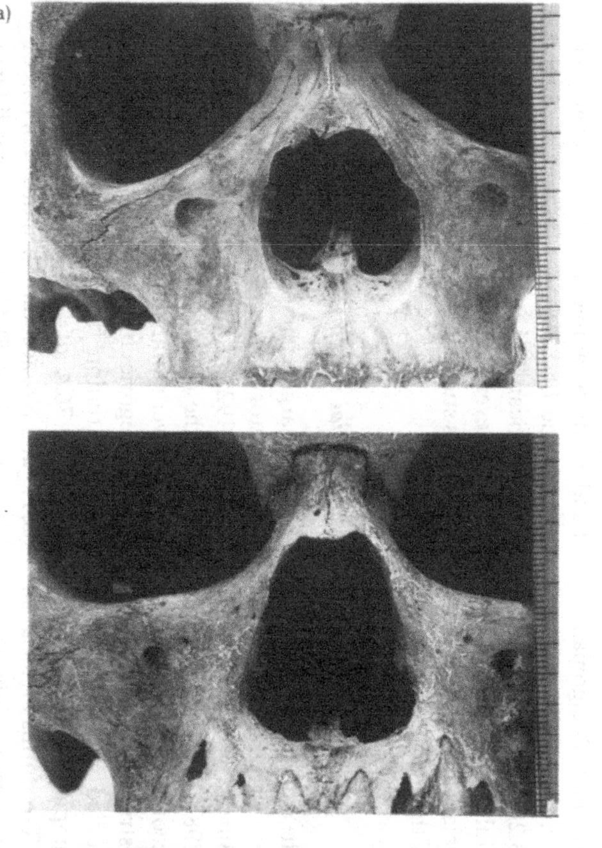

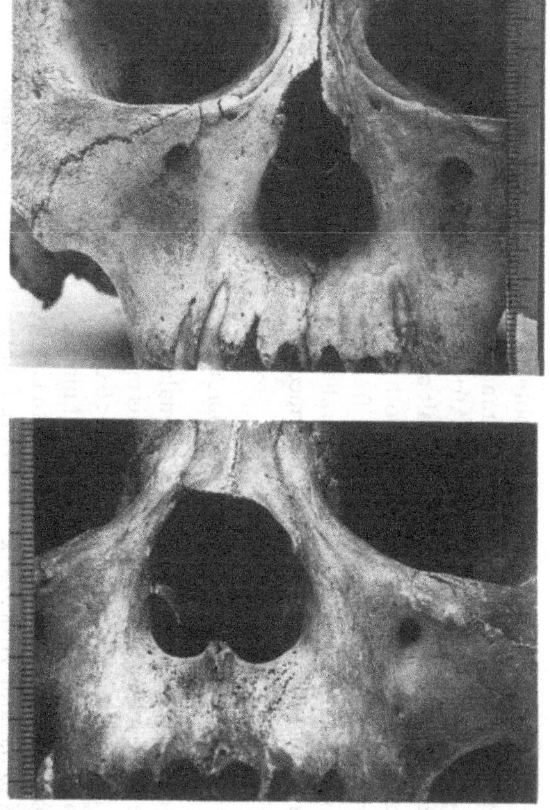

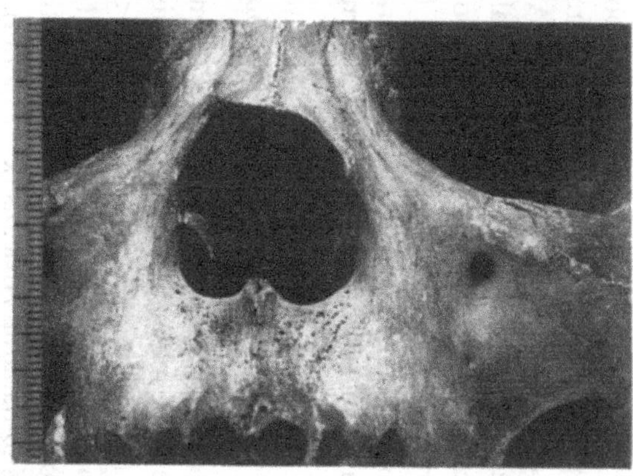

Figure 4.4 Examples of cases of ill-defined (a) and absent (b) lower narial margins.

1882) in N8. They resemble each other in that both show an ill-defined fossa. Narial margins formed by the *crista intermaxillaris*, with an ill-defined fossa (N3) are the most frequent type observed in all populations except the Australians and the sub-Saharan Africans. In both these groups, the most frequent form is the margin formed by the *crista maxillaris* (Holl, 1882), with an ill-defined external fossa (N8). A 'sharp' narial margin (N1), together with a relatively sharp margin in which the *crista maxillaris* and *intermaxillaris* are partially separated (N2) form the next most common types, accounting for 29.8% of cases. The Australians are the only population to lack cases of single/sharp narial margins (N1). The remaining 18.7% of cases show types N6 (5.3%), N5 (4.1%), N9 (4.1%), N4 (2.4%), N7 (2.4%) and N10 (0.4%). Least variation in lower narial margins was observed among the Mediterranean Epi-Palaeolithic samples.

These results indicate that, although lower narial margins are a very variable feature in modern skulls, four types account for 81.3% of 245 cases examined (N3, N8, N1, N2). Furthermore, these types could be observed in all populations, with the exception of the Afalou and Taforalt for type N8 and of the Australians for type N1. It is clear that completely ill-defined narial margins are uncommon, seen in only 4% of cases, while complete absence of lower narial margins is extremely rare, observed in a single African skull (Figure 4.4b). Among the few Inuit skulls studied (not included in the above comparisons because of the small sample size) two other cases of complete absence of lower narial margins were observed. The Australians show the greatest frequency among the populations examined of cases of ill-defined margins (N9) (Figure 4.4a). However, the occurrence of ill-defined margins was not unique to Australians, nor was this the most common Australian condition. It is possible that, because this population lacks cases of single/sharp margins, it reinforces a visual impression of absence of narial margins in this group. However, the uniquely Australian character of this trait is not confirmed.

AUS10: Curvature of the alveolar plane of the maxilla

This trait was measured by projecting the distances from the Frankfort Plane to prosthion and to the posterior-most alveolar surface onto a two dimensional plane. The difference between these two measures was then taken as the curvature of the alveolar plane of the maxilla (Figure 3.6). The variation observed in the five regional samples was statistically significant ($F=4.4$, $p<0.01$) (Figure 4.3i). The populations with the greatest mean curvature of the alveolar plane of the maxilla were the Europeans and East Asians (EU: $x=3.3$ mm; EA: $\bar{x}=3.0$ mm). Both these samples were significantly different ($p<0.05$) from the Australians, who show the least mean alveolar maxillary curvature (AUS: $\bar{x}=0.9$ mm). The Fueguian-Patagonian sample presents values above those of any recent population (F–P: $\bar{x}=4.8$ mm), and are also significantly different from the Australians ($p<0.001$). The African sample is similar to the Australian, showing a relatively horizontal alveolar plane (SSA: $\bar{x}=1.6$ mm), a condition also observed in the Epi-Palaeolithic Afalou and Taforalt groups (AF: $\bar{x}=1.9$ mm; TAF: $\bar{x}=1.3$ mm). Contrasting with these low values, the Natufians show the greatest mean alveolar curvature of all populations examined (NAT: $\bar{x}=5.6$mm). The Australian character of this trait is not confirmed.

AUS11: Pronounced development of the zygomatic trigone

The development of the zygomatic trigone (TR) was scored as four grades (App. I, Figure AI.6), ranging from a completely smooth trigone (TR1) to an inflated surface, with the frontal portion much larger than the frontal process of the zygomatic (TR4). Two intermediate scores were recognised (TR2 and TR3). The distribution of the incidence of this feature among the five regional samples is statistically significant ($x^2=53.5$, df$=12$, $p<0.001$). In the case of the Europeans, Southeast Asians and East Asians, the most frequent condition is that of a slightly developed trigone in which the orbital surface of the frontal portion is raised (TR2). This is also the case in the Natufian and Taforalt samples, while the Afalou show equal numbers of cases of TR2 and TR3. In the case of the sub-Saharan Africans, with their pronounced frontal gracility already seen in the lack of development of the supraorbital ridges, the most frequent condition is that of a completely smooth trigone (TR1). In the Australian sample, the most frequent score is TR4, pronounced development of the zygomatic trigones (Figure 4.3j). This would seem to support the claim that Australian crania have significantly higher frequencies of very large zygomatic trigones than in any other modern population. However, the robust Fueguian-Patagonian group presents an even higher proportion of cases of very large zygomatic trigones (TR4), together with total absence of cases of a smooth trigone (TR1). Therefore, the uniquely Australian character of this feature is not confirmed.

AUS12: Presence of a large suprameatal tegmen

The suprameatal tegmen is the extension of the auditory meatal roof beyond the edge of the tympanic tube. This was measured as the difference between biauricular and inter-tympanic breadths, divided by two to represent the suprameatal tegmen on each side. The variation in this trait among the five regional samples was statistically significant ($F=7, p<0.001$). The European and East Asian samples show the greatest mean values (Figure 4.3k). The Europeans are significantly different from the Africans ($p<0.001$) and the Australians ($p<0.05$), while the East Asians are also significantly different from the Africans ($p<0.05$). The Afalou, Taforalt and Natufian samples show values comparable to the European and East Asian samples, while the Fueguian-Patagonians show the largest mean suprameatal tegmen of all these modern groups. The Australian character of this feature is not confirmed. Again, the common development of supramastoid crests, coupled with small parietal expansion in Australian crania, provides the false visual impression of an overhanging roof to the external auditory meatus.

AUS13: Angling of the petrous to the tympanic in the petro-tympanic axis

The angle formed between the petrous and tympanic bones (Figure 3.7) varied significantly among the five regional samples ($F=6.3, p<0.001$). The Australian and African samples show equally low values in relation to the other populations, and are significantly different from the East and Southeast Asian crania at the 5% level (Figure 4.3l). The Fueguian-Patagonian sample shows the largest mean value, similar to the East Asian crania. The uniquely Australian character of this feature is not confirmed.

AUS14: Presence of an occipital crest

The development of an occipital crest (OCR) was scored as three grades (App. I, Figure AI.7), representing an absent (OCR1), partial (OCR2) or complete crest (OCR3) across the nuchal plane of the occiput. The distribution of these scores among the five regional samples was not statistically significant ($x^2=8.3$, $df=8$, $p>0.05$). All the recent populations examined present cases of absent, partial or complete crests (Figure 4.3m), although East Asians and to a lesser degree the Australians, show proportionally more cases of partial cresting (OCR2). The Afalou and Taforalt fossil samples lack cases of a complete occipital crest, while the Natufians show one. The Fueguian-Patagonian sample differs from the other populations in showing a higher incidence of partial and complete occipital crests than any other population studied. This is consistent with their generally very robust occipital morphology (see below). The Australian character of this feature is not confirmed.

AUS15: Pronounced occipital torus

Scoring the development of an occipital torus (OT) involved seven different categories, some concerned with the configuration of the nuchal lines without development of an occipital torus (App. I, Figure AI.8). These categories are the following:

OT1 – Nuchal lines not visible, external occipital protuberance (EOP) absent.
OT2 – Supreme nuchal line not visible, EOP present.
OT3 – Supreme and superior nuchal lines distinct and separated, EOP present.
OT4 – Supreme and superior nuchal lines joined medially by the EOP.
OT5 – Torus, separated in two lateral portions by the EOP (Tubercle of Hasebe).
OT6 – Torus on which the EOP is visible along the superior toral margin.
OT7 – Torus, EOP absent.

For the purpose of this analysis, the seven original scores were subsumed into three, one representing absence of a torus (OT × 1: OT2 + OT3 + OT4), one development of a torus with EOP present (OT × 2: OT5 + OT6), and one representing development of a torus without an EOP (OT × 3: OT7). Cases of absence of an occipital torus and an external occipital protuberance (OT1) were not observed, and this category was not considered any further. The variation in distribution of these three scores among the five regional samples was significant ($x^2 = 30.4$, df = 8, $p < 0.001$). Only a single case of an occipital torus without an external occipital protuberance (OT7) was recorded among Australians (Figure 4.3n-1,n-2). The most frequent occipital morphology in all populations is not associated with the development of an occipital torus (OT × 1), but Australians, and to a lesser extent East Asians, show a relatively large number of cases of an occipital torus (OT × 2). They differ in that East Asians have similar number of cases of a Tubercle of Hasebe (OT5) and an occipital torus with EOP (OT6) while among Australians a Tubercle of Hasebe is relatively rare (two cases). The Epi-Palaeolithic fossils from Afalou, Taforalt and Natufian sites show relatively gracile occipital morphologies, with frequencies similar to the European, African and Southeast Asian samples. The Fueguian-Patagonian remains are clearly very robust, and they show the largest frequency of cases of an occipital torus with visible EOP of all populations examined (68.9%).

This is the only group in which an occipital torus is the most common morphology. The uniquely Australian character of this trait is not confirmed.

EA/AUS: Facial prognathism and lateral facial flatness

The degree of facial projection was examined as two relatively independent components – facial prognathism in the sagittal plane (EA6/AUS16) and facial flatness in the coronal plane (EA7/AUS17) – and are discussed below. The regional distribution of these two measures are examined independently as the gnathic index and the difference between nasion, subspinale and prosthion radii in the first case, and as the bifrontal subtense and the difference between nasion, frontomalare and ectochonchion radii in the second.

EA6/AUS16: Facial prognathism

Variation in Gnathic Index (GI) among the five regional samples was statistically significant ($F = 18.0$, $p < 0.001$). Greatest prognathism was observed in the African sample (GI: $\bar{x} = 103.7$), followed closely by the Australian (GI: $\bar{x} = 101.3$) (Figure 4.3o-1). The sub-Saharan Africans and Australians are significantly different from the Europeans and East Asians ($p < 0.01$). The Africans are also significantly different from the Southeast Asians ($p < 0.01$), while the Australians only at the 5% level. The Fueguian-Patagonian and Epi-Palaeolithic populations are not very prognathic, and present values similar to the European and Asian groups. Crania from Afalou show the least prognathism of all the recent samples studied (GI: $x = 94.1$), and are significantly different from the Australians and sub-Saharan Africans at the 5% level. Using the difference between nasion and prosthion radii as a measure of total facial prognathism gives very similar results (Figure 4.3o-2). The variation among the five regional samples is significant ($F = 14.9$, $p < 0.001$), with the same pattern of differences between Africans and Australians and the other populations studied.

In order to assess whether the prognathism observed is associated with the whole face or restricted to the alveolar portion, two further measures were examined. The degree of projection of the alveolar portion of the face was calculated as the difference between prosthion and subspinale radii. Variation in this trait among the five regional samples was significant ($F = 7.5$, $p < 0.001$). The Africans clearly show the greatest alveolar prognathism (Figure 4.3o-3), and the Europeans the least. The sub-Saharan Africans are significantly different from the Europeans ($p < 0.001$) and from the East Asians ($p<0.01$), while the Australians are significantly different from the Europeans ($p<0.01$). The North African fossils from Afalou and Taforalt have values below the European mean, showing the degree of sagittal flatness of their lower face. The degree of projection of the nasal portion of the face was calculated as the difference between subspinale and nasion radii. Variation in this trait among the five regional samples was also pronounced ($F=7.0$, $p<0.001$). The sub-Saharan African and Australian samples show exactly the same mean values (Figure 4.3o-4), and are significantly different from the Europeans at the 0.1% level and from the East Asians at the 5% level. The Afalou show the least upper facial prognathism with a mean difference between nasion and subspinale radii of zero. However, all the Epi-Palaeolithic fossils show a relatively large standard deviation and range in comparison to the recent samples.

The uniquely Australian character of pronounced facial prognathism is not confirmed. Furthermore, in terms of sagittal projection, although the East Asians are relatively non-prognathic, they are not the population with most pronounced sagittal facial flatness.

EA7/AUS17: Lateral facial flatness

Facial flatness in the coronal plane was measured as the bifrontal subtense, i.e. the perpendicular distance between nasion and a bifrontal line, and as the difference between nasion and frontomalare radii. Variation in bifrontal subtense among the five regional samples is statistically significant ($F=9.6$, $p<0.001$), with the East Asians, and to a lesser degree the Southeast Asians, showing the smallest mean values, i.e. the laterally flatest faces (Figure 4.3p-1). The East Asians are significantly different from the Australians ($p<0.001$) and from the Africans ($p<0.01$), while the Southeast Asians are also significantly different from these two populations ($p<0.01$, $p<0.05$). The Afalou fossils show mean values close to the Australian mean. The regional variation in the difference between nasion and frontomalare radii is significant but less pronounced than that of the bifrontal subtense ($F=4.5$, $p<0.01$). In this case, it is the Southeast Asians, followed closely by the East Asians, that show the least mean difference, i.e. the flattest face. However, the differences between any two groups are not statistically significant.

The difference between nasion and ectochonchion radii is used as a broader measure of lateral facial flatness, encompassing not only the suborbital area but the frontal process of the zygomatic. The variation in this trait among the five regional samples yields results similar to the above, but more pronounced ($F=12.1$, $p<0.001$). The differences between the

East and Southeast Asians, with their coronally flat faces, and the Australians, with their laterally retracted faces, is greater and significant at the 0.01% level in both cases (Figure 4.3p-2). However, in this case the Europeans are also significantly different from the East Asians ($p<0.05$), and the Australians are also significantly different from the African sample ($p<0.05$). The East Asian character of 'lateral facial flatness' is confirmed.

EA8/AUS18: Profile of the frontal bone

Degree of frontal flatness and its opposite condition, very 'steep' frontal bones, were measured through the frontal angle. Variation in frontal angle among the five regional samples was significant ($F=7.1$, $p<0.001$). The group with the 'steepest' frontals were the sub-Saharan Africans ($\bar{x}=125.3° \pm 4.7$), who are significantly different from all others, the Australians ($p<0.001$), the East Asians ($p<0.01$), the Southeast Asians and the Europeans ($p<0.05$). Among the regional samples, the Australians have, on average, the flattest frontal bones ($\bar{x}=130.8° \pm 4.0$). However, once the specific populations are included in the comparison, this pattern disappears. Both the Afalou and Taforalt samples have on average flatter frontal bones (Afalou: $\bar{x}=132.2° \pm 4.1$; Taforalt: $\bar{x}=132.2° \pm 3.0$) than the Australians, but the flattest of all are to be found in the Fueguian-Patagonian sample. This group has an average frontal angle of 135.7°. The comparatively large standard deviation ($sd=5.5$) belies great levels of within-group variation, but this population is clearly very different from all other modern groups in its supraorbital morphology (Figure 4.3q). The Australian character of flat frontal bones is not confirmed, nor is the opposite East Asian character of rounded frontals.

EA9/AUS19: Rounding of the inferolateral margin of the orbit

Rounding of the inferolateral margin (RO) of the orbit was scored as three grades (App. I, Figure AI.9), ranging from a clear sharp margin (RO1), to a relatively rounded margin (RO2) to a rounded margin levelled with the floor of the orbit (RO3). Variation among the five regional samples was significant ($x^2=48.7$, $df=8$, $p<0.001$). The Australians not only show the highest frequency of cases of RO3 of all populations examined, but it is the most common condition in this group (Figure 4.3r). Contrary to Weidenreich's observations, it is among East Asians and not the Europeans that the highest frequency of 'sharp' borders is found. Only the Australian character of this trait is confirmed.

Table 4.3. *(A) Observed combinations of inclination of the three orbital elements (superior, lateral and inferior margins) and incidence in the samples examined*

Orbital shape		TOT (N=252) %	EU (N=24) %	SSA (N=26) %	SEA (N=25) %	EA (N=42) %	AUS (N=37) %	F/P (N=29) %	AF (N=30) %	TAF (N=17) %	NAT (N=15) %
O1	I/I/II	1.2	–	–	3.8	–	–	–	5.9	–	–
O2	I/II/I	5.9	4.0	–	–	2.3	10.8	3.4	14.7	20.0	–
O3	I/II/II	5.5	–	7.7	11.5	–	13.5	–	5.9	6.7	5.9
O4	I/II/III	3.6	–	–	–	2.3	18.9	–	–	–	5.9
O5	I/III/III	0.4	–	3.8	–	–	–	–	–	–	–
O6	I/IV/I	3.6	–	–	3.8	–	8.1	6.9	5.9	–	5.9
O7	I/IV/II	6.3	4.0	11.5	3.8	2.3	18.9	–	5.9	6.7	–
O8	I/IV/III	0.4	–	–	–	–	2.7	–	–	–	
O9	II/I/I	1.6	–	–	7.7	–	–	–	5.9	–	–
O10	II/I/III	0.4	–	–	–	–	–	–	2.9	–	–
O11	II/II/I	13.1	4.0	–	7.7	18.6	–	48.3	11.8	26.7	–
O12	II/II/II	9.9	24.0	7.7	23.1	7.0	2.7	6.9	8.8	6.7	5.9
O13	II/II/III	13.5	24.0	23.1	7.7	20.9	16.2	10.3	–	–	11.8
O14	II/III/I	1.2	–	–	–	2.3	–	3.4	–	6.7	1.8
O15	II/IV/I	7.9	–	3.8	3.8	7.0	–	13.8	17.6	13.3	17.6
O16	II/IV/II	5.2	–	7.7	7.7	2.3	2.7	–	8.8	6.7	17.6
O17	II/IV/III	2.8	–	23.1	–	2.3	–	–	–	–	–
O18	III/I/II	0.8	–	–	3.8	2.3	–	–	–	–	–
O19	III/II/I	0.8	–	–	–	2.3	–	–	–	6.7	–
O20	III/II/II	7.5	16.0	–	7.7	18.6	2.7	6.9	5.9	–	–
O21	III/II/III	5.2	24.0	11.5	–	7.0	–	–	–	–	5.9
O22	III/III/III	0.4	–	–	–	–	–	–	–	–	5.9
O23	III/IV/I	0.8	–	–	3.8	–	–	–	–	–	5.9
O24	III/IV/II	1.6	–	–	3.8	2.3	–	–	–	–	11.8
O25	IV/II/II	0.4	–	–	–	–	2.7	–	–	–	–

EU: Europe; SSA: sub-Saharan Africa; SEA: Southeast Asia; EA: East Asia; AUS: Australia; F/P: Fueguian-Patagonian; AF: Afalou; TAF: Taforalt; NAT: Natufian.

Table 4.3. *(B) Distribution of categories by region in order of frequency*

	Combination of orbital shape														
EU	O12=	O13=	O21	O20	O2=	O7=	O11								
SSA	O13=	O17	O7=	O21	O3=	O12=	O16	O5=	15						
SEA	O12	O3	O9=	O11=	O13=	O16=	O20	O6=	O7=	O15=	O18=	O23=	O24		
EA	O13	O11=	O20	O12=	O15=	O21	O2=	O4=	O7=	O14=	O16=	O17=	O18=	O19=	O24
AUS	O4=	O7	O13	O3	O2	O6	O8=	O12=	O16=	O20=	O25				
F/P	O11	O15	O13	O6=	O12=	O20	O2=	O14							
AF	O15	O2	O11	O12=	O16	O1=	O3=	O6=	O7=	O9=	O20	O10			
TAF	O11	O2	O15	O3=	O7=	O12=	O14=	O16=	O19						
NAT	O15=	O16	O13=	O24	O3=	O4=	O6=	O12=	O21=	O22=	O23				

EU: Europe; SSA: sub-Saharan Africa; SEA: Southeast Asia; EA: East Asia; AUS: Australia; F/P: Fueguian-Patagonian; AF: Afalou; TAF: Taforalt; NAT: Natufian.

EA10/AUS20: Orbital shape/horizontal superior orbital margin

Two different features are considered here, the first an East Asian trait 'orbital shape' and the other an Australian one, 'horizontal superior orbital border'. To measure these two features, the same categorical classification was used. The superior, lateral and inferior margins of the orbit were each scored in relation to degree of inclination, and the combination of the three scores (superior, lateral and inferior) used as orbital shape. The superior margin scores were used as the orientation of the superior orbital margin.

As it is to be expected, variation in orbital shape among modern populations is enormous. No statistical test of this variation in terms of regional distribution was attempted, since it is clear that all populations vary markedly in what orbital shapes they have, how many and at which frequency. Table 4.3 lists the observed combinations of orbital shape and its incidence in the groups being examined. All the most common orbital shapes involve an oblique superior margin with varying degrees of inclination of the inferior margin (O11, O12, O13, O15). O12 is the only orbital shape observed in all nine samples. In terms of each population, most European crania show four types of orbits (O12, O13, O20, O21), accounting for 88% of cases. These all represent a moderate to pronounced oblique shape. The sub-Saharan Africans are more varied, 69.2% of cases are accounted for by four types (O7, O13, O17, O21), including horizontal superior margins. The Southeast Asian and East Asian samples are so variable that generalisations are difficult. The predominant types in the Southeast Asian crania are O12 (23.1%) and O3 (11.5%), while in East Asian crania 58.1% are accounted for by three types (O11, O13, O20). The Fueguian-Patagonian crania show less variation in forms, with almost half of the sample (48.3%) showing type O11. Among Afalou crania, 44.1% are O2, O11 and O15, while the Taforalt crania show the same pattern but more pronounced. The Natufian are a smaller sample with much variation. The Australian sample differs from most other populations in that although showing a similar level of variability in orbital shape as the other samples, the majority of cases do not have an oblique orbit, but one in which the superior margin is horizontal (72.9% show types O2, O3, O4, O6, O7 and O8). A horizontal superior margin is clearly not unique to Australians, but this population shows a significantly higher incidence. The Afalou and Taforalt samples are the other two populations to show relatively high incidences of a horizontal superior border, but under 40%. While a particular orbital shape associated with the East Asian cranialsamples is not confirmed (on the contrary, this

Table 4.4. *Incidence of significant regional differences of the 'regional continuity traits' in the regions proposed by the Multiregional Model. Asteriks represent level of significance for the regions proposed by the Multiregional Model ($p<0.05$*; $p<0.01$**; $p<0.001$***), while if in brackets they represent significance for a region other than that proposed by the model*

Trait		Region expected	Region found
*EA*1	Sagittal keeling (***)	East Asia	Australia
*EA*2	Course of nasofrontal and frontomaxillary sutures (**)	East Asia	S.S. Africa
*EA*3	Low profile of nasal saddle and nasal roof (***)	East Asia	S.S. Africa
*EA*4	M3 Agenesis *	East Asia	**East Asia**
*EA*5	Reduced posterior dentition	East Asia	none
*EA*6	Sagittal facial flatness (***)	East Asia	Afalou/Europe
*EA*7	Lateral facial flatness***	East Asia	**East Asia**
*EA*8	Rounded frontal bones (***)	East Asia	S.S. Africa
*EA*9	Rounded infero-lateral margin of the orbit*** (=AUS19)	East Asia	Australia
*EA*10	Orbital shape	East Asia	none
*AUS*1	Pronounced development of supraorbital ridges (***)	Australia	T.Fuego/**Australia**
*AUS*2	Lambdoid depression	Australia	none
*AUS*3	Small pterion articulation ***	Australia	**Australia**
*AUS*4	Deep and narrow infraglabellar notch ***	Australia	**Australia**/Afalou
*AUS*5	Backward position of minimum frontal breadth (***)	Australia	Taforalt/Australia
*AUS*6	Low position of maximum parietal breadth (**)	Australia	Europe
*AUS*7	Presence of zygomaxillary tuberosity *	Australia	**Australia**
*AUS*8	Eversion of the lower border of the malars (***)	Australia	Taforalt/T.del Fuego
*AUS*9	Absence of external narial margins (***)	Australia	none
*AUS*10	Curvature of alveolar plane of the maxilla (**)	Australia	Natufian/T.del Fuego
*AUS*11	Developed zygomatic trigone (***)	Australia	T.del Fuego
*AUS*12	Large suprameatal tegmen (***)	Australia	T.del Fuego/Europe
*AUS*13	Angled petrous to tympanic in petro-tympanic axis (***)	Australia	**Australia**/S.S.Africa
*AUS*14	Presence of an occipital crest	Australia	T.del Fuego
*AUS*15	Presence of an occipital torus (***)	Australia	T.del Fuego
*AUS*16	Facial prognathism (***)	Australia	S.S. Africa
*AUS*18	Flat frontals (***)	Australia	T.del Fuego
*AUS*19	Rounded infero-lateral margins of the orbit ***	Australia	**Australia**
*AUS*20	Horizontal superior orbital border ***	Australia	**Australia**

In bold those cases in which the incidence of a feature in the region proposed by the Multiregional Model is confirmed.
S.S. Africa: sub-Saharan Africa; T. del Fuego: Tierra del Fuego.

group is clearly the most variable), the Australian character of a horizontal superior orbital border is shown.

Which regional continuity traits?

The results presented above show that the great majority of the features proposed by the Multiregional Model as stable regional characteristics that reflect regional morphological continuity for a million years, are actually not confined to, or even most common in, the two regions expected, namely East Asia and Australia. Table 4.4 summarises these results. Of the 11 East Asian 'regional continuity traits' examined, only two have their highest incidence among East Asians, M3 agenesis and lateral facial flatness, although M3 agenesis occurs at such low frequencies that it can hardly be considered as characterising the population. Of the 20 Australian 'regional continuity traits', only four exhibit an exclusively high incidence in Australia (small pterion articulation, presence of a zygomaxillary tuberosity, rounding of the inferolateral margin of the orbits and a horizontal superior orbital margin). Two features show highest incidence in Australian crania, but with very similar frequencies in other populations, namely a deep and narrow infraglabellar notch which is also very common in skulls from the site of Afalou, and an angled petrous bone, which is also very common in sub-Saharan African skulls. Lastly, there are two features which show very high frequencies in Australian crania, but are not the highest observed, namely pronounced supraorbital ridges which are most common in Fueguian-Patagonian crania, and a backward position of minimum frontal breadth, which has its most pronounced expression in the Taforalt sample. Two features do not show a particular regional pattern, while the other nine, supposedly Australian regional continuity traits, occurred far more frequently in other populations than in the Australian sample studied.

These results clearly indicate that the claimed regional specificity of these features among present day and recent modern populations is not supported by the data. Most (24 of 30) of the traits examined show significant regional distributions, i.e. occur in modern populations at very different frequencies or with very different values. They just do not occur in the populations claimed by the Multiregional Model. A small number (6 of 30) of the features examined do occur in East Asia and Australia as proposed by the Multiregional Model. Is this proof of regional morphological continuity? It may be argued that if these features, however few, are indeed unique to recent and fossil East Asian and Australian populations, then they may be evidence for regional continuity in morphology. If this is the case, part of the recent geographical diversity in morphology would reflect archaic regional diversity. The next chapter examines the distribution of the supposedly East Asian and Australian regional features in late Middle and Upper Pleistocene hominid fossils.

5 *Temporal distribution of the 'regional continuity traits' in late Pleistocene hominids*

We have seen that a small number of the East Asian and Australian 'regional continuity traits' occur in present Chinese and Australian aborigines differentially from other recent populations. But what is their distribution among late Pleistocene hominids? In other words, do we know that Australians uniquely inherited these particular features from Javanese archaic hominids and not from an early modern ancestor? Here we examine the distribution and incidence of the East Asian and Australian 'regional continuity traits' in individual late Pleistocene fossil specimens. The fossil specimens (listed in Table 5.1) were chosen with the intention of sampling diverse groups that can throw light onto the process of modern differentiation.

If multiregional evolution of modern geographical populations took place, the following predictions should be fulfilled:

(1) The Javanese late *Homo erectus* (Ngandong) fossils should present all or most of the Australian regional continuity traits; the African late Middle to early Upper Pleistocene fossils (Ndutu, Djebel Irhoud, LH18, Omo 1, Florisbad, Klasies River Mouth) and the Middle Eastern early Upper Pleistocene fossils (Qafzeh 9 and 6, Skhūl 5 and 4) should not present any of the regional continuity features, either East Asian or Australian.

(2) Two clear regional lineages should be identified – one in east Asia including the fossils from UC101, UC103, Liujiang, Minatogawa I and IV, Tepexpán, Lagoa Santa 6, and recent East Asians and Fueguians; the other in Australasia, including the fossils from Wajak I and II, Keilor, Kanalda, Kow Swamp 1, 5 and 15, and recent Australians and Southeast Asians.

Table 5.1. *Fossil crania studied for the incidence of the regional continuity traits*

BMNH – NDUTU (cast)	Tanzania	Middle Pleistocene
BMNH – SOLO I (cast)	Java	Early Upp. Ple. ?
BMNH – SOLO IV (cast)	Java	Early Upp. Ple. ?
BMNH – SOLO V (cast)	Java	Early Upp. Ple. ?
BMNH – SOLO VI (cast)	Java	Early Upp. Ple. ?
BMNH – SOLO IX (cast)	Java	Early Upp. Ple. ?
BMNH – SOLO X (cast)	Java	Early Upp. Ple. ?
BMNH – SOLO XI (cast)	Java	Early Upp. Ple. ?
BMNH – DJEBEL IRHOUD 1 (cast)	Morocco	*c.* 190 ka
BMNH – OMO I (cast)	Kenya	*c.* 130 ka (U-S)
BMNH – Klasies River Mouth (casts)	South Africa	*c.* 130 ka
BMNH – FLORISBAD (cast)	South Africa	*c.* 120 ka ?
TASSM – LH18 (cast)	Laetoli, Tanzania	*c.* 120 ka (U-S)
TASSM – QA H6 (cast)	Djebel Qafzeh, Israel	*c.* 90, ka (ESR/TL)
RM-J – QA H9	Djebel Qafzeh, Israel	*c.* 90 ka (ESR/TL)
RM-J – SK IV	Skhūl, Israel	*c* 100 ka yrs (ESR/TL)
TASSM – SK V (cast)	Skhūl, Israel	*c.* 100 ka (ESR/TL)
TASSM – NEG 1	Nahal Ein Gev, Israel	22–24,000 BP (no ^{14}C)
TASSM – OH II H2	Ohalo, Israel	*c.* 19,000 BP (^{14}C)
TASSM – EG 1	Ein Gev, Israel	14–16,000 BP (no ^{14}C)
BMNH – PREDMOST III (cast)	Aurignacian	35–25,000 BP
BMNH – Ckn UC 101 (cast)	Zhouk. Upper Cave, China	24,000 BP (^{14}C)
BMNH – Ckn UC 103 (cast)	Zhouk. Upper Cave, China	24,000 BP (^{14}C)
BMNH – LIUJIANG (cast)	Liujiang, Kwangsi, China	? late Upper Pleistocene
BMNH – MINATOGAWA I (cast)	Japan	17 ka (^{14}C)
BMNH – MINATOGAWA IV (cast)	Japan	17 ka (^{14}C)
BMNH – LAGOA SANTA	Sumidouro Cave, Brasil	*c.* 8 ka (^{14}C)
BMNH – TEPEXPAN (cast)	Tepexpan, Mexico	11 ka (^{14}C)
BMNH – WAJAK I (cast)	Java	20–10 ka?
BMNH – WAJAK II (cast)	Java	20–10 ka?
BMNH – KOW SWAMP 15 (cast)	Australia	16 ka (^{14}C)
BMNH – KEILOR (cast)	Australia	15–10 ka
BMNH – KANALDA	Australia	15–10 ka

BMNH: The Natural History Museum, London; TASSM: Tel Aviv Sackler School of Medicine; RM-J: Rockefeller Museum, Jerusalem; Ple.: Pleistocene.

Temporal distribution of the East Asian regional traits

See Table 5.2

EA1: Sagittal keeling

The distribution of the three grades of sagittal keeling in late Pleistocene hominids shows that sagittal keeling is not an ancestral trait of late African

Table 5.2. *Values and incidence of the East Asian regional continuity traits in the recent samples and some late Pleistocene fossil hominids*

Regional continuity traits	Ngan	Ndutu	Dj Irh	LH18	Omo I	KRM	Flor	*SSA*	Afa	Taf	Q9	Q6	SK5	SK4	EGev	Oha	NEG	Natuf	Pred	*EU*
Sagittal keeling	SK2	–	SK1	SK1	SK1	–	SK1	*SK1*	SK1	SK1/2	SK1	SK1	SK2	SK1	SK1	SK1	SK1	SK1	SK2	*SK1*
Course n-fr./fr.-max. sutures	–	12.0	7.0	–	–	9.0	8.0	*4.4*	5.7	8.4	–	1.0	2.0	–	–	6.0	6.0	6.0	11.0	*6.2*
Flat nasals	NS3	NS4	NS3	–	–	–	NS3	*NS1*	NS3	NS3	–	NS4	NS1	–	–	NS3	NS2	NS2/3	NS3	*NS3*
M3 agenesis	–	–	–	M3	–	–	–	*M3*	M3	M3	M3	M3	M3	M3	–	M3	M3	M3	M3	*M3*
Red. post dentition																				
Area M1 – Area M3	–	–	–	–	–	–	–	*15.4*	21.9	17.8	18.5	–	–	5.7	–	–	–	27.4	11.0	*25.8*
blM1 – blM3	–	–	–	–	–	–	–	*–0.3*	–0.1	–0.1	–0.6	–	–1.0	–	–	–	0.8	0.4	*0.3*	
mdM1 – mdM3	–	–	–	–	–	–	*–1.5*	1.9	1.5	1.9	–	–	2.1	–	–	–	1.6	0.5	*2.0*	
Coronal facial flatness																				
Bifrontal subtense	–	–	9.0	–	6.0	–	25.0	*17.7*	17.8	16.3	–	13.0	10.0	–	–	15.0	13.0	16.1	16.0	*17.0*
NAR – EKR	–	–	26.0	–	20.0	–	–	*22.0*	22.2	19.8	–	20.0	16.0	–	–	18.0	21.0	22.8	26.0	*23.0*
Orbital shape	–	–	O21	–	–	–	O19	*O13/17*	O15	O11	O11	O21	–	–	–	O12	O2	O15/16	O2	*O12/21*

Regional continuity traits	UC101	UC103	Liu	Min. I	Min.IV	*EA*	Tepex	L.Sta6	*FUE*	Keilor	Kan	WaI	Wa II	K.S1	K.S5*	K.S15	*SAUS*	*SEA*
Sagittal keeling	SK2	SK3	SK1	SK1	SK1	*SK1*	SK3	SK2	*SK2*	SK1	SK3	SK1	–	SK2	SK2	–	*SK3*	*SK1/2*
Course n – fr./fr.-max. sutures	7.0	8.0	6.0	5.0	4.0	*6.5*	4.0	–	6.9	4.0	10.0	–	–	9.0	3.0	2.0	*6.4*	*5.2*
Flat nasals	NS3	NS3	–	NS2	NS3	*NS2/3*	NS2	–	*NS3*	NS1	NS3	NS2	–	NS3	NS3	NS3	*NS3*	*NS1/2/3*
M3 agenesis	M3	M3	**M3A**	M3	–	*M3*	M3	M3	*M3*	M3	M3	M3	M3	M3	M3	M3	*M3*	*M3*
Red. post dentition																		
Area M1 – Area M3	–	–	–	–	–	*20.6*	–	–	*8.7*	26.0	14.2	42.2	19.0	–11.9	–	15.4	*22.7*	*25.5*
blM1 – blM3	–	–	–	–	–	*0.3*	–	–	*0.2*	0.7	0.3	1.0	0.0	–0.6	–	0.2	*0.2*	*0.5*
mdM1 – mdM3	–	–	–	–	–	*1.5*	–	–	*1.2*	1.4	0.8	2.4	1.5	–0.4	–	1.0	*1.6*	*1.8*
Coronal facial flatness																		
Bifrontal subtense	15.0	14.0	16.0	9.0	8.0	*14.8*	10.0	–	*17.2*	21.0	19.0	16.0	16.0	21	12	10.0	*17.9*	*15.1*
NAR – EKR	23.0	21.0	20.0	10.0	–	*20.3*	20.0	–	*22.7*	23.0	24.0	20.0	–	–	–	–	*24.6*	*20.8*
Orbital shape	O15	O2	O11	O3	O13	*O13*	O15	O11	*O11*	O16	O6	O16	–	O6	O11	O17	*O4*	*O12*

*Deformed.

archaic hominids or early modern fossils (Figure 5.1a). The Ngandong sample shows moderate keeling, as do Skhūl 5, Predmost, UC101 and Lagoa Santa 6 specimens. Pronounced sagittal keeling is only observed in three of the 30 fossil specimens, UC103, Tepexpán and Kanalda, all chronologically very late. The non-Asiatic nature of this feature has already been discussed, but it is interesting to note that it is also inconsistent among modern East Asian fossils. Pronounced sagittal keeling seems to be a late specialisation of some modern human populations.

EA2: Horizontal course of the naso-frontal and fronto-maxillary sutures

This is clearly not an ancestral African trait, since the greatest inclination was recorded in the Ndutu specimen (Figure 5.1b). The only fossil specimens that present low values of this trait are the early modern fossils of Qafzeh 6 and Skhūl 5. Relatively low values are also observed in Keilor Minatogawa IV and Tepexpán, and as mentioned before, among recent sub-Saharan Africans. This trait does not seem to be geographically restricted at any point in the Upper Pleistocene.

EA3: Flat nasal saddle and nasal roof

This is not an ancestral trait. It is present only in Skhūl 5, Keilor (Figure 5.1c). This trait is very rare throughout the Upper Pleistocene and not geographically restricted at any point. The characteristic recent sub-Saharan African flat nasal saddle appears to be a later specialisation of the group.

EA4: M3 agenesis

This was only observed in the Liujiang specimen. All the other 21 fossils in which this feature could be recorded, including the East Asian Upper Cave and Minatogawa specimens and the East Asian derived Tepexpán and Lagoa Santa, lack this trait. It should be noticed that even though East Asians have a higher frequency of M3 agenesis than other populations, 74% of the East Asian and 96.6% of the Fueguian-Patagonian populations have third molars.

EA5: Reduced posterior dentition

A trend towards relative reduction of the molars in fossil modern humans is not identified. This is consistent with the variation among recent populations,

(a) Sagittal keeling

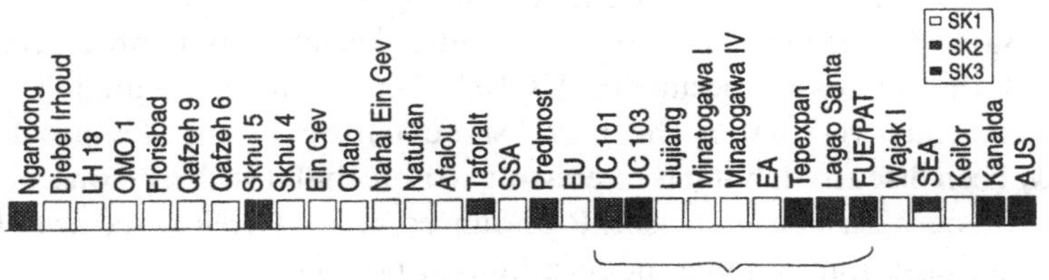

(b) Course of the naso-frontal and fronto-maxillary sutures (either individual or mean values)

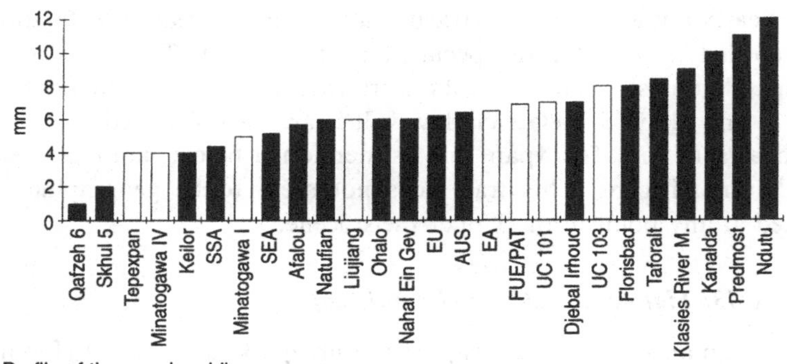

(c) Profile of the nasal saddle

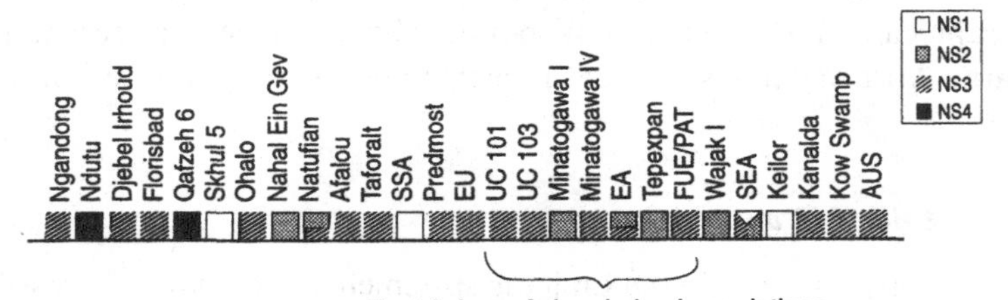

Figure 5.1 Incidence of the East Asian 'regional continuity traits' in some late Pleistocene hominid fossils (either individual values or the most frequent expression within a sample).
(a) Sagittal keeling.
(b) Course of the naso-frontal and fronto-maxillary sutures (in white: East Asian or American crania).
(c) Profile of the nasal saddle.
EU: Europeans; SEA: Southeast Asians; AUS: Australians; SSA: sub-Saharan Africans; FUE/PAT: Fueguian-Patagonians

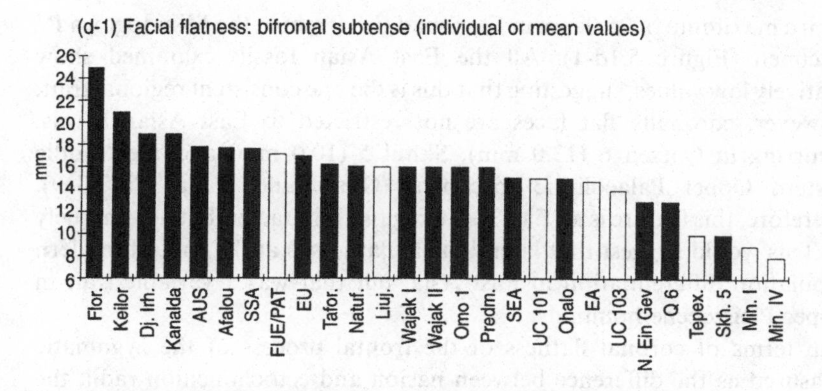

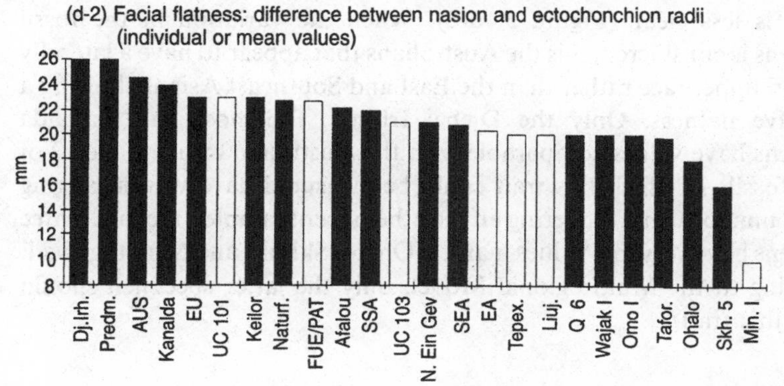

Figure 5.1 (*cont.*) (d) Facial flatness: d1 - bifrontal subtense; d2: difference between nasion and ectochonchion radii (in white East Asian or American crania).

where we observe a wide range of values, from the Fueguians and sub-Saharan Africans in whom first and third molars are of similar size, to the Europeans and Southeast Asians in whom the occlusal area of the third molars are on average 18.6% and 20.1% smaller than that of the first. In the fossils examined these values also ranged from relatively similar sized molars (Predmost) to much reduced third molars (Skhūl 4).

EA7: Coronal facial flatness

This was the second feature to show a distinctive East Asian character among recent populations, in which average bifrontal subtense values vary from 14.8 mm and 15.1 mm in East and Southeast Asians respectively, to 17.0 mm (Europe), 17.2 mm (Fueguians), 17.7 mm (sub-Saharan Africans) and 17.9 mm (Australians). The fossil values of bifrontal subtense range from a maximum of 25.0 mm in Florisbad to 8.0 mm in the Minatogawa IV specimen (Figure 5.1d-1). All the East Asian fossils examined show relatively low values, suggesting that this is the one consistent regional trait. However, coronally flat faces are not restricted to East Asian fossils, occurring in Qafzeh 6 (13.0 mm), Skhūl 5 (10.0 mm) and the Middle Eastern Upper Palaeolithic specimens (Ohalo and Nahal Ein Gev). Therefore, this feature is an East Asian regional character, but not uniquely so. This would suggest that lateral facial flatness is at the root of modern population differentiation in East Asia, but that was a variable trait in Upper Pleistocene hominids.

In terms of coronal flatness of the frontal process of the zygomatic measured as the difference between nasion and ectochonchion radii, the pattern is less clear (Figure 5.1d-2). When the variation in the fossil specimens is considered, it is the Australians that appear to have a laterally retracted upper face rather than the East and Southeast Asians showing a distinctive flatness. Only the Djebel Irhoud, Predmost and Kanalda specimens have values comparable with the Australian mean, while 12 of the 18 fossils in which this trait could be measured have values ranging from 23 mm to 20 mm (covering all the other recent samples' means). Three specimens have very low values, namely Ohalo, Skhūl 5 and Minatogawa I. According to the Multiregional Model, only the latter specimen should exhibit this trait.

EA10: Orbital shape

We have already seen that there is great within-group variation in the shape of the orbits in recent modern human regional populations, specially in the East Asian sample. The fossil specimens studied also show a wide range of forms (Table 5.3), but, similar to the recent samples, moderately oblique orbits are the most common type observed. The East Asian Pleistocene fossils all differ with each other in regard to orbital shape. Both UC101 and Liujiang show a moderately oblique superior border with a horizontal inferior one, but they differ in the inclination of the frontal process of the zygomatic, revealing a broader bifrontal breadth in the Upper Cave specimen. UC103 has the superior and inferior margins horizontal, resembling Australian skulls. The two Minatogawa specimens are different from each other. In Minatogawa I the superior border is horizontal, while it is oblique in Minatogawa IV. It is clear that a particular orbital shape has not been a regional character at any point in the late Pleistocene. This is exactly what would be expected if one considers that the superior, lateral and inferior components which determine the shape of the orbit, actually grow and relate to different aspects of the facial skeleton.

Table 5.3. *Orbital shape among the fossil specimens examined, represented by the combination of the variation in the inclination of the three (superior, inferior and lateral) orbital elements: superior and inferior – I: horizontal, II: oblique, III: very oblique; lateral – I: sharp superior angle, vertical margin, II+III: broad superior angle, margin broadening outwards, either slightly or pronouncedly, IV; margin directed inwards)*

Orbital shape	Fossil specimen
O2: I/II/I	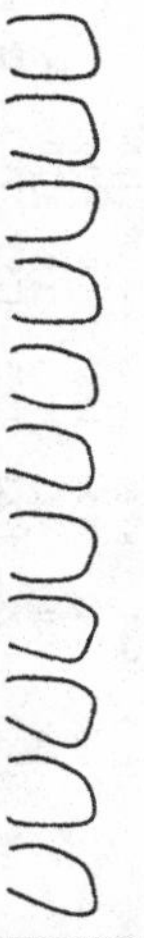 Nahal Ein Gev, Predmost, UC 103
O3: I/II/II	Minatogawa I
O6: I/IV/I	Kanalda, Kow Swamp 1
O11: II/II/I	Qafzeh 9, Liujiang, Lagoa Santa, Kow Swamp 5
O12: II/II/II	Ohalo
O13: II/II/III	Minatogawa IV
O15: II/IV/I	UC101, Tepexpán
O16: II/IV/II	Wajak I, Keilor
O17: II/IV/III	Kow Swamp 15
O19: III/II/I	Florisbad
O21: III/II/III	Djebel Irhoud, Qafzeh 6

Temporal distribution of the Australian regional traits

See Table 5.4.

AUS1: Pronounced development of the supraorbital torus (ST5)

This is clearly an ancestral condition of all archaic hominids, present in most early modern humans. A horizontal continuous torus, in which there is no clear division between the superciliary arches and the trigone is observed in Ngandong, Florisbad, Qafzeh 6, Skhūl 5 (?), Predmost, Kanalda, Kow Swamp 1 and four individuals of the recent Australian sample that also show a torus formation (i.e. no separation of the arches and trigones), but there is a reduction of thickness laterally and the arches are well-defined as such (Figure 5.2a). Clearly, although it is a rare occurrence, supraorbital tori do occur in fossil and recent modern crania. The remains of Omo 1, Qafzeh 9, Skhul 4, Ohalo, Tepexpán, Keilor, Kow Swamp 5 and 15 and 58.6% of Fueguian-Patagonian, 43.6% of

Table 5.4.A. *Values and incidence of the Australian regional continuity traits in the recent samples and some late Pleistocene fossil hominids*

Regional continuity traits	Ngan	Ndutu	Dj Irh	LH18	Omo 1	KRM	Flor	*SSA*	Afa	Taf	Q9	Q6	SK5	SK4	EGev	Oha	NEG	Natuf	Pred	*EU*
Supraorbital torus	ST5	ST5	ST5	ST5	ST4	ST2	ST5	ST1	ST4/2	ST2	ST4	ST5	ST5	ST4	ST3	ST4	ST1	ST2	ST5	*ST3*
Lambdoid depression	(−/+)	(−)	(−)	(−)	(−)	–	(+)	(−)	(−)	(−)	–	(−)	(+)	(−)	(−)	(−)	(−)	(−)	(−)	(−)
Size of pterion articulation	10.3	–	7.0	8.0	–	–	–	10.9	10.0	12.4	7.0	–	10.0	–	8.0	9.0	12.0	10.5	15.0	13.5
Profile of infraglabellar notch	IN2/3	IN3	IN3	–	IN4	IN2	IN3	IN1/2	IN3	IN3	–	IN3	IN4	–	–	IN2	IN1	IN2	IN4	IN3
Position of min. frontal br:																				
WFBZFS	26.8	23.0	25.0	–	23.0	–	22.0	18.6	21.7	24.9	15.0	22.0	22.0	23.0	17.0	13.0	9.0	15.6	19.0	15.6
WFBORB	38.3	28.0	22.0	24.0	26.0	–	24.0	17.2	20.8	22.2	17.0	25.0	24.0	27.0	17.0	14.0	12.0	15.7	18.0	15.6
Position of max. parietal br.	36.7	–	33.0	37.4	–	–	–	42.41	45.7	46.9	24.4	–	27.5	–	45.6	37.4	27.5	45.5	30.3	35.7
Zygomaxillary tuberosity	–	–	–	–	–	ZT1	–	ZT1	ZT1	ZT1	ZT1	ZT4	ZT1	–	ZT2	ZT3	ZT1	ZT1	ZT2	ZT3
Eversion of malars	–	–	–	–	–	–	–	53.9	63.9	73.2	–	65.0	62.0	–	–	63.0	58.0	57.5	62.0	63.0
Narial margins	–	–	N8	N6	–	–	–	N8	N3	N3	N3	N1	–	N7	–	N3	N1	N3	N3	N3
Curvature of alv. pl. maxilla	–	–	–	–	–	–	–	1.6	1.9	1.3	15.0	0.0	5.0	–	–	0.0	9.0	5.6	12.0	3.3
Zygomatic trigone	TR5*	TR4	TR4	TR4	TR4	–	TR4	TR1	TR2/3	TR2	TR1	TR4	TR4	–	TR2	TR3	TR2	TR2	TR4	TR2
Size of suprameatal tegmen	15.3	10.5	–	6.5	–	–	–	7.9	10.6	10.2	10.0	13.5	12.5	10.0	9.0	13.5	9.0	10.6	12.5	10.8
Petro-tympanic angle	42°	–	–	–	–	–	–	31.8°	27.6°	31.2°	–	–	–	–	–	–	–	–	–	29.7°
Occipital crest	OCR2/3	OCR1	OCR1	–	OCR1	–	–	–	OCR1	OCR2	OCR1	–	OCR1	OCR1	–	OCR2	OC1	OCR1	OCR2	OCR2
Occipital torus	OT7	OT6	OT7	OT3	OT6	–	–	OT3	OT3	OT3	OT3	OT4	OT6	–	OT4	OT4	OT3	OT3	OT4	OT4
Facial projection:																				
Gnathic index	–	–	–	–	–	–	–	103.7	94.1	97.6	–	–	15.6	–	–	99.0	101.1	98.3	104.6	95.3
Alveolar: PRR – SSR	–	–	2.0	–	–	–	–	8.5	5.1	4.5	14.0	7.0	10.0	–	–	3.0	8.0	6.0	4.0	5.2
Facial: SSR – NAR	–	–	3.0	–	–	–	–	3.8	0.0	4.0	–	1.0	12.0	–	–	7.0	−1.0	0.7	4.0	0.2
Frontal angle	142.5	–	133.3	141.9	135.4	–	140.0	125.3	132.2	132.2	136.3	125.3	132.1	129.5	131.2	122.7	122.4	127.0	134.7	128.9
Rounding of the orbit i-l. m	–	–	RO1	–	RO3	RO2	RO2	RO2	RO2	RO2	RO3	RO3	RO2	–	RO2	RO2	RO2	RO2	RO2	RO1
Horizontal sup. orbital m	–	–	O'3	–	–	–	O'3	O'2	O'2	O'2	O'2	O'3	O'2	–	–	O'2	O'1	O'2	O'1	O'2

*Deformed.

Table 5.4.B. *Artificial cranial deformation may have affected the data for the Kow Swamp crania (1,5,15), specially in terms of frontal angle*

	UC101	UC103	Liu	Min. I	Min.IV	*EA*	Tepex	L.Sta6	FUE	Keilor	Kan	WaI	Wa II	K.S1*	K.S5*	K.S15	AUS	SEA
Supraorbital torus	ST3	ST2	ST3	ST3	ST3	ST1/2	ST4	ST2	ST4	ST4	ST5	ST3	ST3	ST5	ST4	ST4	ST4	ST2
Lambdoid depression	(+)	(−)	(−)	(−)	(−)	(−)	(+)	(−)	(−)	(−)	(−)	(+)	–	–	(−)	–	(−)	(−)
Size of pterion articulation	(12?)	–	15.0	8.0	–	12.3	14.0	14.0	13.5	–	9.0	12.0	–	–	–	–	8.1	12.9
Profile infraglabellar notch	IN3	IN2	IN2	IN3	IN3	IN1/2	IN2	IN2	IN2	IN2	IN3	IN3	IN3	IN3	IN3	IN3	IN3	IN1/2
Position of min. frontal br:																		
WFBZFS	22.0	16.0	18.0	25.0	–	16.7	27.0	19.0	22.0	30.0	33.0	27.0	18.0	38	22	32.0	23.0	17.1
WFBORB	20.0	16.0	20.0	23.0	16.0	16.0	24.0	16.0	19.7	29.0	33.0	20.0	16.0	38	22	32.0	22.6	15.2
Position of max. par. br.	32.2	38.4	27.0	33.6	–	40.6	–	34.9	37.0	34.4	34.4	29.3	–	48*?	–	–	37.0	42.3
Zygomaxillary tuberosity	ZT3	ZT2	ZT2	ZT2	ZT2	ZT1/2/3	ZT4	ZT1	ZT2/3	ZT4	ZT4	ZT2	–	ZT4	ZT4	ZT4	ZT4/3	ZT3/2
Eversion of malars	64.0	61.0	52.0	62.0	–	63.1	–	–	64.2	–	83.0	67.0	–	–	63	–	57.5	58.9
Narial margins	N8	N8	N3	–	N3	N3	N3	N3	N3	N3	N6	–	N6	N3	N3	N6	N8	N3
Curvature of alv. pl. maxilla	5.0	4.0	3.0	–	–	3.0	–	–	4.8	–	1.0	–	–	-2	–	–	0.9	2.3
Zygomatic trigone	TR4	TR3	TR3	TR3	–	TR2	TR3	TR3	TR3/4	TR4	TR4	TR3	TR3	TR4	TR4	TR4	TR3	TR2
Size of suprameatal tegmen	13.0	13.5	12.0	8.5	–	10.1	12.0	10.0	11.4	10.5	7.5	13.0	–	–	–	–	8.7	9.7
Petro-tymp. angle	26.0°	32.0°	31.0°	–	–	31.1°	–	–	30.2°	–	–	–	–	–	–	–	30.2°	30.1°
Occipital crest	OCR2	OCR1	OCR1	OCR1	OCR2	OCR1	OCR2	OCR2	OCR2	OCR1	OCR1	OCR2	OCR1	OCR2	OCR2	–	OCR1	OCR1
Occipital torus	OT3	OT3	OT3	OT3	OT6	OT3/4	OT6	OT6	OT6	OT6	OT4	OT3	OT4	OT5	Ot3	–	OT6/3	OT3
Facial projection:																		
Gnathic index	100.0	100.9	100.0	101.0	108.4	96.6	–	–	97.6	99.1	100.9	102.9	–	–	–	–	101.3	97.6
Alveolar: PRR – SSR	7.0	6.0	6.0	6.0	–	6.2	–	–	6.8	6.0	9.0	7.0	–	–	–	–	7.7	6.9
Facial: SSR – NAR	1.0	8.0	−2.0	7.0	–	1.1	1.0	–	2.4	4.0	6.0	3.0	–	–	–	–	3.8	1.3
Frontal angle	132.6	132.6	129.1	136.7	–	129.3	131.7	126.9	135.7	134.4	134.0	133.6	–	146.8*?	147.7*?	–	130.8	129.5
Rounding of inf-lat. orbital m	RO2	RO2	RO2	RO3	RO3	RO1	RO2	RO2	RO2	RO2	RO3	RO2	–	RO2	RO2	RO2	RO3	RO1/2
Horizontal sup. orbital m	O'3	O'1	O'2	O'1	O'2	O'2	O'2	O'2	O'2	O'2	O'1	O'2	O'2	O'2	O'2	O'2	O'1	O'2

*Deformed.

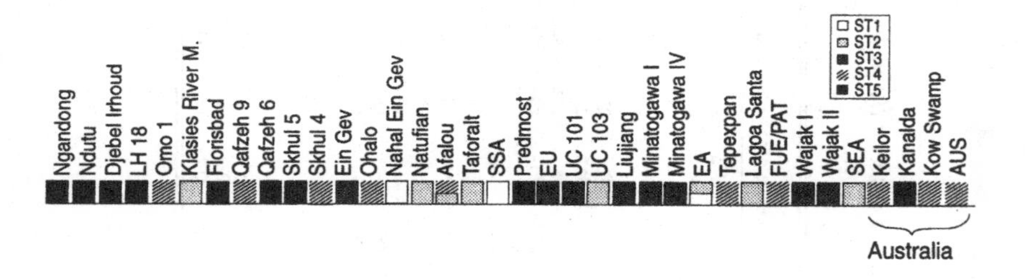

(b) Lambdoid depression

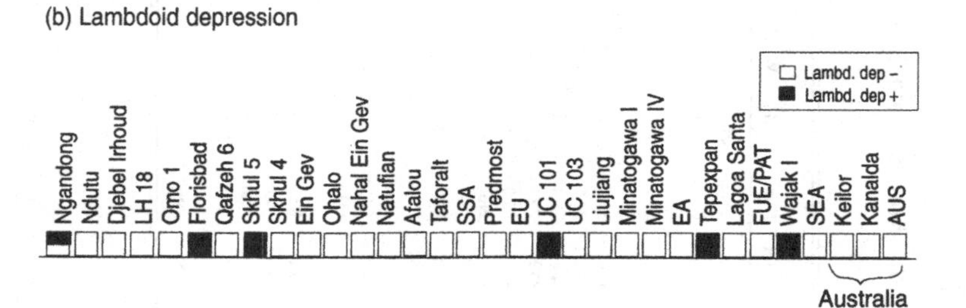

(c) Size of pterion articulation (individual or mean values)

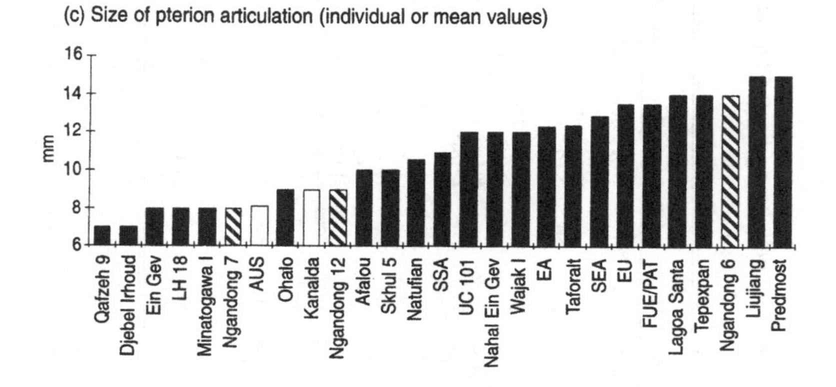

(d) Depth of infraglabellar notch

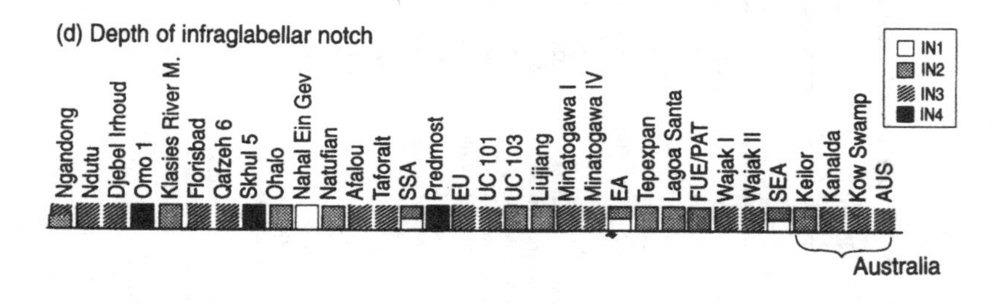

Australian and 37.8% of Afalou crania have very pronounced supraorbital ridges (ST4). The incidence of this feature does not support a particular relationship between Ngandong and recent Australians.

AUS2: A lambdoid depression

This may be observed in half of the Ngandong specimens, in Florisbad, Skhul 5, UC101, Tepexpán and Wajak I. This condition is very variable, and although it is absent in late African archaics and most early modern specimens, its presence in such geographically and temporally variable specimens suggests that its occurrence has always been aleatory (Figure 5.2b). The incidence of this feature does not support a particular relationship between Ngandong and recent Australians.

AUS3: The pterion articulation

This is significantly smaller in the Australian sample than in other recent populations as predicted by the Multiregional Model. However, in this case it is the Ngandong fossils that do not consistently follow the prediction of the Multiregional Model (Figure 5.2c). On the three Ngandong specimens that this measurement could be taken, the values range from 8 mm (similar to the Australian mean) to 14 mm, higher than any recent population average. On the other hand, values smaller than the Australian mean are observed in Djebel Irhoud, Qafzeh 9, LH18, Ein Gev and Minatogawa I, while the values of Kanalda and Ohalo are only slightly higher than the Australian average. With the exception of Ngandong 6, all the earlier specimens have small values, and increased pterion articulations seem to be a feature of later modern humans. The Australians, and to a lesser degree the sub-Saharan Africans, have retained the ancestral condition of a small articulation at pterion. Pterion articulation size relates to vault expansion in terms of breadth (PTE/XPB: $r = 0.282$, $p < 0.001$, $N = 148$); PTE/AUB: $r = 0.228$, $p < 0.001$, $N = 148$) but not length (PTE/GOL: $r = -0.066$,

---

Figure 5.2 Incidence of Australian 'regional continuity traits' in some late Pleistocene hominid fossils (either individual values or the most frequent expression within a sample).
(a) Development of the supraorbital torus-ridges.
(b) Lambdoid depression.
(c) Size of pterion articulation.
(d) Depth of infraglabellar notch.
(See Figure 5.1 for explanation of abbreviations.)

$p>0.05$, $N = 148$) or height (PTE/BBH: $r = 0.079$, $p>0.05$, $N = 143$). This feature does not support a particular relationship between Ngandong and recent Australians.

AUS4: A deep and narrow infraglabellar notch (IN4)

This is very rare, only observed in Omo 1, Skhūl 5 and Predmost (Figure 5.2d). Most importantly, is that it is not the most common condition in either the Australian or Ngandong crania. As mentioned in Chapter 4, the majority of Australian crania (42.1%) actually show a broad deep curve as a sub-glabellar profile (IN3), which is clearly the most common condition in most of the early fossils examined (Ndutu, Djebel Irhoud, Florisbad, Qafzeh 6), also present in some of the later material (UC101, Minatogawa I and IV), and most of the Australian fossils (Kanalda, Wajak I and II, Kow Swamp 1, 5 and 15). Among the Ngandong crania, a broad deep curve (IN3) is observed in three specimens, while another two present only slight infraglabellar depression (IN2). This feature does not support a particular relationship between Ngandong and recent Australians.

AUS5: Position of minimum frontal breadth

It is clear that the Ngandong hominids and several Australian fossils (Kanalda, Kow Swamp 1, 5 and 15, Keilor) show comparatively large distances from the point of minimum frontal breadth to both the zygomatico-frontal suture and the orbital margin (Figures 5.3a,b). However, they are not unique in these values. Wajak I, Tepexpán, Djebel Irhoud, Minatogawa I, Taforalt, Ndutu, Omo 1 and Skhūl 4 have the distance from minimum frontal breadth to the zygomatico-frontal suture larger than the recent Australian mean; and Ndutu, Skhul 4, Omo 1, Tepexpán, Florisbad, Skhūl 5 and Minatogawa I have the distance from minimum frontal breadth to the orbital margin larger than the recent Australian mean. But more importantly, the difference between the two values, the distances minimum frontal breadth to the zygomatico-frontal suture and minimum frontal breadth to the orbital margin (Figure 5.3c), shows that the Ngandong and Australian similarities do not reflect the same structure. The Ngandong hominids, and to a lesser degree Ndutu and early modern humans (Omo 1, Florisbad, Qafzeh 6 and 9, Skhūl 4 and 5), are characterised by very large zygomatic trigones that are not only pronounced in vertical thickness, but in antero-posterior width. Consequently, the insertion of the *temporalis* onto the anterior frontal bone is placed backwards in relation to recent modern crania, and thus the great

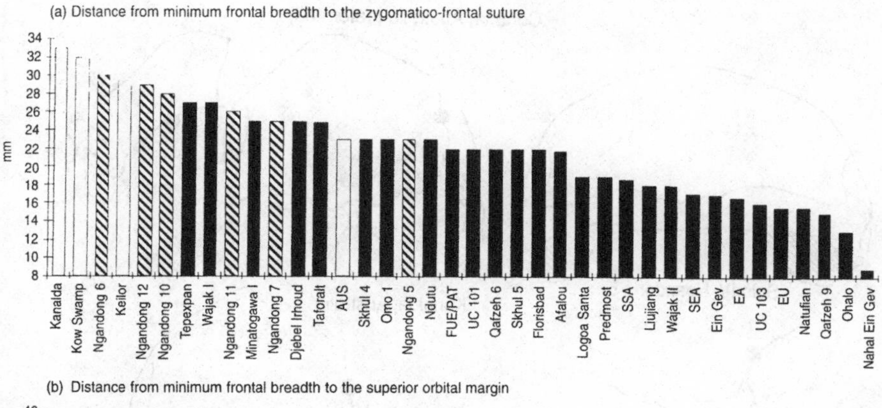

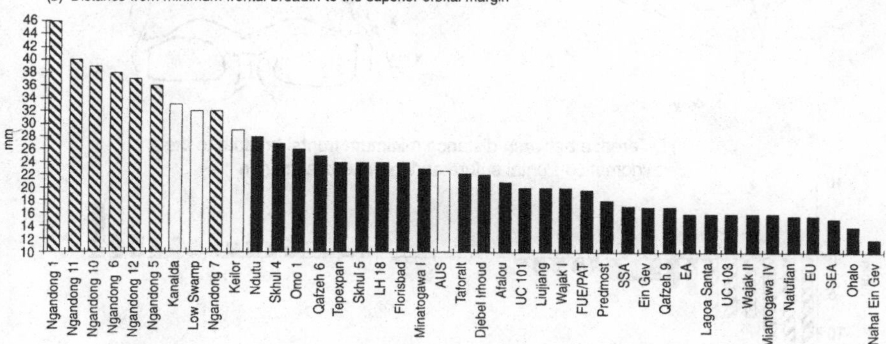

Figure 5.3 Relative position of minimum frontal breadth in some late Pleistocene hominid fossils (in white: Australian crania; hatched: the Ngandong specimens). (a) Distance between minimum frontal breadth and the zygomatico-frontal suture. (b) Distance between minimum frontal breadth and the superior orbital margin. (See Figure 5.1 for explanation of abbreviations.)

difference between the two measures of position of minimum frontal breadth. This pattern is markedly different from that observed in all recent modern human crania, including the Australians, in which the difference is either absent or positive, i.e. the distance from minimum frontal breadth to the zygomatico-frontal suture is actually larger than the distance from minimum frontal breadth to the orbital margin. Although there is a clear temporal trend to this feature, with the early modern crania presenting small negative values, the Ngandong hominids' lateral supraorbital area stands out as a different structure. The position of minimum frontal breadth does not support a particular relationship between Ngandong and recent Australians.

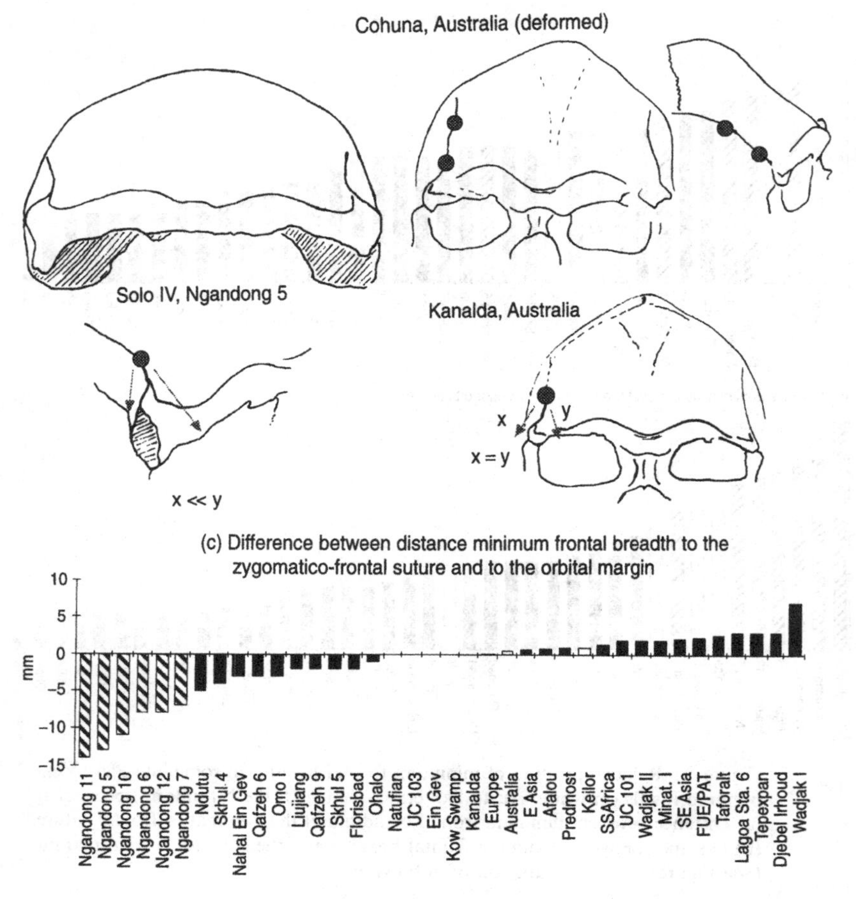

Figure 5.3 (*cont.*) (c) Difference between the distance from minimum frontal breadth to the zygomatico-frontal suture and to the orbital margin, resulting in a different morphological pattern in archaic and modern hominids as seen in the comparison between Ngandong and Australian crania.

AUS6: Position of maximum parietal breadth

Among recent regional populations this is lowest in European but not in Australian crania. Furthermore, only one of the three Ngandong specimens in which this measure could be taken shows a low position of maximum parietal breadth (Ngandong 12), so that this trait does not seem to follow the prediction of the Multiregional Model in either its present or archaic incidence (Figure 5.4a).

It was mentioned in Chapter 4 that a number of recent skulls have

Table 5.5. *Differences between maximum parietal breadth and breadth on the temporals (A) and on the supramastoid crests (B) in some late Middle and Upper Pleistocene hominids*

Fossil specimen	(A) Max. parietal breadth – max. temporal breadth	(B) Max. parietal breadth – max. breadth at the supramastoid crests
Ngandong 6	−1	−6
Ngandong 7	−1	−7
Ngandong 10	−5	−11
Ngandong 11	−7	−11
Ngandong 12	−7	−12
Ndutu	–	−6
Djebel Irhoud	−2	2
LH 18	–	1
Qafzeh 9	–	6
Skhūl 5	0	3
Ein Gev	–	17
Ohalo	−1	8
Nahal Ein Gev	–	10
Natufian	4.4 ± 4 (n=8)	5.9 ± 5.5 (n=15)
Predmost	−3	−5
UC 101	−1	0
UC 103	−5	−3
Liujiang	3	6
Minatogawa I	–	3
Tepexpán	0	5
Lagoa Santa 6	0	−3
Keilor	0	0
Kanalda	1	0
Wadjak I	–	4
Kow Swamp 5*	7	12

*Deformed.

maximum cranial breadth not on the parietal bones but on the temporals, and a few on the supramastoid crests. Maximum cranial breadth placed on the temporal bones rather than on the parietals is by far the most common condition in Upper Pleistocene modern fossil crania (Table 5.5). On the other hand, it remains true that the majority of modern skulls have poorly developed supramastoid crests, and only in very few cases (Predmost, UC103 and Lagoa Santa 6) is maximum cranial breadth placed at the supramastoid crests. This clearly separates archaic skulls like Ndutu and Ngandong, which show a large difference between maximum breadth at the supramastoid crests and the smaller vault width, ranging from 6 to 12 mm (Table 5.5).

AUS7: Pronounced development of a zygomaxillary tuberosity (ZT4)

This is one of the features that differentiates most recent Australian crania from other geographical populations. It is not possible to assess the occurrence of this trait in the Ngandong fossils because these are represented only by vaults, so that the Australasian archaic regional incidence cannot be corroborated. Among the Upper Pleistocene fossils, this feature is rare. Most specimens have no zygomaxillary tuberosity (ZT1: KRM, Qafzeh 9, Skhūl 5, Nahal Ein Gev, Afalou, Taforalt, Natufians, Lagoa Santa 6 – also the most common condition among recent sub-Saharan Africans), some have a slight tubercle on their malar surfaces (ZT2: UC103, Liujiang, Minatogawa I and IV, Ein Gev, Wajak I – the most common among recent Fueguians-Patagonians), and very few have a pronounced tubercle (ZT3: Ohalo, UC101 – the most common among recent Europeans, East Asians, Southeast Asians, some Fueguian-Patagonians and some Australians). Only Qafzeh 6, Tepexpán and all the Australian fossils (Keilor, Kanalda, Kow Swamp 1, 5 and 15) show a pronounced zygomaxillary tuberosity, also the most common condition among recent Australians (Figure 5.4b). Given that there is disagreement about whether this feature, and certainly its degree of expression, occurs in Sangiran 17 (Groves, 1989b; Habgood, 1989; Wolpoff, 1989a), and that all other Javanese archaic hominids lack the facial skeleton, the regional archaic incidence of this trait is unclear. A pronounced zygomaxillary tuberosity in the form of a ridge running parallel to the lower border of the malar is clearly an Australian regional

---

Figure 5.4 Incidence of the Australian regional continuity traits in some late Pleistocene hominid fossils (either individual values or the most frequent expression within a sample).
(a) Relative position of maximum parietal breadth (in white: Australian crania; hatched: the Ngandong specimens).
(b) Development of the zygomaxillary tuberosities.
(c) Eversion of the lower border of the malars (in white: Australian crania).
(d) Curvature of the alveolar plane of the maxilla (in white: Australian crania).

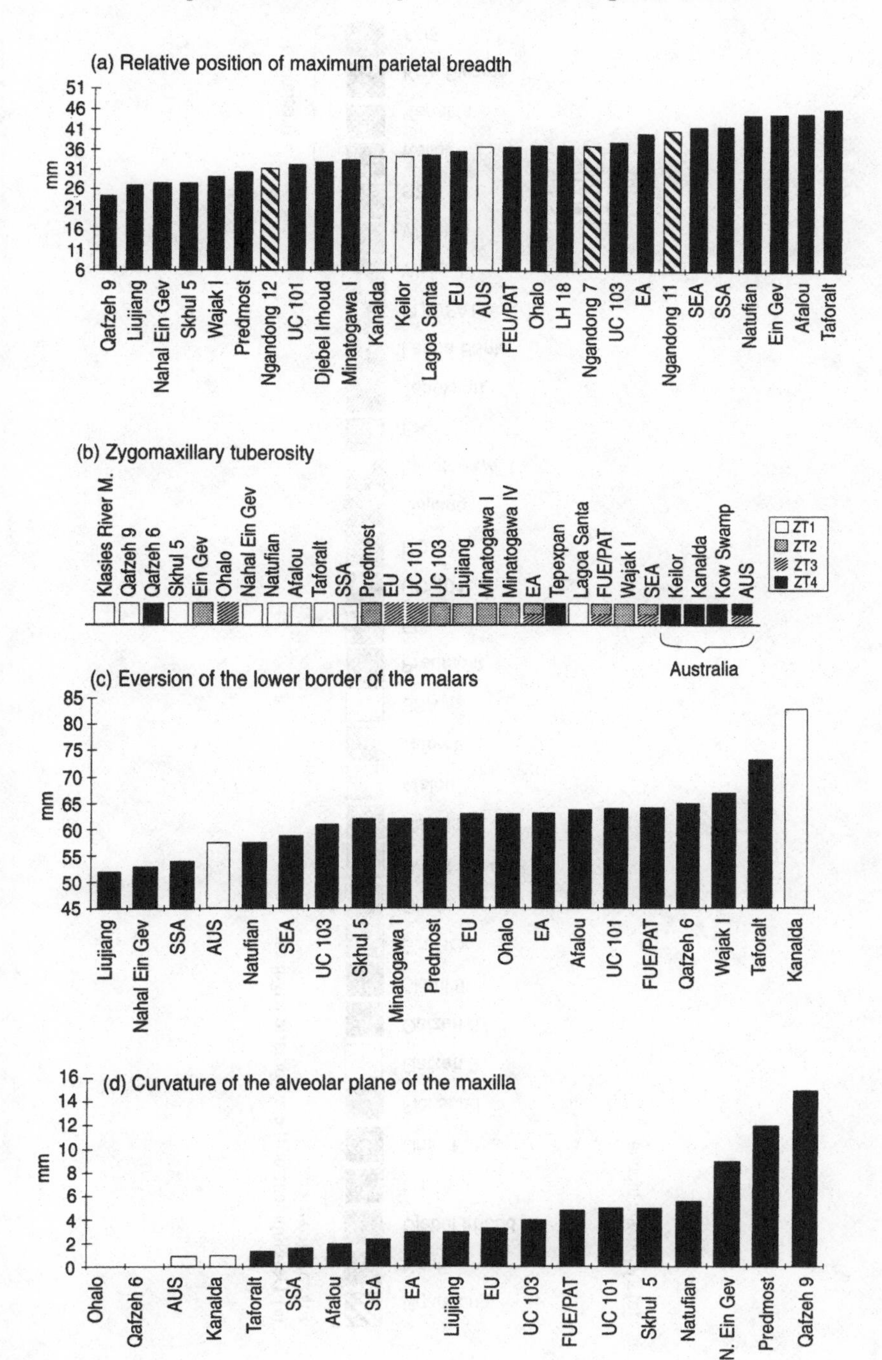
(a) Relative position of maximum parietal breadth
mm
51
46
41
36
31
26
21
16
11
6
Qafzeh 9
Liujiang
Nahal Ein Gev
Skhul 5
Wajak I
Predmost
Ngandong 12
UC 101
Djebel Irhoud
Minatogawa I
Kanalda
Keilor
Lagoa Santa
EU
AUS
FEU/PAT
Ohalo
LH 18
Ngandong 7
UC 103
EA
Ngandong 11
SEA
SSA
Natufian
Ein Gev
Afalou
Taforalt
(b) Zygomaxillary tuberosity
Klasies River M.
Qafzeh 9
Qafzeh 6
Skhul 5
Ein Gev
Ohalo
Nahal Ein Gev
Natufian
Afalou
Taforalt
SSA
Predmost
EU
UC 101
UC 103
Liujiang
Minatogawa I
Minatogawa IV
EA
Tepexpan
Lagoa Santa
FUE/PAT
Wajak I
SEA
Keilor
Kanalda
Kow Swamp
AUS
ZT1
ZT2
ZT3
ZT4
Australia
(c) Eversion of the lower border of the malars
mm
85
80
75
70
65
60
55
50
45
Liujiang
Nahal Ein Gev
SSA
AUS
Natufian
SEA
UC 103
Skhul 5
Minatogawa I
Predmost
EU
Ohalo
EA
Afalou
UC 101
FUE/PAT
Qafzeh 6
Wajak I
Taforalt
Kanalda
(d) Curvature of the alveolar plane of the maxilla
mm
16
14
12
10
8
6
4
2
0
Ohalo
Qafzeh 6
AUS
Kanalda
Taforalt
SSA
Afalou
SEA
EA
Liujiang
EU
UC 103
FUE/PAT
UC 101
Skhul 5
Natufian
N. Ein Gev
Predmost
Qafzeh 9

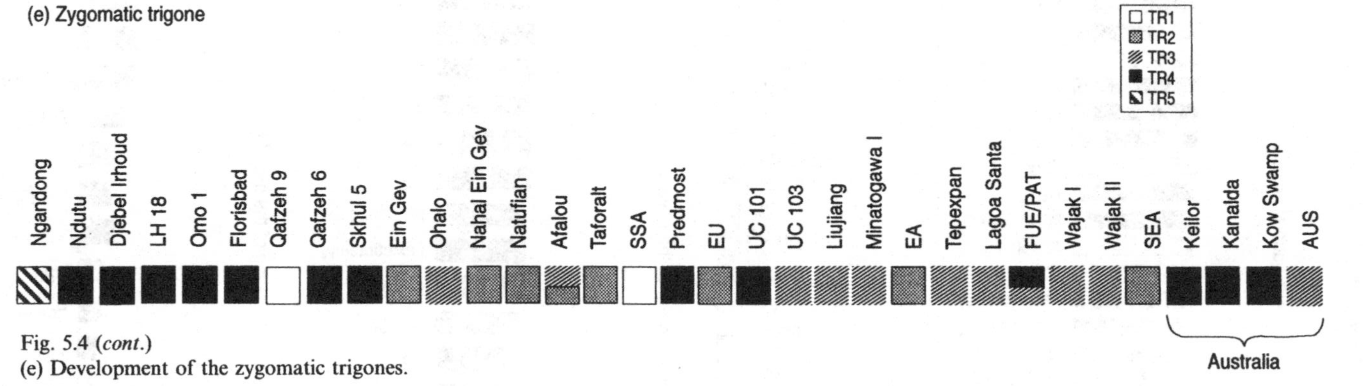

Fig. 5.4 (*cont.*)
(e) Development of the zygomatic trigones.

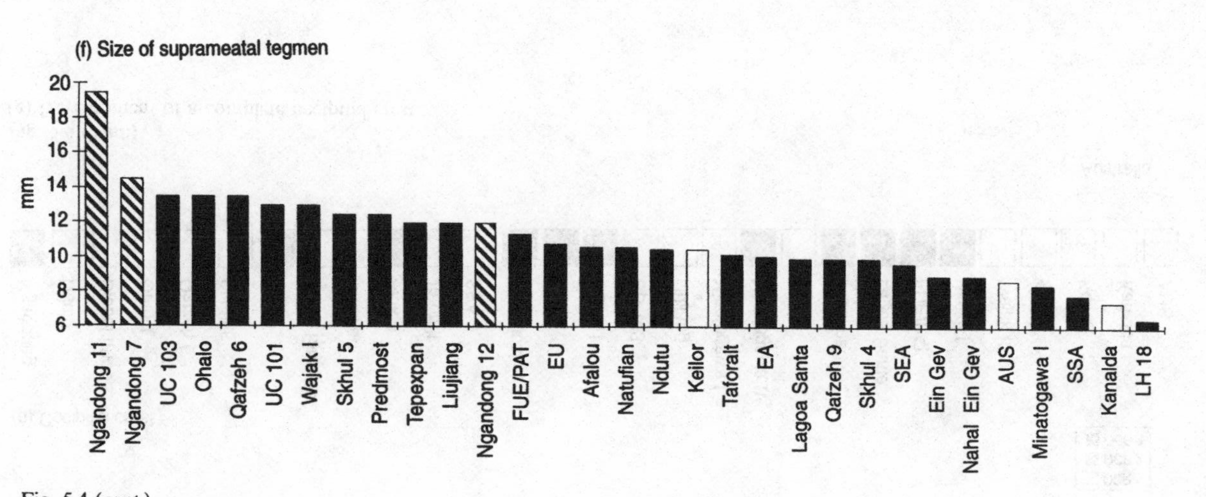

Fig. 5.4 (*cont.*)
(f) Size of the suprameatal tegmen (in white: Australian crania; hatched the Ngandong specimens).

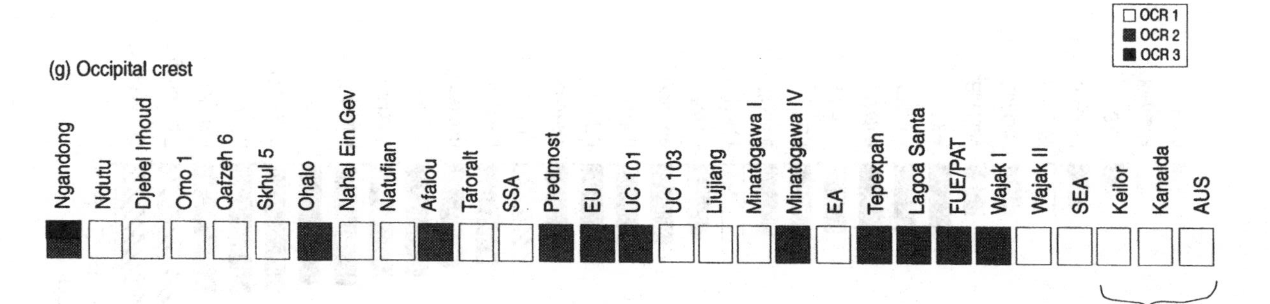

Fig. 5.4 (*cont.*)
(g) Development of a complete occipital crest.

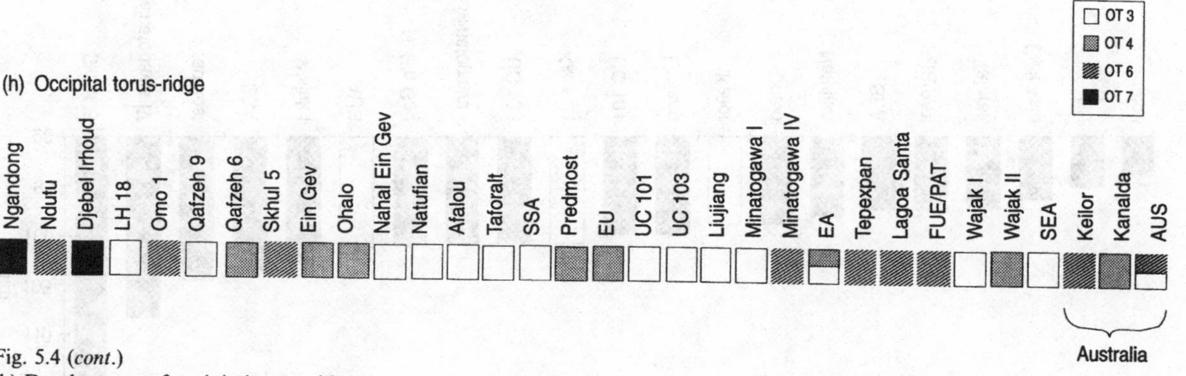

Fig. 5.4 (*cont.*)
(h) Development of occipital torus-ridges.

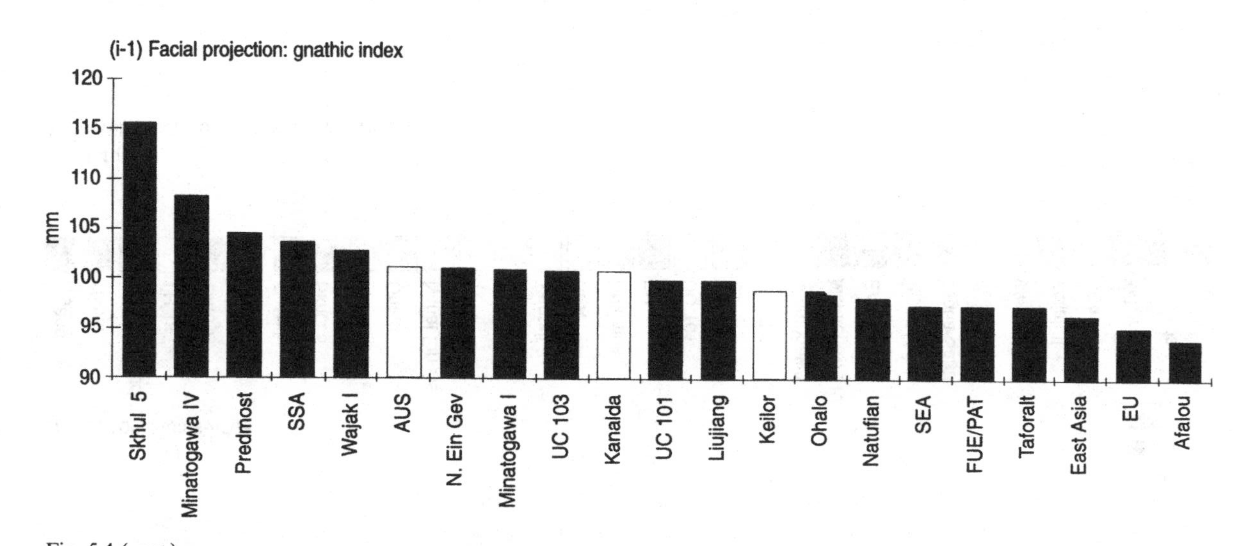

Fig. 5.4 (*cont.*)
(i) Facial prognathism (in white: Australian crania).

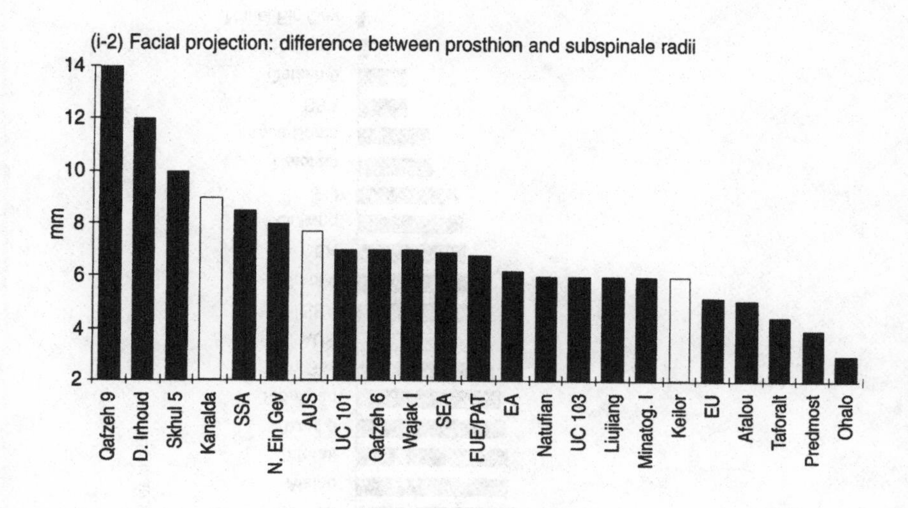

(i-2) Facial projection: difference between prosthion and subspinale radii
mm
14
12
10
8
6
4
2
Qafzeh 9
D. Irhoud
Skhul 5
Kanalda
SSA
N. Ein Gev
AUS
UC 101
Qafzeh 6
Wajak I
SEA
FUE/PAT
EA
Natufian
UC 103
Liujiang
Minatog. I
Keilor
EU
Afalou
Taforalt
Predmost
Ohalo

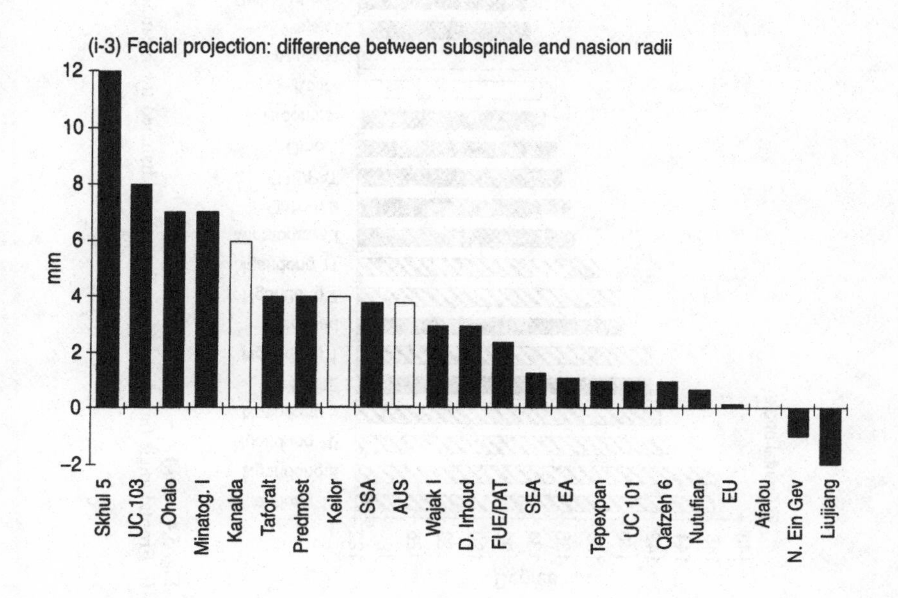

(i-3) Facial projection: difference between subspinale and nasion radii
mm
12
10
8
6
4
2
0
-2
Skhul 5
UC 103
Ohalo
Minatog. I
Kanalda
Taforalt
Predmost
Keilor
SSA
AUS
Wajak I
D. Irhoud
FUE/PAT
SEA
EA
Tepexpan
UC 101
Qafzeh 6
Nutufian
EU
Afalou
N. Ein Gev
Liujiang

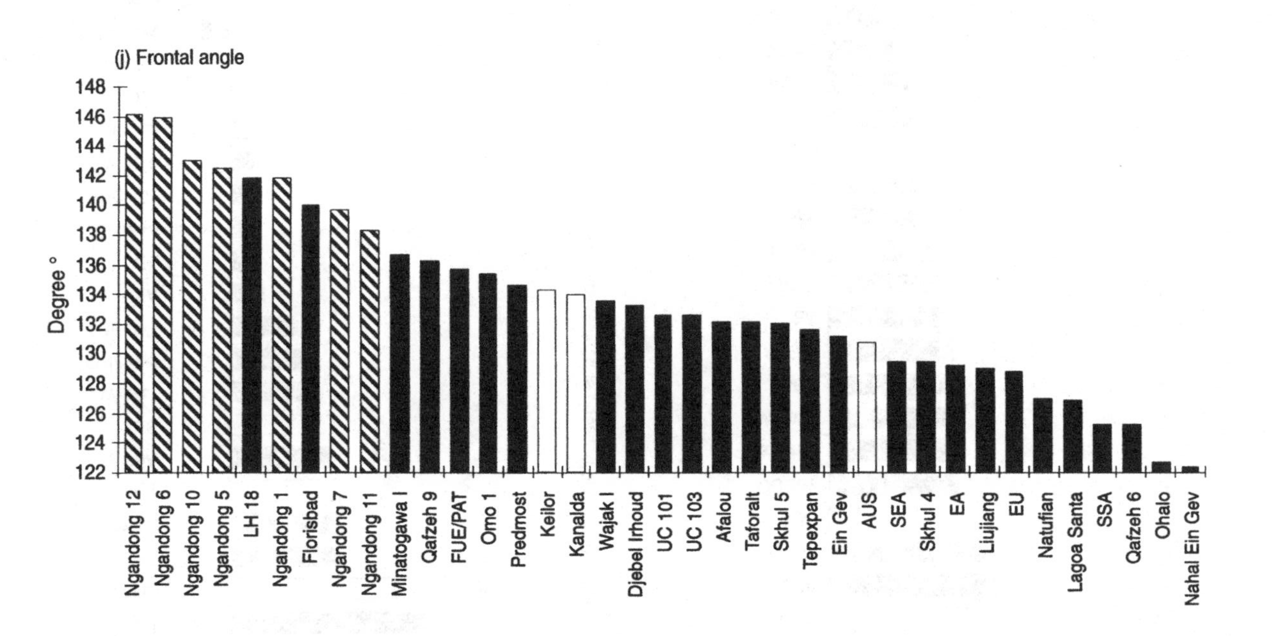

Fig. 5.4 (*cont.*)
(j) Frontal angle (in white: Australian crania; hatched the Ngandong specimens).

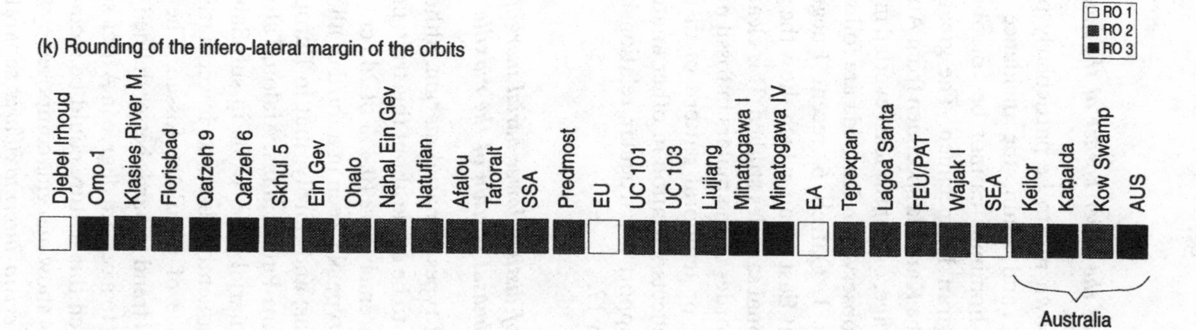

Fig. 5.4 (*cont.*)
(k) Rounding of the infero-lateral margin of the orbits. (See Figure 5.1 for explanation of abbreviations.)

trait. It occurs in Qafzeh 6 and possibly Sangiran 17, so that its ancestral source remains open.

AUS8: Eversion of the lower border of the malars

This has already been shown not to be particularly pronounced among recent Australian crania, and again, the incidence of this feature in Southeast Asian archaic hominids cannot be corroborated but for the single reconstructed Sangiran 17 specimen. The greatest value of malar eversion was recorded in the Kanalda specimen from Australia while recent Australians show, on average, one of the least everted malar configurations (Figure 5.4c). The values observed in Kanalda are followed by the average of Taforalt, then Wajak I, Qafzeh 6, recent Fueguians-Patagonians, UC101, Afalou, and recent East Asian, which show the greatest eversion of the five broad recent regional samples studied. It is clear that the incidence of eversion of the lower border of the malars in fossil modern populations shows no distinct spatial or temporal pattern of distribution, and it is probable that this trait reflects variation in other anatomical parameters. This feature does not support a particular relationship between recent Australians and Sangiran 17.

AUS9: Absence of distinct lower narial margins dividing the nasal floor from the subnasal portion of the maxilla

The recent distribution of types of narial margin, although very variable, allowed three conclusions to be made about this trait: firstly, that complete absence of a margin is extremely rare (0.4% of 245 crania); secondly, that there is one predominant type N3, seen in a third of all modern crania and observed in all populations; and thirdly, that in Europeans, East Asians, Southeast Asians, Fueguians-Patagonians, Afalou, Taforalt and Natufians, N3 is the most common narial type, while in sub-Saharan Africans and Australians, the most frequently observed lower narial margins are less well-defined, with presence of maxillary fossae. The fossil incidence is consistent with a temporal trend towards better defined narial margins, not observed in Australia, but to a certain degree in Africa south of the Sahara. Of the early fossils in which this feature could be recorded clearly, Djebel Irhoud, LH 18 and Skhūl 4 show margins composed by the *crista maxillaris* of Holl (1882) with the *crista intermaxillaris* separated internally, and formation of maxillary fossae (N6, N7, N8). The Upper Cave specimens, all the Australian fossils, recent Australians and Africans show similarly formed narial margins (N6 and N8). The two Qafzeh specimens examined

(6 and 9) do not follow this pattern, and show very clear narial margins (N1 and N3), formed by the *crista intermaxillaris*. These distinct types, that together account for more than 50% of the types observed in recent crania, are the ones observed in all the other fossil specimens. It is possible to hypothesise that margins formed by the *crista maxillaris*, which are also less well defined, were the ancestral condition for modern humans, retained to a certain degree in Africans, early fossil Asians (UC and Wajak) and fossil and recent Australians. This feature does not support a particular relationship between recent Australians and Sangiran 17.

AUS10: Curvature of the posterior alveolar plane of the maxilla

The recent distribution of this trait was found to be the opposite of that expected by the Multiregional Model, since the Australians show the least curvature of the posterior alveolar plane of the maxilla. The Southeast Asian character of this trait (again, observable only in Sangiran 17) has also been called into question recently by a new reconstruction of Sangiran 17 by Baba *et al.* (1993). Therefore, it is possible that this trait is not regionally characteristic of Australasia in either its recent or archaic distribution. Among the fossils examined, maximum curvature was observed in Qafzeh 9, while Qafzeh 6 shows none. Similarly, Nahal Ein Gev shows a very pronounced curvature, while Ohalo shows none (Figure 5.4d). This inconsistent pattern is true for most regional fossils, and as it is the case of eversion of the lower border of the malars, this feature is probably varying in terms of other anatomical parameters.

AUS11: Pronounced development of a zygomatic trigone (TR4)

This is not the most common condition in recent Australian crania, although it is observed in more than a third of the Australian sample (TR4: 35.9%). Among recent populations, its highest incidence was found in Fueguian-Patagonian crania (TR4: 41.4%). All the late African archaic and early modern hominids examined (Ndutu, Djebel Irhoud, LH18, Omo 1, Florisbad) show a pronounced development of the zygomatic trigones (TR4), as do Qafzeh 6, Skhūl 5, Predmost, UC101, Keilor, Kanalda and Kow Swamp (Figure 5.4e). With the exception of Qafzeh 9, which shows an extraordinarily gracile trigone, there is a clear temporal trend to this feature, and a pronounced zygomatic trigone may be considered the ancestral condition for modern humans.

The Ngandong hominids clearly have very pronounced zygomatic trigones, but as was discussed above in relation to the position of minimum frontal breadth, this structure is very different from that observed in modern crania. The trigone is deep antero-posteriorly, forming part of the main body of the frontal bone, rather than a lateral extension of it as seen in modern skulls. Consequently, it causes the frontal bone of these hominids not to angle downwards laterally, resulting in a supraorbital structure that forms a horizontal bar over the face. The different structural characteristics of the zygomatic trigone in the Ngandong hominids guarant the attribution of a separate grade in our scoring system, TR5, representing an archaic structure, so far not observed in any fossil or recent modern skull. The temporal and spatial distribution of this trait not only fails to support a particular relationship between recent Australians and the Ngandong hominids, but the fact that Australians together with all other recent populations share a modern trigone structure with late African archaic and early modern fossils absent in the Ngandong hominids, supports an African morphological ancestry for the modern expression of this trait.

AUS12: The suprameatal tegmen

For Australian skulls this was not the largest as proposed by the Multiregional Model. On the contrary, Australians show a relatively small suprameatal tegmen in relation to other recent populations. Australian fossils vary markedly in the expression of this feature (Figure 5.4f). Kanalda shows a suprameatal tegmen even smaller than the recent Australian average, Keilor is larger, similar to the Mediterranean Epi-Palaeolithic samples, while the Javanese fossil of Wajak I has a comparatively large tegmen. Of the Ngandong sample, Ngandong 11 shows a very large suprameatal tegmen, much larger than all the other fossils examined, including the other two Ngandong specimens in which this feature could be observed (7 and 12). This feature does not show any spatial or temporal trend, and does not support a particular relationship between recent Australians and the Ngandong hominids.

AUS13: Angling of the petrous to the tympanic in the petrous-tympanic axis

The angle formed between the petrous and the tympanic bones in the petro-tympanic axis is more acute in Australian and sub-Saharan African crania than in the other recent populations studied. This measure could be taken with certainty in only three fossils, UC101 and two Ngandong specimens, 7 and 12. The UC101 individual shows a more acute angling of the petrous towards the tympanic than that observed in Australians and sub-Saharan Africans (160°), but well within the recent human range (148°–176°). Of the two Ngandong specimens, Ngandong 7 has a petro-tympanic angle of 163°, greater than the Australian and African averages and similar to the Afalou and Taforalt means. Ngandong 12 has a value of 143°, a very angled petrous in relation to the tympanic, with a value not observed in any of the recent crania studied. The lack of Australian uniqueness in this trait, together with the variation within the Ngandong hominids, shows that this feature is too undifferentiated between groups to support a particular phylogenetic relationship.

However, what is interesting in the Ngandong crania is not their petro-tympanic angle, but rather the actual angles of the petrous and tympanic in relation to the coronal plane of the skull. Both the Ngandong hominids in which these measures could be taken show a markedly more oblique (less coronally oriented) tympanic bone, with angles outside the modern human range (Ngandong 7: 42°, Ngandong 12: 38°). In the case of the petrous, Ngandong 7 shows a very pronounced angle (59°), resulting in a petro-tympanic axis comparable to recent humans (163°). Ngandong 12 on the other hand, has a petrous angle of 47°, similar to recent humans, but which combined with the large tympanic angle results in a small petro-tympanic angle (143°). Therefore, what differentiates these archaic hominids most is the inclination of the tympanic bone. This is all the more interesting considering that, while the petrous angle varies markedly among the five recent regional populations ($F = 5.5$, $p < 0.001$), the variation in tympanic angle is not significant ($F = 0.9$, $p > 0.05$). This point highlights the difference between the Ngandong hominids and all recent populations, including the Australians.

In his *Sinanthropus* monograph, Weidenreich states that the angle of the pyramid axis (the sagittal instead of coronal angle of the petrous bone) in *Sinanthropus* (40°) differs from that of Europeans (63°), while both differ from the apes (15° in gorillas and 30° in orangs) (Weidenreich, 1943a:p.57, plates XLII). However, the results found in this study show that modern human values of the pyramid angle have a wide range (26° to 60°) and that all recent populations have an average between 41° and 48° which is close to the *Sinanthropus* value, quite different from the 63° reported by Weidenreich for Europeans. This is consistent with the discussion above in terms of the variation observed in petrous angles in modern and archaic populations.

AUS14: Occipital crest

In all recent populations, absence of an occipital crest is the most frequent condition observed, with the exception of the Fueguians-Patagonians and

Afalou crania that show more cases with a partial occipital crest than without. The frequency of cases of partial occipital crests in the Australian, East Asian, Taforalt and Natufian samples are also relatively high. This is consistent with the more robust occipital morphology of these populations while in the occipitally gracile sub-Saharan Africans, Southeast Asians and Europeans this feature is much rarer. No recent population was characterised by the common occurrence of a complete occipital crest (Figure 5.4g). This is also true for all the fossils examined except the Ngandong hominids. The late African archaic (Ndutu, Djebel Irhoud) and the African, (Omo 1) and Middle Eastern (Qafzeh 6, Skhūl 5) early modern hominids all lack an occipital crest. So do the fossils of Nahal Ein Gev, UC103, Liujiang, Minatogawa I, Keilor, Kanalda and Wajak II. The fossils of Ohalo, Predmost, UC101, Minatogawa IV, Tepexpán, Lagoa Santa 6 and Wajak I have partial occipital crests. Among the Ngandong hominids, Ngandong 7, 10, 11 and 12 show partial occipital crests, and Ngandong 1 and 6 complete ones. In Ngandong 1 this feature is very pronounced. This feature does not support a special relationship between recent Australians and the Ngandong hominids.

AUS15: Occipital torus

An occipital torus was observed in comparatively more Australians skulls than in any of the other regional samples (44.7% in comparison to the minimum of 0% in Europeans and the maximum of 27.9% in East Asians). However, it was not the most common Australian occipital morphology, nor the highest frequency observed in modern groups (Figure 5.4h). Of the Fueguian–Patagonian crania examined, 69% have an occipital torus. In the fossils studied, occipital tori occur in Ngandong, Ndutu, Djebel Irhoud, Omo 1, Skhūl 5, Minatogawa IV, Tepexpán, Lagoa Santa 6 and Keilor. Therefore, the presence of this feature is certainly not confined to Southeast Asia in either past or present times.

It could be argued that occipital tori are a not so rare feature of modern skulls. Of the 187 recent skulls examined (Europeans, sub-Saharan Africans, Southeast Asians, East Asians, Australians and Fueguian-Patagonians), 27.8% show an occipital torus. Furthermore, if the Europeans are not included, the frequency increases to 32.1%. Approximately a third of recent skulls have an occipital torus. However, even the cases in which an occipital torus is present they differ from an archaic morphology in that an external occipital protuberance is visible along the superior toral margin. Hublin (1978a) has argued that the presence of an external occipital protuberance is one of the few modern human apomorphies. This observation is fully supported by the results in this study. All the modern fossils that show an occipital torus show an external occipital protuberance. A single Australian skull was found in which an external occipital protuberance could not be identified. On the other hand, in all the six Ngandong specimens in which an occipital torus could be observed, an external occipital protuberance is absent, as it is also in the Djebel Irhoud cranium. Not only does the temporal and spatial distribution of this feature not support a special relationship between the Ngandong hominids and recent Australians, but the latter share with late African archaic and early modern fossils the presence of an external occipital protuberance along the superior toral margin.

AUS16: Pronounced facial prognathism

As in the case of eversion of the malars, curvature of the maxilla , zygomaxillary tuberosity, narial margins and rounding of the infero-lateral margin of the orbits, facial prognathism cannot be measured on the Ngandong hominids, and the past Southeast Asian characterisation is based on a single reconstructed specimen (Sangiran 17). The recent regional distribution shows that African crania, and to a slightly lesser degree Australian, are prognathic (in terms of the Gnathic Index) in relation to other recent populations. This prognathism is surpassed by the fossils of Skhūl 5, Minatogawa IV and Predmost, and in the case of the Australians, also by Wajak I (Figure 5.4i-1). Furthermore, it is clear that orthognathous faces are a relatively late specialisation of both European and Asian populations, since virtually all the modern fossils examined are more prognathic than these recent groups. Recent Australians do not show any special prognathism that particularly differentiates them from other modern crania and guarantees a special relationship with an early Javanese *Homo erectus*. Interestingly, there is a lot of variation as to whether the prognathism is mainly alveolar or nasal in both the fossil and recent samples (Figure 5.4i-2,3; Table 5.6). If prognathic, the great majority of modern skulls exhibit alveolar prognathism. However, in a small number of fossil specimens (Skhūl 5, Ohalo, UC103 and Minatogawa IV) there is more nasal than alveolar prognathism, while in Predmost, the proportions are equal.

AUS18: A low (flat) frontal angle

This is an archaic characteristic for all hominids. Accordingly, the late *Homo erectus* Ngandong specimens indeed show very flat frontal bones.

Table 5.6. *Relative percentages of nasal and alveolar prognathism in total facial prognathism*

	Facial Prognathism (PRR – NAR)	Nasal Prognathism (SSR – NAR)	Alveolar Prognathism (PRR – SSR)
Djebel Irhoud	15.0	3.0	12.0 (80.0%)
Qafzeh 6	8.0	1.0	7.0 (87.5%)
Skhūl 5	22.0	12.0	10.0 (45.4%)
Ohalo	10.0	7.0	3.0 (30.0%)
Nahal Ein Gev	7.0	−1.0	8.0 (100.0%)
Natufians	6.7	0.7	6.0 (89.6%)
Predmost	8.0	4.0	4.0 (50.0%)
Europeans	5.5	0.2	5.3 (96.4%)
Afalou	5.2	0.0	5.2 (100.0%)
Taforalt	8.5	4.0	4.5 (52.9%)
Sub-Saharan Africans	12.5	3.8	8.5 (68.0%)
Wajak I	10.0	3.0	7.0 (70.0%)
Southeast Asians	8.2	1.3	6.9 (84.1%)
UC 101	8.0	1.0	7.0 (87.5%)
UC 103	14.0	8.0	6.0 (42.9%)
Liujiang	4.0	−2.0	6.0 (100.0%)
Minatogawa I	13.0	7.0	6.0 (46.1%)
East Asians	7.3	1.1	6.2 (84.9%)
Fueguians/Patag.	9.2	2.4	6.8 (73.9%)
Keilor	10.0	4.0	6.0 (60.0%)
Kanalda	15.0	6.0	9.0 (60.0%)
Australians	11.5	3.8	7.7 (67.0%)

However, the fossils of LH 18 and Florisbad show similarly flat frontals, and 14 modern fossil specimens, as well as the archaic fossil of Djebel Irhoud and the recent Fueguian-Patagonian sample all have very flat frontals (Figure 5.4j). All these specimens have flatter frontal bones than the recent Australians. It is impossible to argue for the regional Southeast Asian character for this trait across the Upper Pleistocene.

AUS19: Rounding of the infero-lateral margin of the orbit (RO3)

Among recent modern humans this occurs with a uniquely high incidence in Australian skulls. This feature is variable in the late archaic and early modern fossils examined. A rounded infero-lateral orbital margin, levelled with the floor of the orbit, is found in Omo 1, Qafzeh 9, Qafzeh 6, Minatogawa I and IV, and Kanalda (Figure 5.4k). The presence of this feature in three out of six African and Middle Eastern early modern fossils examined shows that it was part of the morphological variation of this early modern population. Therefore, recent Australian crania could have retained this feature from a Javanese *Homo erectus* ancestor (even though the incidence of this feature in Javanese archaic hominids for the past 700 ka is untestable) or an African early modern population. As it is the case for the zygomaxillary tuberosity, the ancestral source of this Australian feature remains uncertain.

AUS20: A horizontal superior orbital margin

This is observed in the majority of Australian crania (73%). Its incidence outside Australia is much less common, varying from 7% in East Asians to 38% in the Afalou sample. A horizontal superior orbital margin is observed in Predmost, Nahal Ein Gev, UC103, Minatogawa I, Kow Swamp 1 and Kanalda fossils, while it is absent in Keilor, Kow Swamp 5 and 15 and the Wajak I and II specimens. Among the Ngandong hominids, the bar-like supraorbital torus and trigonid structure reinforce the impression of a horizontal superior orbital margin. However, with the exception of Ngandong 6, in the Ngandong 1, 5 and 12 specimens in which the superior orbital margin can be seen completely, the margin slopes laterally in relation to the coronal plane. The Ngandong-Australian relationship is not confirmed.

Ancestral or derived?

At the beginning of this chapter, I set out to question the temporal distribution of the East Asian and Australian continuity traits predicted by the Multiregional Model. Those questions can now be answered.

How many of the Australasian continuity traits are present in the Ngandong hominids and how many are present in the late African archaic and early modern fossils? The first point to notice is that of the 19 regional continuity traits studied, six cannot be observed in the preserved portions of the Ngandong hominids (Table 5.7). Of the remaining 13, two are present in the Ngandong sample, three are present but not homologous with the modern form, five have a variable incidence, and three are absent. If the features with variable incidence are taken as showing that that particular trait was part of the morphological variability of that population, we may say that seven of the 19 claimed Australian regional continuity traits are present in the Ngandong sample. In the late African archaic

Table 5.7. *Incidence of the Australian regional continuity traits in the Ngandong hominids, late African archaic, early modern and Australian crania*

Regional community traits	Ngandong	Djebel Irhoud Ndutu+Lh 18	Omo 1, Florisbad, KRM, Qafzeh 6+9, Skhul 4+5	AUS
Supraorbital torus (ST5)	+	+	±	±
Lambdoid depression	±	−	±	−
Small pterion	−	+	±	+
Deep infraglabellar notch (IN4)	−	−	±	−
Backward position of min. fr. br.	(+)	+	±	±
Low position of max. par. br.	−	±	±	−
Zygomaxillary tuberosity (ZT4)			±	±
Eversion of malars			±	−
Absence of narial margins (N9/10)		−	−	−
Curvature of alveolar plane of max.			±	−
Zygomatic trigone (TR4)	(+)	+	±	−
Large suprameatal tegmen	±	±	±	−
Low petro-tympanic angle	±			−
Occipital crest (OCR3)	±	−	−	−
Occipital torus (OT6/7)	(+)	±	±	+
Facial prognathism		+?	±	+
Flat frontal bone	+	+	±	±
Rounding of i-l margin of orbit (RO3)		−	±	+
Horiz. sup. orbital margin (O'1)	±	−	−	+

−: absent; +: present; ±: variable expression in the sample; (+): present as a different structure; ?: unclear.

hominids studied (Ndutu, Djebel Irhoud and LH 18) 15 of the 19 Australian regional continuity traits could be recorded, of which six are present, three have variable incidence and six are absent. In the early modern sample examined (Omo 1, Florisbad, KRM, Qafzeh 6, Qafzeh 9, Skhūl 4 and Skhūl 5), 18 of the 19 features could be recorded. Of these, 15 have a variable expression and three are absent (also absent in the recent Australian sample). Only nine of the 19 features are clearly or variably present in recent Australians, all of which occur at least in some specimens of the early modern human fossil sample (Table 5.7). These results show that the early modern African and Middle Eastern fossils are actually a better model for the ancestral population of recent Australians than the Ngandong hominids in the very features that are claimed to characterise the Javanese archaic-Australian line of descent. The first two of the three predictions of the Multiregional Model, that the Ngandong hominids should show all the Australian regional continuity traits, and that these traits should not occur in the late African archaic and early modern fossils, are not fulfilled.

The second question relates to the presence of East Asian and Australian regional lineages in the Upper Pleistocene. Five East Asian (UC101, UC103, Liujiang, Minatogawa I, Minatogawa IV) and two Amerindian (Tepexpán, Lagoa Santa 6) Pleistocene fossils were examined. Table 5.8 summarises the incidence of the East Asian regional continuity traits in these specimens and in the recent East Asian and Fueguian-Patagonian samples. These data show that the features being studied can hardly be called regional even in this all modern East Asian sample. The only possible exception is coronal facial flatness, which as mentioned before, seems to characterise to some degree all the specimens except the recent Fueguian–Patagonian crania. Independent of the archaic variation, the claimed East Asian regional continuity traits do not identify an East Asian regional lineage in the Upper Pleistocene. To examine the reality of a Southeast Asia-Australian regional lineage, two Southeast Asian (Wajak I and II) and three Australian (Keilor, Kanalda, Kow Swamp 1, 5 and 15) Pleistocene–Holocene fossils were examined. Table 5.9 summarises the incidence of the Australian regional continuity traits in these specimens and in the recent Southeast Asian and Australian samples. The first point to notice from these data is that the Southeast Asian and Australian samples clearly do not follow the same pattern. If only the Australian fossils are considered, four features show consistent regional incidences: large supraorbital ridges (ST4), relatively backward position of minimum frontal breadth, zygomaxillary tuberosities and relatively large zygomatic trigones (TR4 – this degree of development of zygomatic trigone is not the most common in recent Australians, but it does occur with a relatively high frequency). These four features relate to the robust supraorbital morphology of Australian crania and are clearly regional in their expression. As we have seen before, this regionality does not extend to the Ngandong hominids, since the three comparable features – supraorbital tori, trigones and distance from minimum frontal breadth to the face – in these archaic hominids are not homologous to the modern Australian morphology. All the other claimed Australian regional continuity traits fail to show a consistent regional incidence that can be interpreted as representing an evolutionary lineage. Therefore, we can safely say that the second prediction of the Multiregional Model, that there should be clear regional lineages in East Asia and Australia identifiable through the variation in the proposed continuity traits, is not fulfilled.

We may conclude that most of the regional features of continuity used as

Table 5.8. *East Asian regional lineage*

Regional community traits	UC101	UC103	LIUJ.	MIN I	MIN IV	TEP	L.STA.	EA	FUE/PAT
Sagittal keeling	±	+	−	−	−	+	±	−	±
Horiz. course of n-fr. and fr.max. sut	−	−	±	±	+	+		±	−
Flat nasals	−	−	±	−	±	−		−	
M3 agenesis	−	−	+	−	−	−		±	−
Coronal facial flatness	±	+	±	+	+	+		±	−
Orbital shape	O15 II/IV/I	O2 I/II/I	O11 II/II/I	O3 I/II/II	O13 II/II/III	O15 II/IV/I	O11 II/II/I	O13: IIII/III O11: II/II/I O20: III/II/II	O11 II/II/I

−: absent; +: present; ±: variable or intermediate expression.

Table 5.9. *Australian regional lineage*

Regional community traits	Wajak I	Wajak II	SEA	Keilor	Kanalda	K.Swamp	AUS
Supraorbital torus ST5/ST4	−	−	−	+	+	+	+
Lambdoid depr.	+		−	−	−	−	−
Small pterion	−		−		+		+
Deep infra-glab. (IN4)	−	−	−	−	−	−	−
Back. pos. WFB	+	−	−	+	+	+	±
Low pos. of XPB	+		−	±	±		±
Pron. zygom. tub. (ZT4)	−		−	+	+	+	+
Eversion of malars	+		−		+!	+?	−
Narial margins		N6	N3	N3	N6	N3/N6	N8
Curvature of maxilla			±		−	−?	−
Pron. zyg. trigone (TR4)	−	−	−	+	+	+	−
Large supram. tegmen	±		−	±	−		−
Occipital crest (OCR3)	−	−	−	−	−	−	−
Occipital torus (OT6)	−	−	−	+	−	±	+
Facial prognathism	+		−	−	±		±
Flat frontal bone	+		−	+	+	+??	±
I-l margin of orbit (RO3)	−		−	−	+	−	+
Horiz. sup. orb. border	−	−	−	−	+	−	+

−: absent; +: present; ±: variable expression in the sample; +!: very pronounced; ?: uncertainty because of the effects of artificial cranial deformation.

evidence of multiregional evolution are actually ancestral for all modern humans, and that the variable incidence observed in the Australian fossil and recent samples reflect the differential retention of ancestral traits. In the case of the East Asians, not only do they show almost none of the expected morphological differentiation, but they also show a comparatively different pattern from the early modern sample. This would suggest that the modern East Asian populations are relatively derived.

6 *The independence of expression of the 'regional continuity traits'*

The results of the previous chapters on the spatial and temporal comparisons show that only a few of the East Asian and Australian 'regional continuity traits' proposed by the Multiregional Model exhibit a significant greater incidence in those regions. However, it has been argued that it is not just the presence of these features in Australian skulls that carries phylogenetic weight, but their combined incidence (Habgood, 1989). This would only be true if it could be shown that these traits occur in combination in Australian skulls more frequently than in other populations and, most importantly, that their expression is independent of each other and other cranial dimensions. Only if these features are shown to be independent phylogenetic markers, is their combined incidence along evolutionary lines phylogenetically important.

Most of the features that were uniquely or highly 'Australian' in the regional comparisons (sagittal keeling, pronounced supraorbital ridges, deep infraglabellar notch, zygomaxillary tuberosities, rounding of the inferolateral margin of the orbit, and horizontal superior orbital margin) are those traits that were measured as grades or categories in the scoring system described in this study. These represent mostly features of robusticity. Three other features were found to have a high incidence in Australian skulls, a small articulation at pterion, position of minimum frontal breadth and an acute petro-tympanic angle, but these were already discussed extensively. Therefore, this chapter will focus on the features of robusticity measured as degrees of development or categories.

Combined occurrence

As observed by Habgood (1989), a large proportion of Australian crania show the concurrent presence of several pronounced tori and ridges. An Australian skull that has very pronounced supraorbital ridges is also likely to have quite pronounced zygomatic trigones, infraglabellar notches, zygomaxillary tuberosities, a broad nasal saddle, and even a pronounced sagittal keel or an occipital torus (Table 6.1). Although these features also show a high frequency of combined occurrence in other populations, the

Table 6.1. *Proportion of (A) Australian and (B) other recent crania with differing degrees of development of supraorbital ridges that have another pronounced feature of robusticity*

(A)	TR4/3 (%)	IN4/3 (%)	ZT4/3 (%)	NS3 (%)	RO3 (%)	SK3 (%)	OT5/6 (%)	TR3/4+IN3/4 (%)
ST1	–	–	–	–	–	–	–	–
ST2	25.0	12.5	12.5	25	50.0	25.0	25.0	–
ST3	80.0	60.0	55.5	50	44.4	30.0	33.3	50.0
ST4	94.1	93.8	87.5	75	75.0	64.7	52.9	87.5
ST5	100.0	75.0	75.0	100	50.0	50.0	50.0	75.0
(B)	**TR4/3 (%)**	**IN4/3 (%)**	**ZT4/3 (%)**	**NS3 (%)**	**RO3 (%)**	**SK3 (%)**	**OT5/6 (%)**	**TR3/4+IN3/4 (%)**
ST1	4.8	4.8	12.7	18.8	8.3	–	12.5	–
ST2	20.4	9.3	46.3	23.7	17.9	–	15.4	5.2
ST3	51.9	40.8	42.3	72.4	13.8	3.4	10.3	24.1
ST4	80.0	32.0	68.0	75.0	50.0	–	–	100.0
ST5	–	–	–	–	–	–	–	–

TR: zygomatic trigone; IN: infraglabellar notch; ZT: zygomaxillary tuberosity; NS: profile of the nasal saddle; RO: rounding of the infero–lateral margin of the orbit; SK: sagittal keeling; OT: occipital torus-ridges; ST: supraorbital ridges.

Australian incidence is slightly higher. Among Australian skulls, 87.5% of those with pronounced supraorbital ridges have both well-developed zygomatic trigones and infraglabellar notches, while no skull with slightly developed supraorbital ridges will show these two other features to such a degree of development. However, this pattern of co-variation is not true for all the traits examined. Pronounced rounding of the infero-lateral margin of the orbit is present in most Australian crania independent of the degree of development of the supraorbital ridges. Type of narial margins also seems to be independent, in this case with a variable rather than constant expression. Lastly, it is clear from Table 6.1 that the patterns of combination of degree of expression observed between the development of the supraorbital ridges and the other features of robusticity do not apply to supraorbital torus formation. It is possible that this grade, which represents not an extreme expression of the supraorbital ridges, but a different supraorbital morphology, relates differently to the other cranial superstructures.

Should the high frequency of a group of morphological features in combination in Australian crania be taken as evidence of morphological continuity? For a morphological feature to be considered a phylogenetic trait, i.e. a trait that characterises a taxonomic unit and can be attributed a cladistic value or character-state, two conditions should be satisfied:

(1) To have a stronger genetic than environmental determination.
(2) To be independent of other morphological characters.

The first point implies that those traits that occur as the result of mechanical or other external forces during development are not suitable for the characterisation of evolutionary relationships. This is in spite of the fact that the functional or external origin may hide an underlying genetic propensity, since in this case the anatomical conditions under which the environmental force may induce a feature to occur should be investigated instead. These will present a more stable characterisation of the populations under study. The second point implies that if two traits can be shown to be different phenotypical manifestations of the same morphological variable or part of a functional complex that constrains its expression, to use them independently gives undue weight to this variable in any phylogenetic study (Szalay, 1981).

We have already seen that there is a certain degree of co-variance with the features of robusticity, but we have not established whether this pattern of co-variance is related to the genetic background of the population as argued by Habgood (1989) or because these features are jointly responding to variation in other anatomical parameters. The most obvious alternative to Habgood's view is that these traits are actually under similar functional constraints, and thus show a similar pattern of expression. But are there any reasons to suppose that the development of the morphological features under study may be related to functional mechanisms? One way of assessing this possibility is by examining whether there is a significant relationship between more pronounced development of these features and the dimensions of the masticatory apparatus.

Dental dimensions and size of the temporal muscle (measured as the height, length and minimum bilateral distance of the superior temporal lines) are larger in crania with more pronounced development of the supraorbital ridges and sagittal keeling, while the development of the occipital torus seems to be related to temporal muscle size but not the dental dimensions (Table 6.2). This variation in dental and temporal muscle size in crania with different degrees of robusticity is highly significant, and suggests that there is at least a partial functional component in the aetiology of these morphological features. Therefore, the independence of expression of these traits is examined here by testing whether and to what extent it is possible to predict the development of these features on the basis of metrical dimensions of the human skull.

Table 6.2. *One way analyses of variance of first molar bucco-lingual and mesiodistal dimensions (BLM1, MDM1), height and length of the temporal lines (STLHCH, STLLCH), and distance between bilateral superior temporal lines across the vault (SILLPA, SILLOC) by grades of supraorbital ridge/torus, sagittal keeling and occipital torus development (all metrical values in mm)*

	ST1	ST2	ST3	ST4	ST5	
BLM1	11.8 (42)	11.9 (56)	12.0 (44)	12.5 (23)	13.3 (4)	F= 7.3***
MDM1	10.6 (42)	10.6 (56)	10.8 (45)	11.0 (23)	11.6 (4)	F= 3.7**
STLHCH	85.5 (66)	89.8 (94)	93.3 (90)	96.0 (63)	87.4 (5)	F= 17.1***
STLLCH	130.1 (67)	136.6 (92)	141.7 (91)	144.9 (63)	139.8 (5)	F= 22.2***
SILLPA	102.1 (64)	100.1 (91)	95.5 (90)	90.3 (62)	93.2 (5)	F= 6.4***

	SK1	SK2	SK3	
BLM1	11.9 (99)	12.0 (52)	12.7 (15)	F= 9.0***
MDM1	10.6 (100)	10.8 (50)	11.3 (17)	F= 8.4***
STLHCH	88.8 (180)	93.0 (101)	96.8 (39)	F= 20.9***
STLLCH	134.8 (176)	141.0 (101)	147.8 (39)	F= 29.6***
SILLPA	102.9 (175)	93.3 (101)	81.5 (39)	F= 46.7***

	OT2	OT3	OT4	OT5	OT6	OT7	
BLM1	11.4 (2)	12.0 (86)	12.0 (53)	11.3 (3)	12.1 (24)	–	F= 1.1
MDM1	9.9 (2)	10.7 (89)	10.6 (52)	10.3 (2)	11.0 (24)	–	F= 2.0
STLHCH	80.3 (3)	89.1 (147)	91.9 (103)	94.2 (10)	94.9 (54)	88.0 (1)	F= 5.7***
STLLCH	128.2 (4)	134.8 (142)	139.9 (103)	141.2 (10)	144.9 (55)	139.0 (1)	F= 8.7***
SILLOC	110.7 (3)	101.2 (140)	100.9 (99)	91.4 (10)	88.3 (54)	113.0 (1)	F= 9.4***

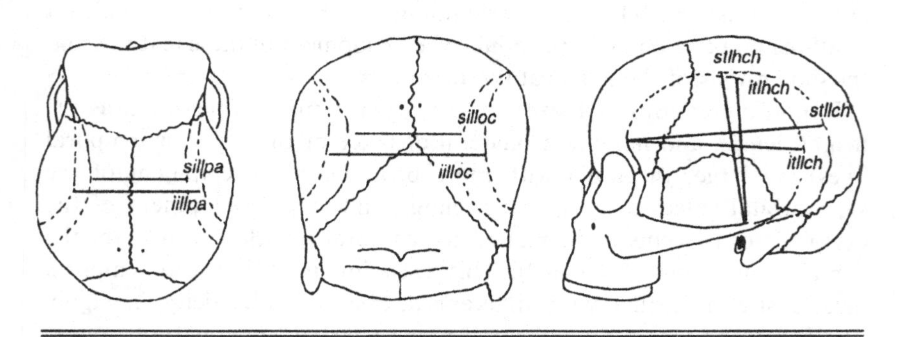

**: $p<0.01$; ***: $p<0.001$.

The relationship between the degree of development of the morphological features and cranio-dental dimensions

Statistical analysis

The association between degree of expression in superstructure morphology and cranio-dental dimensions was investigated through discriminant analyses. Discriminant analysis examines whether it is possible to classify a number of cases into specified groups or categories on the basis of a group of independent variables, and compares this classification with the actual group membership of those cases. Therefore, this analysis tests whether independent (cranio-dental) variables correctly predict the degree of development of the morphological features scored as grades in this study. Standard entry/removal criteria were used (FIN = 1.0, FOUT = 1.0), but to avoid high levels of multicolinearity among the independent variables, the tolerance level was set at 0.1. The relative importance of the discriminating variables was established by a stepwise method of variable selection through minimisation of Wilk's λ, but the functions were built using a direct entry method.

The samples used for these analyses consisted of 135 skulls from various parts of the world. They included most of the European, African, Southeast Asian, East Asian and Australian samples used in the previous chapters, with the addition of a few crania from Egypt, Melanesia and Polynesia and some Inuit crania (Table 6.3). By their distinctiveness, these crania contribute to the variation observed. Because the discriminant analysis automatically removes all cases with missing values in any of the dependent or independent variables, the sample size unavoidably decreased. While most skulls included in the study were relatively complete, the dental sample was between 70 and 80 cases for molar teeth. The selection of dental variables in all the analyses determined that sample sizes of 60 to 75 were used. Furthermore, anterior dental variables had to be excluded altogether, since the number of crania in the sample with anterior teeth was below 25. Although the importance of tooth size in predicting the development of these features is apparent by the presence of various molar dimensions among the discriminating variables in all the analyses, the full extent and character of the relationship between these features and dental dimensions could not be examined in the present study. Therefore, the results obtained here examine the significance of the posterior dentition in predicting the expression of these features, but the effect of the anterior dentition remains unknown.

Treatment of the independent variables

The metrical cranial variables listed in Appendix II were used as 'independent' variables in the discriminant analyses. In order to increase the dental sample size, the mean values of left and right individual posterior tooth dimensions were used. This was justified by the high levels of bilateral tooth size correlations (Lahr, 1992). From these values, the dental area of each posterior maxillary tooth (buccolingual × mesiodistal) was calculated, and these were also used as independent variables in the analyses.

The facial features

Supraorbital torus

The results of a discriminant analysis of the five grades of supraorbital ridge/torus show that the degree of development of this feature can be correctly discriminated in 98.48 % (65/66) of crania using 22 cranio-dental metrical variables (Table 6.4a). Four discriminant functions were obtained, the first three of which are significant:

Function 1: Eig = 4.64, % Var = 51.4, λ = 0.014, x^2 = 218.2, $p<0.001$
Function 2: Eig = 2.66, % Var = 29.4, λ = 0.082, x^2 = 129.1, $p<0.001$
Function 3: Eig = 1.22, % Var = 13.6, λ = 0.298, x^2 = 62.3, $p<0.05$

The distribution of mean values of the most important predicting variables by the five grades of supraorbital ridges/torus shows that the degree of development of the supraorbital ridges increases with an increase in M1 size, an increase in lateral facial retraction, an increase in occipital length, nasal breadth, malar length, cranial length and distance between position of minimum frontal breadth to the zygomatico-frontal suture. The relationship to temporal fossa breadth and length of temporal muscle varies between the grades, as does length and breadth of the foramen magnum and interorbital width. The relationship between development of the supraorbital ridge and its predicting variables is not a simple linear one, and this may indicate that pattern as well as size, may be represented in the grade discrimination. The maximum grade scored, a supraorbital torus continuous with the zygomatic trigone by a thickening of the orbital roof, showed a different relationship to some of the predicting variables away from the trend observed for the other four grades, such as minimum values of foramen magnum breadth and minimum frontal breadth positioned closer to the face. However, two facts make this point lose significance: first, only two skulls with this grade were included in the analysis, and

Table 6.3. *Recent crania used in the discriminant analyses*

Europe (N=25)	*Africa (N=25)*
DC Eu.1.5.109 – Lanhill, Scotland; Neolithic	DC Af.11.5.235 – Naqada, Egypt
DC Eu.1.4.96 – Isle of Harris, Scotland; Bronze Age	DC Af.11.5.238 – Naqada, Egypt
DC Eu.1.5.103/JT 278 – Gatcombe, UK; Neolithic	DC Af.11.5.7/B7 5101 – Badari; Pre-Dynastic, Egypt
DC Eu.1.4.97 – Aldwincle, UK; Bronze Age	DC SUD 62 – Kerma, Sudan; 12–13th Dinasty, Egypt
DC Eu.1.3.379 – Wandlebury, UK; Iron Age	DC E 121 – Gizeh; 19–26th Dinasty, Egypt
DC Eu.1.3.178 – Cambridge, UK; Romano-British	DC E 864 – Gizeh; 19–26th Dinasty, Egypt
DC Eu.1.3.112 – Maiden Castle, UK; Romano-British	DC EDN 332 – Egypt
DC NNI – Sedgeford, UK; Saxon 4–7th C A.D.	DC 4161 – Egypt
DC Eu.1.0.1043 – Whitechapell, UK; L. Medieval	DC 695 NU GR 85 112/119x – Nubia, ASN
DC Eu.1.00.75 – Stumps Cross # 2,UK	DC Af.23.0.23/ K 9 – Haya Tribe, Musira Is., Tanzania
DC TEAN V – Cornwall, UK	DC Af.23.0.22/ K 8, East Africa
DC 1209/1177 – Finland	DC Af.21.0.4 – Turkana
DC 1171 – Lapp, from Paddeby	DC 1774 – Yao Tribe, Uganda
DC Eu.31.00.1/4612–3 – Russia	DC 1777 – M'Zadi, Upper Congo River, Congo
DC Eu.31.00.2/5498–4 – Russia	DC 6096 – Yola, North Nigeria
DC 1205/1178 – Sweden or Denmark	DC 5060 – Osheba, French Congo
DC Eu.26.00.2/JT 220 – Germany	DC 1758 – Bushman, Cape Town
DC Eu.26.00.1/JT 219–1133 – Germany	DC 1738 – Bushman
DC Eu.42.00.5/JT 244 – Ostia, Rome, Italy	DC 6110 – Bushman, South Africa, ?Botswana
DC Eu.42.00.2 – Italy	DC 6109 – Amaxosa, South Africa
DC 1114 – Paestum, Italy	DC 5323 – Africa
DC 1119 – Sardinia	DC xxx1 – Africa
DC 1170 – Austria	DC 4205 – black mulatto from Surinam
DC 1150 – Serb/Austrian, Balkans	DC 4208 – mulatto carbouger from Surinam
DC 2332 – France	DC 4201 – negress from Surinam

Table 6.3. *(cont.)*

Southeast Asia (N=21+4 from Andaman Is.)
DC BU 14 – Burma
DC BU 19 – Burma
DC BU 23 – Burma
DC BU 26 – As.57.0.26, Burma
DC BU 38 – Burma
DC BU 41 – Moulmein, South Burma
DC BU 53 – Moulmein, South Burma
DC BU 71 – Burma
DC BU 76 – As.57.0.76, Burma
DC BU 97– Burma
DC BU 139 – Moulmein, South Burma
DC Pangan Sakei– Skeat Expedition
DC As.44.0.9 – Sarawaak
DC As.44.0.43 – Mulu Cave, Sarawak
DC 1787 – North Borneo
DC 4516 – Lepuasing, Sarawak
DC 4546 – Orang Bukit, Sarawak
DC 4556 – Murut, Sarawak
DC 5936 – Mabuiag, Sarawak
DC As.33.6.11 – Kelantan, Malaysia; Mesolithic
DC 1756/1173C – Malay mulatto from Diepo Negoro
DC 5w – Andaman Islands
DC 9b – Brown Coll., 1909, Andaman Islands
DC 11 – Brown Coll., 1909, Andaman Islands
DC As.55.0.1 – Andaman Islands

India (N=3)
DC 1212 – Punjab, India, from Hindostan
DC 1225 – Bihar, India, Rajput from Behari Patna
DC xxx2 – Pathan,India

East Asia (N=14)
DC As.21.0.3 – China Kowloon, Urn Burials
DC As.21.0.4 – China Kowloon, Urn Burials
DC As.21.0.5 – China Kowloon, Urn Burials
DC As.21.0.7 – China Kowloon, Urn Burials
DC 1760 – China, Macassar
DC 1761 – China
DC 1762 – China, Macassar
DC 4211 – Chinese from Surinam
DC 4213 – Chinese from Surinam
DC 5932 – Formosa, Taiwan, Atayal (?)
DC As.17.0.2 – Tibet
DC 5334 – Dokihol, Tibet
DC 2377 – Japan
DC 1763 – Kamchatka

Americas (N=11)
DC Inuit I – Pend Is., Inuit
DC Inuit IV – Cape Simpson, Inuit
DC Inuit C – Angmacsalik, Inuit
DC Inuit II – Inuit
DC Am.1.0.3 – Inuit

Australia (N = 17)
DC 3 – Australia
DC 4 – Australia
DC 5 – 'Berida', Australia
DC 33 – Australia
DC 42 – West Australia
DC 44 – Australia ? Torres Strait ?
DC 51 – Mackay, Queensland, Australia
DC 'A' – Australia
DC 2161 – Australia
DC 2164 – Australia
DC 2165 – Australia
DC 2133/1168 – New South Wales, Australia
DC 6081 – New South Wales, Australia
DC Oc.4.0.4 – Queensland, Australia
DC Oc.0.0.1 – Australia
DC 2112 – Henley Beach, Sandhills, Australia
DC 1215 – Australia

Tasmania (N = 2)
DC 2100 – Tasmania
DC 2096 – Tasmania

DC Am.0.0.1 – Inuit
DC 1832 – Inuit
DC 1873 – Inuit
DC 1842 – Vancouver
DC xx 5928 – Iroquois, Ontario
DC 4498 – Tierra del Fuego, Patagonia

Polynesia (N = 6)
DC 2131/272 – Polynesian
DC 47 – North Is., New Zealand
DC 4496 – Ngapuhi Tribe, North Is. New Zealand
DC 6099 – 'ancient' Maori, New Zealand
DC xxxx – Moriori
DC Fol. 35, Acc. Book, Easter Island

Melanesia (N = 6)
DC Oc.11.0.28 – Papua New Guinea
DC Oc.11.0.45 – Papua New Guinea
DC Oc.11.0.56 – Papua New Guinea
DC Oc.12.0.1 – New Britain
DC Oc.13.0.1/54 – Solomon Is.
DC Oc.14.0.1 – Sta. Cruz, Ovaova Is.

DC: Duckworth Collection

therefore the range of variation and mean values of the related variables is not correctly represented; and second, the two most important predicting variables, molar size and lateral facial retraction which together account for 51.56% of total correct classification, show maximum observed values for this last grade. Still, the possibility that a different pattern of relationships may exist with the formation of a torus cannot be discarded.

Zygomatic trigone

A discriminant analysis of the four grades of zygomatic trigone shows that the degree of expression of this feature was correctly predicted in 95.7% (67/70) of cases on the basis of the multivariate variation of 20 cranio-dental measurements (Table 6.4b). Three canonical discriminant functions were obtained, all of which are statistically significant:

Function 1: Eig = 4.74, % Var = 72.3, $\lambda = 0.049$, $x^2 = 172.3$, $p<0.001$
Function 2: Eig = 1.31, % Var = 17.3, $\lambda = 0.279$, $x^2 = 72.7$, $p<0.001$
Function 3: Eig = 0.68, % Var = 10.4, $\lambda = 0.595$, $x^2 = 29.6$, $p<0.05$

The distribution of predicting variables by grade of zygomatic trigone shows that the development of the lateral element of the supraorbital torus is directly related to the development of the superciliary arches measured as supraorbital projection. This variable is responsible for almost half of the total discrimination achieved. An increase in the size of the trigone is also associated with wide upper faces, large teeth, broader skulls at the level of the root of the zygomatic arches, with wide noses, long occipital surfaces with short nuchal planes of the occipital bone. Furthermore, the lack of a trigone is associated with absence of malar eversion. Like in the case of the supraorbital torus, not all related variables show a linear relationship with trigone development.

Infraglabellar notch

A discriminant analysis of the four grades representing depth of the infraglabellar notch obtained a correct discrimination in 96.8% (61/63) of skulls on the basis of 19 cranio-dental measurements (Table 6.4c). Three canonical discriminant functions were obtained, the first two statistically significant:

Function 1: Eig = 3.06, % Var = 48.8, $\lambda = 0.0425$, $x^2 = 159.4$, $p<0.001$
Function 2: Eig = 2.62, % Var = 41.7, $\lambda = 0.1728$, $x^2 = 88.7$, $p<0.001$

The depth of nasion in relation to both glabella and the nasal bones increases with supraorbital projection, a lower position of maximum

Table 6.4. *Discriminating variables in order of importnce for the facial features of robusticity: (a) supraorbital torus; (b) zygomatic trigone; (c) depth of the infraglabellar notch; (d) zygomaxillary tuberosity; (e) rounding of the inferolateral margin of the orbit; (f) profile of the nasal saddle; and (g) lower narial margins*

(a) ST: 22 discriminating vars.	(b) TR: 20 discriminating vars.	(c) IN: 19 discriminating vars.	(d) ZT: 21 discriminating vars.
1: AM1 – Area maxillary M1	1: SOS – Supraorbital projection	1: SOS – Supraorbital projection	1: SOS – Supraorbital projection
2: NBS – Biorbital-nasion subtense	2: FMB – Bifrontal breadth	2: PMPB – Relative position of maximum parietal breadth	2: AM2 – Area maxillary M2
3: TFB – Temporal fossa breadth	3: MDM2 – Mesio-distal maxillary M2	3: MDB – Mastoid width	3: NBS – Biorbital-nasion subtense
4: STLLCH – Superior temporal line length chord	4: PALD – Palatal depth	4: PTYMX – Pyramid angle of Weidenreich	4: PTEAST – Pterion-asterion chord
5: LFOB – Foramen magnum breadth	5: STLHTA – Superior temporal line height tape	5: OCF – Lambda-subtense fraction	5: PAF – Bregma-subtense fraction
6: OARCH – Occipital arch	6: TFL – Temporal fossa length	6: BLM2 – Bucco-lingual maxillary M2	6: WFB – Minimum frontal breadth
7: NLB – Nasal breadth	7: AUB – Biauricular breadth	7: BLM1 – Bucco-lingual maxillary M1	7: MDH – Mastoid height
8: XML – Maximum malar length	8: WNB – Minimum nasal breadth	8: GOL – Glabello-occipital length	8: XFB – Maximum frontal breadth
9: WIOB – Minimum interorbital breadth	9: NFRB – Naso-frontal breadth	9: WFB – Minimum frontal breadth	9: DIFAL – Curvature of the posterior alveolar plane of the maxilla
10: GOL – Glabello-occipital length	10: DIFE – Eversion of the lower border of the malars	10:PTEAST – Pterion-asterion chord	10: XML – Maximum malar length
11: WFBZFS – Distance between position of WFB and the zygomatico-frontal suture	11: OCF – Lambda-subtense fraction	11: OCS – Lambda-opisthion subtense	11: WCB – Minimum cranial breadth
12: FOL – Foramen magnum length	12: OARCH – Occipital arch	12: NFFMS – Course of the naso-frontal/ fronto-maxillary sutures	12: FP–AL – Distance between the posterior alveolar plane of the maxilla and the FP
13: FRAN – Frontal angle	13: OCANG – Occipital angle	13: NLB – Nasal breadth	13: INIOPIS – Inion-opistocranium length

Table 6.4. *(cont.)*

(a) ST: 22 discriminating vars.	(b) TR: 20 discriminating vars.	(c) IN: 19 discriminating vars.	(d) ZT: 21 discriminating vars.
14: NPH – Nasion–prosthion height	14: PMPB – Relative position of maximum parietal breadth	14: SSR – Subspinale radius	14: D2 – Palatal–opisthion subtense
15: NLH – Nasal height	15: TFB – Temporal fossa breadth	15: ASB – Asterionic breadth	15: FRF – Nasion subtense fraction
16: OCF – Lambda-subtense fraction	16: PTEAST – Pterion-asterion chord	16: PALL – Palatal length	16: AVR – Molar alveolar radius
17: BNL – Basion-nasion length	17: BNL – Basion-nasion length	17: PTE1 – Size of pterion articulation	17: AM1 – Area maxillary M1
18: ALVOC – Palate-basion length	18: PROG3 – Difference between prosthion and nasion radii	18: ITLLTA – Inf. temporal line length tape	18: FOB – Foramen magnum breadth
19: PROG1 – Difference between prosthion and subspinale radii	19: NAS – Naso-frontal subtense	19: DIFE – Eversion of the lower border of the malars	19: WNB – Minimum nasal breadth
20: DIFAL – Curvature of the posterior alveolar plane of the maxilla	20: XFB – Maximum frontal breadth		20: NFRB – Naso-frontal breadth
21: FP-PR – Distance prosthion and the FP			21: WIOB – Minimum interorbital breadth
22: AM2 – Area maxillary M2			

Table 6.4. *(cont.)*

(e) RO: 23 discriminating vars.	(f) NS: 21 discriminating vars.	(g) N: 22 discriminating vars.
1: FMB – Bifrontal breadth	1: SILLPA – Distance between superior temporal lines across the parietals	1: BLM1 – Bucco-lingual M1
2: BAH – Basion-hormion length	2: WIOB – Min. interorbital breadth	2: PTEAST – Pterion-asterion chrod
3: PARCH – Parietal arch	3: NOL – Nasion-occipital length	3: PROG1 – Difference between prosthion and subspinale radii
4: IML – Inferior malar length	4: WNB – Minimum nasal breadth	4: FOB – Foramen magnum breadth
5: SILLPA – Distance between superior temporal lines across the parietals	5: ALVOC – Palate-basion length	5: ZOR – Zygoorbitale radius
6: INOP – Inion-opisthion length	6: XML – Max. malar length	6: LAMOPIS – Lambda-opistocranium chord
7: GI – Gnathic Index	7: BAH – Basion-hormion length	7: SSS – Bimaxillary subtense
8: NLH – Nasal height	8: MDH – Mastoid height	8: OBH – Orbital height
9: HPAL – Palate-hormion height	9: PAF – Bregma-subtense fraction	9: MDM1 – Mesiodistal M1
10: ITLHCH – Inf. temporal line height chord	10: IILLPA – Distance between inferior temporal lines across the parietals	10: AUB – Auricular breadth
11: OCF – Lambda-subtense fraction	11: PTYMC – Petro-tympanic angle	11: SOS – Supraorbital projection
12: FOL – Foramen magnum length	12: SSS – Bimaxillary subtense	12: INOP – Inion-opisthion length
13: ZM-ZM – Distance between the orbital edge on one zygomaxillary suture to another	13: D2 – Palate-opisthion subtense	13: LAMINI – Lambda-inion length
14: XFB – Max. frontal breadth	14: WFB – Min. frontal breadth	14: PAF – Bregma-subtense fraction
15: WIOB – Min. interorbital breadth	15: ZM-ZM – Distance between the orbital edge on one zygomaxillary suture to another	15: PALD – Palatal depth
16: PTYMC – Petro-tympanic angle	16: PTE1 – Size of pterion articulation	16: FP-AL – Distance between the posterior alveolar plane of the maxilla and FP
17: TYMPA – Tympanic angle	17: STLHCH – Superior temporal line height chord	17: HPAL – Hormion-palate height
18: AVR – Molar alveolar radius	18: VRR – Vertex radius	18: D2 – Palate-opisthion subtense
19: PTELAM – Pterion-lambda chord	19: ITLLTA – Inferior temporal line length tape	19: NLB – Nasal breadth
20: DIFAL – Curvature of the posterior alveolar plane of the maxilla	20: TFB – Temporal fossa breadth	20: PROG3 – Difference between prosthion and nasion radii
21: AUB – Biauricular breadth	21: MDM1 – Mesiodistal M1	21: PTYMA – Petrous angle
22: BLM2 – Bucco-lingual M2		22: MDB Mastoid width
23: BLM1 – Bucco-lingual M1		

parietal breadth, wider mastoids, smaller angle between the petrous bone and the cranial midline, and variable values for the position of the occipital angle – shortest occipital plane for maximum depth of nasion. The first two variables account for 60.32% of correct grade prediction. A deep and narrow infraglabellar notch is also related to large teeth, long skulls, reflected in total cranial length and pterion-asterion length, wide frontal bones, angled occipitals, and un-levelled naso-frontal and fronto-maxillary sutures.

Zygomaxillary tuberosity

The results of a discriminant analysis of the four grades of zygomaxillary tuberosity show that it is possible to predict the degree of development of this feature in 98.4% (60/61) of skulls on the basis of 21 cranio-dental variables (Table 6.4d). The first two canonical discriminant functions are statistically significant:

Function 1: Eig = 14.12, % Var = 75.4, $\lambda = 0.008$, $x^2 = 229.5$, $p<0.001$
Function 2: Eig = 3.91, % Var = 20.9, $\lambda = 0.121$, $x^2 = 100.5$, $p<0.001$

An examination of the most important predicting variables shows that the degree of development of zygomaxillary tuberosity increases with an increase in supraorbital projection, area of the second molar, and lateral facial retraction. Mean values of pterion–asterion length and bregma–subtense fraction show a non-linear variation in the four grades of zygomaxillary tuberosity – the highest mean values of these two variables are found in grade 3. The degree of development of a zygomaxillary tuberosity is also related to an increase in minimum frontal breadth, an increase in mastoid height (with the highest mean value in grade 3), a smaller maximum frontal breadth associated with grade 4, a decrease in the curvature of the posterior alveolar plane of the maxilla, and an increase in malar length.

Rounding of the infero-lateral margin of the orbit

A discriminant analysis of the three grades of degree of rounding or the infero-lateral margin of the orbit shows that correct discrimination was obtained in 95.4% (63/66) of skulls on the basis of 23 cranio-dental variables (Table 6.4e). The two canonical discriminant functions obtained are statistically significant:

Function 1: Eig = 3.58, % Var = 76.8, $\lambda = 0.1051$, $x^2 = 117.1$, $p<0.001$
Function 2: Eig = 1.08, % Var = 23.2, $\lambda = 0.4811$, $x^2 = 38.0$, $p<0.05$

An examination of the average values of the most important discriminating variables for each grade shows that a rounded infero-lateral orbital margin levelled with the floor of the orbit is found associated with broad upper faces, short distances for hormion–basion, short parietals, long malars, proximity of left and right temporal muscles across the parietals, prognathic faces, long noses and short occipital planes of the occipital bone. A rounded, but not levelled infero-lateral orbital margin (grade 2) is particularly associated with large nuchal planes, minimum values for nasal height and foramen magnum length.

Nasal saddle

The results of a discriminant analysis of the four categories of nasal saddle show that the curvature of the nasal bones may be correctly predicted in 90.5% (67/74) of skulls on the basis of 21 cranio-dental measurements (Table 6.4f). The three canonical discriminant functions obtained are statistically significant:

Function 1: Eig = 1.86, % Var = 40.6, $\lambda = 0.0634$, $x^2 = 166.9$, $p<0.001$
Function 2: Eig = 1.60, % Var = 35.0, $\lambda = 0.1813$, $x^2 = 103.3$, $p<0.001$
Function 3: Eig = 1.12, % Var = 24.4, $\lambda = 0.4719$, $x^2 = 45.4$, $p<0.001$

The most important predicting variable, proximity of left and right temporal muscles across the vault, does not vary linearly with categories of nasal saddle. A low profile of the nasal saddle, interpreted as a flat angle between the nasal bones and characteristic of sub-Saharan African crania, is found in skulls with left and right temporal muscles set apart, large interorbital breadths, short nasion-occipital lengths, long distances from the posterior border of the palate to basion, short malars, short mastoids, and parietal angles positioned away from bregma. Its opposite expression, broadly curved nasal bones, characteristic of European and Australian crania, is found in skulls with relatively shorter distances between the left and right temporal muscles across the vault, more narrow interorbital widths, longer crania, a greater minimum nasal breadth, shorter distances from the posterior border of the palate to basion, short malars, large mastoids and parietal angles positioned away from bregma. Spatially close temporal muscles are related to 'pinched noses' and distant ones to flat nasal bones. However, the classification percentages by this variable alone show that its discrimination is mainly for grades 2 and 3. These distributions show that the relationship between nasal saddle and its predicting variables is not a simple one – different variables relating to the four grades in different patterns. This was expected, because the scoring of nasal saddle involved not just degree of development, such as sagittal keeling or zygomaxillary tuberosities, but also a categorical classification.

Lower narial margins

A discriminant analysis of the 10 categories of lower narial margins shows that it is possible to discriminate the configuration of the lower narial margin in 93.3% (72/77) of crania on the basis of 22 cranio-dental variables (Table 6.4g). The first five discriminant functions obtained are statistically significant:

Function 1: Eig = 4.40, % Var = 36.0, $\lambda = 0.0012$, $x^2 = 390.6$, $p<0.001$
Function 2: Eig = 2.34, % Var = 19.2, $\lambda = 0.0064$, $x^2 = 292.8$, $p<0.001$
Function 3: Eig = 1.53, % Var = 12.5, $\lambda = 0.0215$, $x^2 = 222.7$, $p<0.001$
Function 4: Eig = 1.23, % Var = 10.0, $\lambda = 0.0544$, $x^2 = 168.9$, $p<0.001$
Function 5: Eig = 1.05, % Var = 8.6, $\lambda = 0.1212$, $x^2 = 122.4$, $p<0.05$

The distribution of mean values of the predicting variables by the 10 grades of narial margins is not simple, since the scoring performed reflected categories rather than degree of development. However, it is possible to draw some conclusions. The complex relationship between the categories of narial margins and the predicting variables reflects the difficulty in determining these categories in the first place. However, some general categories, like a distinct margin (N1, N6, N5), presence of a pre-nasal fossa (N7, N9, N4, N6), with or without an external margin, and complete absence of external margins (N10) showed some distinct relationships to the predicting variables. From these, we find that large teeth are associated with the development of a prenasal fossa, that crania with a prenasal fossa but no external margin are associated with very pronounced superciliary ridges, and that crania with strong alveolar prognathism and small superciliary ridges are associated with complete absence of narial margins. Although some of these relationships between type of narial margin, tooth size and alveolar prognathism had been proposed before (Burkitt & Lightoller, 1923; Gower, 1923; Johnson, 1937; Macalister, 1898), it should be emphasised that the feature proposed by the Multiregional Model and usually associated with Australians, i.e. lack of external margins, was only observed in African and Inuit crania that presented very prognathic faces with smooth supraorbital regions, while the Australian sample characterised by large teeth and strong superciliary ridges, was associated with the development of prenasal fossae, with or without clear external margins.

Do the facial features of robusticity follow similar functional constraints?

There have been several studies that examined the relationship between cranial morphology and function. In the case of the cranial superstructures, most of this work has been focused on the aetiology of the supraorbital torus. Endo (1965, 1966, 1970) carried out a number of experiments on the skull by applying stressing forces, and concluded that the supraorbital area, especially the glabellar region, is mechanically deformed during chewing. He proposed a model by which the facial skeleton functions as a rigid beam structure through which the stress created during mastication is dissipated. He identified anterior dental loading as the main stressing factor. Russell (1983, 1985) developed this model further, suggesting that intermittent masticatory stress stimulates cortical bone remodelling to resist the central upward forces and the downward pull of both *temporalis*. The bone remodelling would cease when the supraorbital area achieves a shape that counteracts the stress by minimising bending and torsion effects. Endo (1970) and Russell (1985) suggest that these vertical anterior forces can be resisted either by buttressing (supraorbital torus) or by a vertical frontal, which would pose more resistance to the maxillo-dental pressures, the latter accounting for the relatively smaller supraorbital ridges found in Africans. However, Russell (1985) stresses the importance of circumstantial effects, i.e. the combination of the various determining factors, for the development of a supraorbital torus. The circumstantial effects proposed by Russell (1985) may explain why, although frontal angle was among the discriminating variables in the analysis of supraorbital torus, it was not one of the most important ones. Also associating supraorbital morphology with masticatory stress, Hilloowala & Trent (1988a,b) found that the formation of the supraorbital torus in humans and other primates is more closely related to mandibular length than to dental size, as the result of the traction of the masticatory force exerted by the anterior part of the temporal muscle. The smaller the ratio of the power arm to the load arm of the mandible, the greater the anterior part of the temporalis and the larger the supraorbital ridges. However, Demes & Creel (1988) found a consistent relationship between bite force generating the necessary muscle pull and dental crown area. The small number of mandibles available in this study did not allow corroboration of Hilloowala & Trent's findings, but the discriminating results obtained on the basis of tooth size and upper facial structure suggest that these variables might be just as important as mandibular dimensions.

However, alternative models for the role of the supraorbital torus have been proposed. It has been argued that browridge development is not associated with mechanical factors of mastication, but with the spatial relationship between the cranium and the face (Biegert, 1963; Moss & Young, 1960; Shea, 1986). This hypothesis has received strong support in recent years by Ravosa (1988, 1989, 1991a,b) and the *in vivo* bone strain analyses in primates carried out by Hylander and co-workers (Hylander, 1979, 1984; Hylander *et al.*, 1991; Picq & Hylander, 1989). These analyses show that the glabellar region is indeed strained during mastication (equally by biting and chewing), but that the stress created by this mechanical loading is very small, even smaller than that measured in other, more fragile areas of the face (Hylander *et al.*, 1991). These authors conclude that this level of stress cannot account for supraorbital ridge development. Although they argue that all bones require certain repeated levels of strain to complete their development, as shown by Rubin *et al.* (1990), they suggest that brow ridges do not develop as a structural adaptation to counteract masticatory stress, but instead that they serve an important spatial role in joining the face (position in front of the vault) and the braincase.

The results of the discriminant analyses support aspects of both alternative hypotheses. On the one hand, the importance of dental size and temporal muscle size as discriminating variables supports the view that the size of the facial superstructures respond to the pressures of mastication transmitted to the vault through the two pillars of Strasser — the frontal processes of the maxilla and the zygomatic – by means of a local reinforcement of the supraorbital area, and suggest that a similar effect may be associated with the development of other facial superstructures. However, these results also suggest that the supraorbital torus and other facial superstructures are not only ontogenetically developed to respond to chewing forces, as proposed by Endo and Russell. Furthermore, the expression of the superstructures is very strongly associated with specific cranial dimensions, and therefore related to spatial aspects of cranial size and shape. The combination of specific cranial dimensions characterises different populations, and as shown by studies of recent population relationships through multivariate analyses using cranial measurements, these dimensions largely reflect the paths of population differentiation (Brace *et al.*, 1992; Howells, 1973, 1989; Pietrusewsky, 1984, 1992; Wright, 1992). Therefore, we may conclude that the facial superstructures have a genetic basis for their development associated with specific cranial constraints, although it cannot be ruled out that their degree of expression may also be affected during life by the effects of the forces created during chewing on specific cranio-facial forms.

Keeling of the vault

A discriminant analysis of the three degrees of sagittal keeling shows that it is possible to discriminate correctly the degree of keeling in 95.4% (62/65) of skulls on the basis of 21 cranio-dental variables (Tables 6.5a). The two canonical discriminant functions obtained are statistically significant:

Function 1: E = 8.12, % Var = 89.4, $\lambda = 0.056$, $x^2 = 150.0$, $p<0.001$
Function 2: E = 0.96, % Var = 10.6, $\lambda = 0.509$, $x^2 = 35.1$, $p<0.05$

An examination of the 12 most important predicting variables shows that sagittal keeling increases with an increase in area of the second molars, an increase in the proximity of left and right temporal muscles across the vault, an increase in frontal bone surface length, with varying degrees of total facial prognathism, with an increase in the distance from minimum frontal breadth to the face, an increase in temporal muscle and cranial height, a decrease in inion–opistocranium length, an increase in minimum frontal breadth and basion–nasion length, large supraorbital ridges and short pterion articulations. Molar size and proximity of temporal muscles across the parietals account for 67.2% of correct grade prediction.

Is cranial keeling mechanically determined?

In an extensive study of Inuit morphology, Hrdlička (1910) identified a relationship between temporal muscle size and sagittal keeling. He noted that keeling of the vault was absent in children and less developed in females than males, and proposed that it is formed by a vertical upward expansion of the skull, together with a strengthening of the bone along the medial line, due to the compression effect of extraordinarily developed temporal muscles on lateral cranial growth during development, drawing parallels with the effects of artificial cranial deformation practices. He extended this explanation to account for the differences in cranial shape between Inuit infants and adults. However, Washburn (1947) observed that, by unilaterally removing the temporal muscle of newborn rats, there was no significant asymmetry in adult cranial shape, but a complete unilateral absence of temporal lines and resorption of the coronoid process (Rogers & Applebaum, 1941, also observed reduced coronoid processes in edentulous people).

Washburn (1947) recognised three classes of features in relation to their response to mechanical forces: (1) those which never appear unless the muscles are present (like temporal lines, associated crests and ridges), for which there is a direct causation; (2) those which are self-differentiated, but

Table 6.5. *Discriminating variables in order of importance for (a) sagittal keeling, and the features of occipital robusticity: (b) occipital torus and (c) occipital crest*

(a) SK: 21 discriminating vars.	(b) OT: 21 discriminating vars.	(c) OCR: 14 discriminating vars.
1: AM2 – Area maxillary M2	1: DIFAL – Curvature of the post. alveolar plane of the maxilla	1: NFRB – Naso-frontal breadth
2: SILLPA – Distance between superior temporal lines across the parietals	2: MDB – Mastoid width	2: MDM2 – Mesiodistal M2
3: FARCH – Frontal arch	3: PALD – Palatal depth	3: PALD – Palatal depth
4: PROG3 – Difference between prosthion and nasion radii	4: LAMOPIS – Lambda-opistocranium chord	4: OCC – Lambda-opisthion chord
5: WFBORB – Distance between WFB and orbital margin	5: PAF – Bregma-subtense fraction	5: BAH – Basion-hormion length
6: ITLHTA – Inf. temporal line height tape	6: NBS – Biorbital-nasion subtense	6: DIFE – Eversion of the lower border of the malars
7: BBH – Basi-bregmatic height	7: D1 – Palatal-basion subtense	7: D1 – Palate-basion subtense
8: INIOPIS – Inion-opistocranium length	8: FOL – Foramen magnum length	8: FRC – Nasion-bregma chord
9: WFB – Min. frontal breadth	9: MAB – External palatal breadth	9: DIFAL – Curvature of the posterior alveolar plane of the maxilla
10: BNL – Basion-nasion length	10: XFB – Max. frontal breadth	10: PARCH – Parietal arch
11: SOS – Supraorbital projection	11: FARCH – Frontal arch	11: INOP – Inion-opisthion length
12: PTE1 – Size of pteron articulation	12: FOB – Foramen magnum breadth	12: WIOB – Interorbital breadth
13: PAA – Parietal angle	13: TFL – Temporal fossa length	13: ZMB – Bimaxillary breadth
14: FRF – Nasion-subtense fraction	14: BPL – Basion-prosthion length	14: PALL – Palatal length
15: LM-LM – Distance between malar midpoint on its inf. border	15: ASB – Biasterionic breadth	
16: PAC – Bregma-lambda chord	16: WFBORB – Distance between WFB and orbital margin	
17: FMR – Frontomalare radius	17: SOS – Supraorbital projection	
18: NFFMS – Course of the naso-frontal and fronto-maxillary sutures	18: PALL – Palatal length	
19: PALD – Palatal depth	19: NPH – Nasion-prosthion length	
20: MDB – Mastoid width	20: IILLOC – Distance between the inf. temporal lines across the occipital	
21: FRS – Nasion-bregma subtense	21: BLM1 – Bucco-lingual M1	

require the presence of the muscles to remain or develop fully (like the coronoid process); and (3) those which are mainly independent of muscle development (like brain case). He interpreted the absence of any plastic effect on cranial size and shape by the temporal muscles in humans by the fact that 90% of brain growth is reached by age five, while the muscles of mastication have reached only 40% of adult size at the same age. Therefore, by the time the temporal muscle is exerting its full force, cranial size and shape have already been established.

However, Hrdlička (1910) also suggested that the bilateral pull of both *temporalis* during contraction could also result in strengthening of the sagittal sutural area, and it has been shown that sagittal keeling is partly caused by extra bone deposition in the midline region of Eskimo skulls (Hylander, 1977). The results from this study lend support to this interpretation of the origin of sagittal keeling. Although Washburn's argument that 90% of cranial growth has been achieved by age five is true, the cranium does not acquire its adult shape until much later, under the influence of masticatory and muscle bilateral contraction forces creating tension along the midline. These muscle forces depend on bite force, associated to dental and muscle size, which have strong underlying genetic parameters. It is possible that such stress as created by *temporalis* contraction on the mid-line of the vault has an ontogenetic response in the development of sagittal keeling. However, it is also possible that sagittal keeling is partly genetically determined as an adaptation in those populations whose cranial and muscular parameters habitually cause stress along the sagittal suture. This is consistent with the results obtained by Ossenberg and co-workers (1995) in a study of the effects of masticatory stress on certain facial/mandibular traits in two populations of different subsistence (Eskimo hunter–gatherers and E. Indian agriculturists). They conclude that the observed skeletal morphology of the Eskimo masticatory apparatus can be best explained as a plastic response of the skull during life and a long-term selection process favouring those individuals whose craniofacial proportions best respond to the Arctic adaptation in subsistence strategy.

As discussed in the case of the facial superstructures, the effect of the muscle forces on the cranium depends on the anatomical circumstances, i.e. size and shape constraints. Accordingly, what keeled Australian and Inuit crania share is the proximity of the left and right temporal muscles. This small distance results from narrow vaults in both Australian and Inuit crania, but in the latter it is also related to absolutely large muscles, while in the case of the Australians it results from the pronounced narrow breadth of the vault (Figure 6.1). Hrdlička's functional explanation was strongly refuted by Weidenreich, who believed that the development of a mid-sagittal

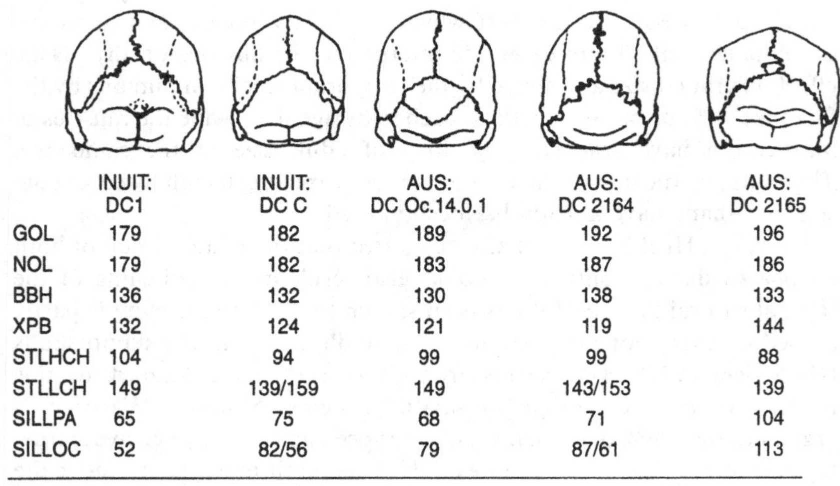

	INUIT: DC1	INUIT: DC C	AUS: DC Oc.14.0.1	AUS: DC 2164	AUS: DC 2165
GOL	179	182	189	192	196
NOL	179	182	183	187	186
BBH	136	132	130	138	133
XPB	132	124	121	119	144
STLHCH	104	94	99	99	88
STLLCH	149	139/159	149	143/153	139
SILLPA	65	75	68	71	104
SILLOC	52	82/56	79	87/61	113

GOL: maximum length
NOL: nasion-occipital length
BBH: basi-bregmatic height
XPB: maximum parietal breadth
STLHCH: superior temporal line height chord
STLLCH: superior temporal line length chord
SILLPA: minimum distance between left and right superior temporal lines across the parietals
SILLOC: minimum distance between left and right superior temporal lines across the occipital

Figure 6.1 Comparison of similar (first four crania) and different (last cranium) vault shapes resulting from the combined effects of cranial breadth (XPB), proximity of left and right temporal muscles across the vault (SILLPA, SILLOC) and sagittal keeling (all values in mm). AUS: Australian

crest, as well as all other tori and ridges of the skull, to be independent from any muscular causal factors (Weidenreich, 1940).

The occipital features of robusticity

Occipital torus

A discriminant analysis of the six grades of occipital torus (the first grade, absence of an external occipital protuberance without torus formation, was never observed) shows that correct discrimination was achieved in 97.3% (73/75) of crania on the basis of 21 cranio-dental variables (Tables 6.5b). Three significant canonical discriminant functions were obtained:

Function 1: Eig = 3.81, % Var = 64.0, $\lambda = 0.042$, $x^2 = 193.2$, $p < 0.001$
Function 2: Eig = 0.88, % Var = 14.8, $\lambda = 0.203$, $x^2 = 97.3$, $p < 0.01$
Function 3: Eig = 0.82, % Var = 13.8, $\lambda = 0.381$, $x^2 = 58.8$, $p < 0.05$

The mean values of the most important discriminating variables by the six grades of occipital torus show that, like the other traits that represent categories rather than degree of development, the relationship is not a simple one. Furthermore, the interpretation of these values should take into account that there is only one case in category 5. The most important discriminating variable is the curvature of the posterior alveolar plane of the maxilla. The development of an occipital torus, on which an external occipital protuberance is visible as an indent along the superior margin or sulcus, is found in skulls with maxillas that do not decrease in size posteriorly (no curvature of the posterior alveolar plane), wide mastoids, small palatal depths, long distances lambda–opistocranium, parietal angles positioned away from bregma, strong basicranial flexion, small lengths of the foramen magnum, wide and long palates, small values of maximum frontal breadth, long frontal bone surfaces, long basion–prosthion distances, pronounced supraorbital ridges and large teeth.

Occipital crest

The results of a discriminant analysis of the three grades of occipital crest development show that it is possible to discriminate correctly 97.2% (69/71) of crania on the basis of 14 cranio-dental variables (Tables 6.5c). The two canonical discriminant functions obtained are statistically significant:

Function 1: Eig = 1.00, % Var = 53.6, $\lambda = 0.2680$, $x^2 = 81.0$, $p<0.001$
Function 2: Eig = 0.86, % Var = 46.4, $\lambda = 0.5361$, $x^2 = 38.3$, $p<0.001$

The mean values of the most important predicting variables by grades of occipital crest show that two main relationships can be identified: (1) those variables that relate linearly, either increasing or decreasing with development of an occipital crest; and (2) those variables for which minimum or maximum mean values are associated with grade 2, i.e. partial development of an occipital crest. Therefore, a complete occipital crest running from the superior nuchal lines to the foramen magnum, is associated with small naso-frontal lengths, large teeth, deep palates, a long basion–hormion distance, malar eversion, small basicranial flexion, long frontal bones, maxillas that decrease in size posteriorly, short parietals, and short nuchal planes of the occipital bone. A partial occipital crest, from either the superior nuchal lines to the inferior nuchal lines, or from the latter to the foramen magnum, is associated with large nasofrontal distances, small palatal depths, long occipitals, small basion–hormion distances, short frontal bones, maxillas that do not decrease in size posteriorly and long parietals.

What are the underlying parameters for occipital robusticity?

Although occipital tori and occipital crests relate differently to cranio-dental dimensions, as also do partial and complete occipital crests, both the occipital superstructures share six important discriminating parameters: palato-dental size, basicranial size and flexion, frontal length, supraorbital protrusion, the relative size of the occipital and nuchal planes, and parietal flexion. It is clear that occipital robusticity, as the other cranial superstructures examined in this chapter, is associated both with dental size and specific cranial dimensions. An interesting aspect of these relationships is the association of occipital features with cranio-facial variables, and the inclusion of basicranial dimensions among the important discriminating features.

The statistical relationships of the cranial superstructures

The results of the discriminant analyses show that the incidence and grade distribution of these 'regional continuity traits' can be correctly predicted on the basis of other features of the skull with 90% accuracy or more. This trait prediction is characterised by the following: (1) in all cases, the relationship is not simple, and involves some 20 discriminating variables; (2) the percentage of correct classification by the first variable alone (maintaining a constant sample size) varies between 30% and 80%, but even in those cases where the total correct classification is high, this classification is very unequal among the grades; the percentage of correct classification by the first two variables varies between 50% and 80%, and in 6 of the 10 analyses, the distribution of cases correctly classified, although not completely homogeneous among the grades, achieves comparable levels; on the basis of the first 10 variables, all features (except categories of narial margins) are correctly classified with more than 70% accuracy; (3) dental dimensions are among the predicting variables in all 10 analyses, and a measure of temporal muscle size and proximity in eight, giving support to the hypotheses of some functional relationship between these features and the masticatory apparatus; and (4) very important spatial constraints, as shown by the importance of overall cranial dimensions in the discriminant functions.

An examination of the described relationships and the distribution of the predicting variables among all 11 features studied (Tables 6.4 and 6.5), indicates that certain traits are amongst the most important discriminating variables in several analyses:

(1) First and second molar dimensions, either as bucco-lingual/mesiodistal values or area, appear as predicting variables in all the analyses; in the case of five features (sagittal keeling, supraorbital torus, zygomaxillary tuberosity, categories of narial margins and occipital crest), they are either the first or second most important variables, for the zygomatic trigone the third, and in the case of depth of the infraglabellar notch, they are amongst the first 10 most important variables.
(2) Relative measures of temporal muscle size (ITLHCH, STLHCH, ITLHTA, STLHTA, ITLLCH, STLLCH, ITLLTA, STLLTA) and proximity across the cranial vault (IILLPA, SILLPA, IILLOC) appear as predicting variables in 9 of 10 analyses, and for six features (sagittal keeling, nasal saddle, supraorbital torus, zygomatic trigone and rounding of the orbit), they are amongst the first five most important variables.
(3) Measures of upper facial structure, either as total upper facial breadth (FMB, EKB), or as interorbital and upper nasal breadth and position (WIOB, WNB, NFRB, NFFMS), or as nasion depth in relation to the zygomatico-frontal process (NAS, NBS), appear as predicting variables in all 10 analyses; for six features (nasal saddle, supraorbital torus, zygomaxillary tuberosity, zygomatic trigone, occipital crest and rounding of the orbit) they are amongst the first three most important variables.

This pattern supports the view that several of the features studied show similar functional responses to these aspects of cranial morphology. The most striking of these is the association of several superstructures to the size of the supraorbital ridges. A measure of development of the superciliary ridges (SOS) appears in 6 of the 10 discriminant analyses (infraglabellar notch, zygomaxillary tuberosity, zygomatic trigone, sagittal keeling, categories of narial margins and occipital torus), in the case of the first three it is the most important discriminating variable. This suggests that these first three features are dependent upon the development of the superciliary ridges for their own expression. Furthermore, all three features show other discriminating variables in common, most notably dental dimensions. However, the overlap of predicting variables does not encompass many other cranial dimensions. Although the maximum expression of these traits will occur always in skulls with large superciliary arches and large teeth, the other determining characteristics vary from one to the other. It can be argued that, both the depth of the infraglabellar notch and the development of the zygomatic trigone are just an extension of the expression of the supraorbital torus, thus explaining the close overall distribution (Figure 6.2), and the complete absence of the minimum grade of all three features in the Australian sample as seen in Table 6.1. However, it is the zygomaxillary tuberosity that shows the closest similarity in combination of the first three predicting variables to that of the supraorbital torus, which together account for 71.4% of ZT grade 4 discrimination, and consequently, the closest regional distribution in its maximum degree of expression.

Weidenreich (1940) recognised that the ridges and tori of the skull were related among themselves, resulting from bony reinforcements to resist the pressures caused by the mandible, dentition and masticatory muscles, forming the cranial superstructures (Figure 6.3). However, contrary to most other authors, in Weidenreich's conception of these reinforcements, the muscular force and dental pressure were not even partially causative, and therefore, he did not consider that the development of these superstructures could be functionally influenced by muscular force during an individual's life.

Hauser & De Stefano (1989) observe that traits deriving from a common fundamental process are not necessarily associated in their occurrence. We have disclosed the cranial dimensions that constrain the expression of these features in spite of the stressing (dento-muscular) factors. This explains why not all skulls with large superciliary arches and large teeth exhibit other reinforcements of the face. Furthermore, this complex relationship between functional stress and anatomical dimensions may explain the variation in regional incidence of the four features studied. As mentioned before, several studies have shown that cranial dimensions can be accurately used to examine genetic distances between populations. Therefore, the anatomical circumstances that interact with the biological and non-biological stresses produced through mastication are largely non-functional in origin.

The morphology of the skull is the result of integrated growth of all its components, there being a general 'timetable' for its growth and development, shown by a common response of all dimensions to a slowing down of growth rates and the existence of specific growth rates of each structure (Moss *et al.*, 1956). The results obtained in this study, show that the various morphological features studied are strongly related to dental, muscle and cranial size, suggesting complex relationships in which both genetic

---

Figure 6.2 Relationship between the incidence of the facial superstructures showing that large supraorbital ridges are associated with a pronounced expression in the other three facial superstructures.
(a) Relationship between the depth of the infraglabellar notch (IN) and supraorbital ridge (ST) development.
(b) Relationship between the development of the zygomatic trigone (TR) and supraorbital ridge (ST) development.
(c) Relationship between the development of the zygomaxillary tuberosities (ZT) and supraorbital ridge (ST) development.

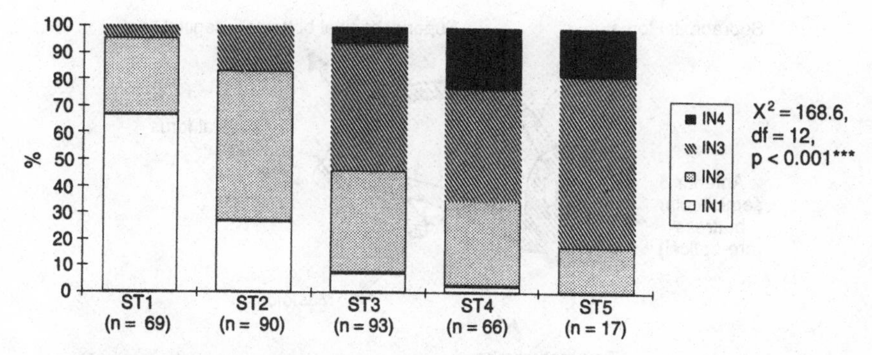

(a) Depth of infraglabellar notch by grade of supraorbital torus-ridges

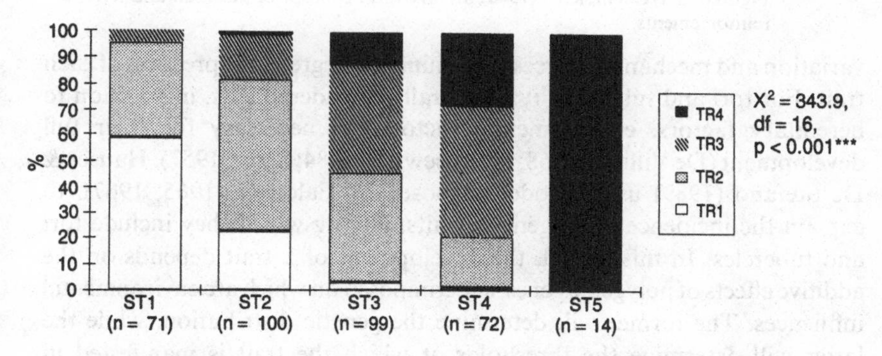

(b) Grades of development of zygomatic trigone by grade of supraorbital torus-ridges

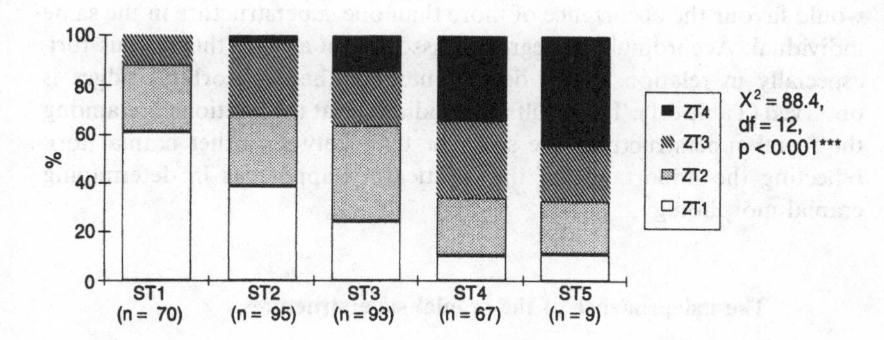

(c) Grades of development of zygomatic trigone by grade of supraorbital torus-ridges

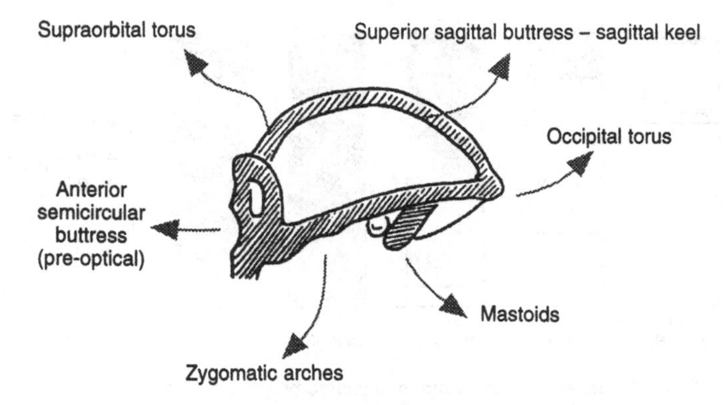

Figure 6.3 Weidenreich's (1940) architectural concept of the skull and its reinforcements.

variation and mechanical forces determine the degree of expression of each trait. For tori and tubercles, it is generally considered that, in addition to hereditary factors, environmental factors are necessary for their full development (De Villiers, 1968; Pietrusewsky, 1984; Scott, 1957). Hauser & De Stefano (1989) use a model proposed by Falconer (1965, 1967), to explain the incidence of epigenetic traits, among which they include tori and tubercles. In this model, the development of a trait depends on the additive effects of polygenic genes, superimposed on which are environmental influences. The former will determine the genetic distribution, while the latter will determine the thresholds at which the trait is manifested in various degrees in different populations. Furthermore, the existence of some correlation among hyperostotic characters (Ossenberg, 1969, 1970) would favour the occurrence of more than one superstructure in the same individual. Accordingly, a degree of association among the various tori, especially in relation to the development of the supraorbital ridges is observed (Table 6.1). The results also indicate that the relationships among the facial superstructures are stronger than between other cranial tori, reflecting the importance of the masticatory apparatus in determining cranial morphology.

The independence of the cranial superstructures

This chapter set out to test whether the cranial superstructures, claimed to represent evidence of multiregional evolution in their high combined occurrence in Australian crania (Habgood, 1989), were related to other aspects of cranial morphology as well as to each other. The results obtained through discriminant analyses are very strong, and show that all 10 features analysed could be predicted to a large extent by other cranio-metrical variables. This prediction/classification was robust, with results beyond 90% accuracy, although the number of discriminating variables, and therefore the complexity of the relationships, was relatively large (some 20 variables). Furthermore, although showing some individual variation in regional and overall distribution, the four facial superstructures (supraorbital torus, zygomatic trigone, depth of the infraglabellar notch and zygomaxillary tuberosity) are closely related. The last three features are dependent upon the degree of development of the superciliary ridges for their own expression. The development of the supraorbital torus also appears to influence the degree of sagittal keeling, categories of narial margins and occipital torus. Lastly, some of the most important aspects of cranial morphology determining these various traits are dental dimensions, temporal muscle size and proximity across the vault, and upper facial architecture.

For the Multiregional Model of modern human origins, these results have strong implications. By being largely an expression of other craniometric variables, these 'regional continuity traits' cannot be used as simple phylogenetic markers in evolutionary studies, for they represent the manifestation of a common functional complex. At the beginning of this chapter we stated that Habgood's (1989) argument for regional continuity in Australasia based on the combined occurrence of cranial superstructures would only be true if these features were shown to be both more commonly associated in Australian crania and independent of each other in their underlying determination. The results obtained show that these features occur in combination in all populations, but that this combination of traits is slightly more frequent in Australian crania. However, the discriminant analyses also show that the expression of these superstructures is strongly related to the actual metrical dimensions of different anatomical parts and masticatory forces acting on the skull. In line with research on the mechanical response of the supraorbital area (Demes & Creel, 1988; Endo, 1970; Hilloowala & Trent, 1988a,b; Russell, 1985), these results show that mechanical stress may be associated with the development of these features. Nevertheless, the further association of the superstructures with cranial dimensions suggests that the actual mechanical effect is not the single most important determining factor, and varies according to the dimensions of the skull, which may enhance or diminish the result of the bite or muscular forces acting upon a particular cranial structure. This is consistent with the view that the spatial relationship between the face and the vault is the main constraining factor in developing some superstructures like the supraorbital ridges, which in turn seem to influence the expression of other tori and ridges. Australian individuals can clearly produce strong bite and muscular forces through their large teeth and musculature, which combined with their particular cranial dimensions, like a narrow vault (resulting in proximate temporal muscles), a broad upper face (increasing the moment arm of the downward pulling force of the *temporalis*), and a prognathic face positioned relatively in front of the brain case, create less resistance to the masticatory stress than the cranial dimensions of other populations, favouring buttressing of the cranial vault. This relationship between superstructures, masticatory power and anatomical constraints is a more consistent explanation for the combined incidence of cranial superstructures in Australian skulls than the inheritance of a number of individual features (which in fact are not independent) from a Ngandong ancestor.

7 *Multiregional evolution as the source of recent regional cranial diversity: A review*

The Multiregional Model of modern human origins proposes that regional populations of *Homo erectus* gradually evolved into modern regional peoples, thus recent cranial diversity would reflect archaic cranial diversity. This model suggests that after an original spread of *H. erectus* out of Africa, regional archaic populations throughout the Old World maintained the right balance between geographic differentiation and gene flow to allow absence of speciation and the acquisition of regional characters (the 'regional continuity traits') stable throughout one million years (Frayer *et al.*, 1993; Wolpoff, 1989a; Wolpoff *et al.*, 1984). Besides representing a theoretical position within studies of human evolution that does not see the evolution of *H. sapiens* as a significant event in relation to the appearance of *H. erectus*, this model claims strong morphological evidence in its support. This morphological evidence relates both to fossil and modern populations, and proposes that regional traits that characterise some modern groups can be identified in the fossils found in those same regions, and therefore, reflect an ancestral-descendant relationship (Aigner, 1976; Weidenreich, 1943a; Wolpoff, 1989a). A discussion of the evidence in support of the Multiregional Model should, therefore, include both a critical review of the fossil data and an assessment of the character and validity of the modern regional characterisations used.

The fossil data has been extensively studied and analysed, and within the past few years several books and papers have described the information available (Bräuer & Smith, 1992; Hublin & Tillier, 1992a; Mellars & Stringer, 1989; Smith & Spencer, 1984; Trinkaus, 1989a). A review of this information, indicates that the development of new dating techniques, such as Electron Spin Resonance (ESR) and Thermoluminescence (TL), together with the discovery of new late Pleistocene fossils during the past decade, have shown that a complete rearrangement of the relative chronology of findings in relation to that existing 15 years ago, was necessary (Foley & Lahr, 1992; Grün & Stringer, 1991). This new chronology, in turn, shows that the appearance of new morphological groups, either 'archaic' *H.*

sapiens, Neanderthals or modern humans, was not synchronous throughout the world, and therefore, the evolution of *H. sapiens* was not characterised by 'grade phases' as postulated before (Brace, 1967). This fact poses an evolutionary problem, since it becomes unclear whether the appearance of these groups in the various regions was the result of local evolution or migration from areas where they already existed. Most authors now accept that after the geographical expansion of *H. erectus*, the regional groups undergo a certain amount of differentiation, represented by the differences between the African and Asian fossils (Andrews, 1984). Furthermore, 'archaic' *H. sapiens* appears most probably in Africa and itself expands into Europe, where it has been suggested that it gradually evolved into the Neanderthals during the Middle Pleistocene (Vandermeersch, 1985). Whether the Chinese fossils currently recognised as 'archaic' *H. sapiens* are a somewhat later part of this geographic expansion (Stringer, 1988) or the result of local evolution (Wolpoff, 1992a) is also a matter of controversy. Although by the end of the Middle Pleistocene there are populations of 'archaic' *H. sapiens* in at least four regions of the world (Africa, Europe, Middle East and China), fossils that are considered 'anatomically modern humans' appear earliest in Africa (Bräuer, 1992a; Deacon, 1992; Grün & Stringer, 1991), where also transitional forms between archaics and moderns are to be found (Bräuer, 1992a; Stringer & Andrews, 1988).

This evidence is currently interpreted as indicating a localised evolution of these modern forms, most probably in the African continent, followed by geographic expansion into other parts of the world. Whether the establishment of modern populations in the different areas was the result of population replacement or admixture (and in the latter case to what extent) is still disputed. The position taken by the Afro-European Hypothesis (Bräuer, 1984a,b) is that a certain amount of hybridisation took place between modern humans and Neanderthals, and that this contact is reflected in the transitional character of some Central European fossils. The Assimilation Model (Smith, 1992; Smith *et al.*, 1989) differs from the above in the importance attributed to the regional archaic's genetic contribution. According to this hypothesis, gene flow between Neanderthals and moderns had only the effect of accelerating a local evolutionary process. The most extreme position is that taken by the Out of Africa Model (Stringer, 1989a, 1992a,b; Stringer *et al.*, 1984) that considers hybridisation between archaics and moderns to have been rare and evolutionary irrelevant, and therefore, the gene pool of all modern populations to have originated from a single African source.

To this chronological and morphological information, the evidence from archaeology and molecular genetics should be added. In Europe, the archaeological evidence indicates a distinct difference between archaic and modern technology and cognition, suggesting a replacement event of the archaic populations (Klein, 1992; Mellars, 1989; White, 1989b). There also seems to be an east to west temporal gradient in the settlement of Aurignacian groups in Europe, supporting the idea of an exogenous origin of the first modern populations in the area (Mellars, 1989, 1991). In Africa, although there is some indication of derived features in the Howieson's Poort MSA industries (Clark, 1982; Deacon, 1989), the pattern is not maintained through time or geographically extensive, and in the Middle East, where there is apparent temporal alternation of modern and archaic groups, the association of different groups with different stone-tool industries is only tentative (Bar-Yosef, 1989b; 1992a). The lack of material and objective information about East Asian and Javanese technologies does not allow the inclusion of these important regions in the discussion, although the appearance of upper Palaeolithic industries in Siberia approximately 35 ka suggests population movements into this area (Klein, 1992). The genetic evidence, although less clear and assertive than previously believed (Maddison, 1991; Templeton, 1992), indicates that there is an original division between African and non-African populations as shown by nuclear studies (Cavalli-Sforza *et al,* 1988; Tishkoff *et al.*, unpub. data) and that, in terms of mtDNA lineages, the Africans are the most diverse group (Cann *et al.*, 1987; Stoneking *et al.*, 1992; Vigilant *et al.*, 1991) and appear to expand before other populations (Harpending *et al.*, 1993), implying a longer evolutionary history of modern peoples in this continent. The genetic data, together with the archaeological evidence in Europe and Africa, strongly support the hypothesis of a single origin for all modern populations in Africa, with a subsequent geographic expansion.

Although the chronological, archaeological and genetic evidence supports a single, localised origin of all modern humans, there are several aspects of this evidence that, as mentioned above, are not completely clear. Furthermore, it is from the fossil and modern morphologies that the Multiregional Model draws its support. In terms of the fossil evidence, both the Single Origin and Multiregional models emphasise different aspects of the existing morphological patterns.

Based on the evidence discussed above, the Single Origin Model claims three main supporting morphological arguments (Stringer & Andrews, 1988): (1) the presence of archaic–modern transitional fossils in the African late Middle Pleistocene, represented by the group of Ngaloba, Irhoud, Omo 2, Florisbad and Eliye Springs; (2) the modern appearance of early Upper Pleistocene fossils in Africa and the Middle East, like Klasies River Mouth, Omo 1, Border Cave, Singa, Mumba, Qafzeh and Skhūl; and (3) the morphological discontinuity between Neanderthals and moderns observed in Europe.

Proponents of the Multiregional Model also put forth three morphological arguments (Wolpoff, 1989a; Wolpoff *et al.*, 1984): (1) the recognition of transitional forms outside Africa, represented by the Ngandong fossils, the Dali and Yingkou crania, and some central European material, like the Vindija remains; (2) the recognition of primitive characteristics in the early modern African and Middle Eastern fossils mentioned above, and therefore, not considering these fossils as 'anatomically modern humans'; and (3) the recognition of a group of features that characterise both archaic and modern populations in at least two regions (East Asia and Java–Australia), reflecting regional continuity through most of the Pleistocene.

The first two morphological arguments of the Multiregional Model centre on the problem of population description and characterisation, in terms of what should be considered 'archaic' *H. sapiens,* whether and to what extent there was regional differentiation of these groups, and what are the limits of variability that should be accepted within modern humans that allow the recognition of a fossil as transitional or fully modern. Although these issues are not solved, it is certain that most students of later hominid evolution recognise the Ngandong sample as an archaic hominid, as suggested by Santa-Luca (1980). Besides the enlarged cranial capacity and accompanying cranial modifications, the Ngandong fossils do not show any of the specialisation's towards the modern morphotype observed in the African transitional forms or even forms of 'archaic' *H. sapiens* (Santa-Luca, 1980). The results reported in Chapter 5 confirm this view by showing the distinctive morphology of the Ngandong hominids, a morphology which in many cases is not-homologous with that of modern humans. Furthermore, in all cases in which the Ngandong hominids present a *H. erectus* structure, the early modern sample from Africa and the Middle East shows the modern morphology. The Chinese archaic hominids, besides lacking strict chronological control, can hardly be said to represent a consistent evolutionary lineage with stable morphological features throughout the million years of hominid occupation in East Asia. Pope (1992) clearly shows that among the Chinese 'archaic' *H. sapiens* material there is much variation, some of which falls well-outside the lineage identified by Wolpoff and co-workers (Groves & Lahr, 1994). Lastly, recent research by Waddle (1994) has shown that the best interpretation for the early modern Central European material is not one of a local evolution from Neanderthals, or even of Neanderthal–modern genetic admixture but an African–Middle Eastern early modern ancestry.

Two points remain, however, that allow some workers in the field to claim a multiregional origin of modern human populations. These are the regional features of East Asian and Australian populations, which are considered to reflect evolutionary continuity from *H. erectus* to modern Chinese and Australian aborigines, and the lack of a consistent characterisation of what are modern humans in terms of cranial morphology, which in turn creates ambiguity in the interpretation of existing data.

The Multiregional Model: A reassessment of its morphological basis

In order to assess the morphological evidence for regional evolutionary continuity used by the Multiregional Model, the preceding chapters examined four points: (1) the description of the traits proposed by the model to reflect continuity and their means of measurement; (2) the degree of regionality of these 'continuity traits' in five recent regional populations and four samples representing specific spatial or temporal groups; (3) the incidence of these 'continuity traits' in late Middle and Upper Pleistocene fossil specimens from East Asia, Australia and other regions of the world; and (4) the independence of expression of these 'continuity traits' from each other and from other cranio-dental dimensions.

The evidence for continuity I: The definition of the 'regional continuity traits'

A review of the historical usage of the 'regional continuity traits' as characterising the evolution of *Homo erectus* forms into modern Chinese and Australians (Chapter 3), shows that many of the traits lack a formal description, resulting in the ambiguous use evident in the literature. Furthermore, with very few exceptions, no standardised means of measurement have been established for these traits, adding to the overall subjectivity of their use. This lack of clear definition also allowed for redundant features to be quoted as independent traits, falsely increasing the number of features of continuity, while some of the traits that can be scored as present or absent have been shown not to present the regionality claimed by the model by studies of the incidence of non-metrical traits in modern populations. Therefore, the number of features suggested to reflect continuity has been unnecessarily inflated. Some traits can be merged and others excluded. The list of 'regional continuity traits' (23 East Asian and 23 Australian) currently proposed by the Multiregional Model was here reduced to 10 East Asian and 20 Australian 'continuity traits'. In order to study these features, objective ways of measuring were determined, either metrically or through the development of a scoring system of grades. The latter represents categorical or grade classifications (minimum, maximum and intermediate degrees of development) of 10 morphological features as observed in a sample of 100 recent crania from varied geographical origins (Appendix I).

The process of developing the scoring system revealed two points: (1) that none of the features being studied show a strict regional incidence; and (2) that most of these traits have a gradient of expression. Therefore, unless clear parameters are used, considering a certain feature as present or absent would only be consistent in extreme cases, leaving all intermediate expressions unclear.

The evidence for continuity II: The 'regional continuity traits' in recent populations

In order to corroborate the regional character of these features, a comprehensive quantified analysis of the incidence in five recent regional populations and four narrowly defined samples was carried out (Chapter 4). The recent regional populations were samples of Europeans, sub-Saharan Africans, Southeast Asians, East Asians and Australians. The sub-Saharan African and East Asian samples include within them what would usually be considered as several populations, such as East, West and South (San/Khoi) Africans in the former, and Chinese, Tibetan, Ainu, Japanese and eastern Siberian in the latter. This choice of crania was intentional in order to eliminate the problem of sampling a single spatial or temporal group as representing African or Mongoloid variation. In any case, the broad composition of the regional samples should not affect the evolutionary interpretation of the analysis, since according to the Multiregional Model all populations within these broad regional groupings share a common regional *Homo erectus* ancestor, while according to the Single Origin Model, all populations had a relatively recent common ancestor.

The results of the regional analyses show that most of the traits studied exhibit geographical patterning, but that only in eight does the observed pattern correspond to the one proposed by the Multiregional Model. Among these eight features, only two belong to the East Asian list (M3 agenesis, coronal facial flatness). Therefore, according to these results, this group lacks most of the specialisations supposed to characterise its evolution from a local *H. erectus* ancestor. It could be argued that the sampling of the Mongoloid population used in this study has affected the results by increasing the within-group variance, but as discussed above, if these features were consistent temporal characteristics that maintained their regional character throughout most of the Pleistocene, we would not expect their incidence to be affected by the recent differentiation and expansion of Mongoloid groups. Six of the 20 features proposed to characterise recent Australians were indeed found at a higher incidence or with a distinctive value in that population (small pterion, deep and narrow infraglabellar notch, zygomaxillary tuberosities, acute petro-tympanic angle, rounded infero-lateral orbital margins and horizontal superior orbital margins). Another two features had their highest incidence in another population, but shared with the Australians a distinctive pattern (pronounced supraorbital ridges and backward position of minimum frontal breadth).

It could be argued that the fossil record never represents regional samples on the scale used for the comparison between recent populations. However, these broadly defined geographical samples serve to show how little inclusive the 'regional continuity traits' are across the breadth of descendants from the proposed East Asian and Australian lineages. This point is further reinforced by the incidence of these same features in four samples that, either because of their chronology or geography, should exhibit most of the East Asian features (the case of the South American Fueguian-Patagonian sample), or none of the Australian regional continuity traits (Afalou, Taforalt, Natufian). The former serves as a test of regional stability, while the latter as an outgroup comparison. The Fueguian-Patagonian sample shows none of the East Asian features of continuity, while exhibiting eight of the Australian ones (the same number as the Australian sample). The Mediterranean Epi-Palaeolithic samples show one of the East Asian and four of the Australian features. These results further show that the regional continuity traits are not exclusive to the regions in which they are supposed to reflect continuity from a *H. erectus* ancestral population, even in their most pronounced expression.

Two main conclusions may be drawn from the incidence of the 'regional continuity traits' among recent populations:

(1) The great majority of the regional features claimed by the Multiregional Model to characterise the East Asian and Australasian evolutionary lines occur outside these areas with a more pronounced expression than in the populations they are supposed to characterise, and therefore cannot be used as regional clades.
(2) The fact that the Afalou, Taforalt and Fueguian-Patagonian populations do not show particular similarities to the African and East Asian

samples used in this study, suggests lack of regional morphological stability in these particular features through time.

The evidence for continuity III: The spatial and temporal distribution of the continuity traits in the Upper Pleistocene

So far, the tests performed have only concentrated on modern samples, without examining the archaic regional incidence of these features in East Asia and Australia. Wolpoff *et al.* (1984) and Frayer *et al.* (1993) show that these features are present in Asian archaic hominids, but from the studies carried out by Groves (1989b) and Habgood (1989) we know that the incidence of these traits in Javanese and Chinese *Homo erectus* is also not exclusive. Many of these features are clearly plesiomorphic for all later hominids, and have been differentially retained and lost by various archaic hominid populations. However, in order to argue that the Multiregional Model is incorrect in its assumption that the few 'continuity' traits that present the proposed recent regional expression reflect morphological continuity from archaic regional populations, we should be able to derive the regional patterns from an alternative single ancestral group. We have seen that it is mainly in Australian crania that we observe a partial regional incidence of the proposed continuity traits. Therefore, the incidence of the Australian continuity traits in both the Ngandong hominids, late African Middle Pleistocene fossils, and the early modern sample from Africa and the Middle East was examined.

In order to do this, 26 Upper Pleistocene hominids from various parts of the world (Table 5.1), six specimens from the Ngandong sample, the Epi-Palaeolithic samples from the sites of Afalou and Taforalt, and a Natufian sample from Israel were examined for the presence and degree of development of the 30 East Asian and Australian 'regional continuity traits' (Chapter 5). In terms of the existence of regional patterns or lineages in East Asia and Australasia during the Upper Pleistocene, it was found that among the five East Asian fossils examined (UC101 and 103, Liujiang, Minatogawa I and IV) only coronal facial flatness was a consistent characteristic of this sample. In relation to all the other 10 East Asian continuity traits, they clearly cannot be called 'regional' for East Asia at any point in the Upper Pleistocene. The Upper Pleistocene–Holocene Australasian sample (Wajak I and II, Keilor, Kanalda, Kow Swamp 1, 5 and 15) does not behave uniformly – the two Wajak specimens are quite different from the Australian fossils. If only the Australian crania are taken into account, four features are consistently present: pronounced supraorbital ridges, backward position of minimum frontal breadth, zygomaxillary tuberosities and large zygomatic trigones. Three of these features relate to the supraorbital area, and are clearly regional in their expression. However, an examination of these traits in the Ngandong hominids reveals a markedly different morphology of an archaic character. The 'regional' Australian supraorbital morphology seems to characterise only fossil and recent Australians, a characterisation they do not share with any other archaic or modern group. The incidence of zygomaxillary tuberosities also characterises many Australian crania, but like all other facial features, the assumption of whether they are shared with Javanese archaic hominids rests only on the incidence in a single specimen (Sangiran 17).

A comparison of the incidence of the 'regional continuity traits' between the early modern humans of Africa and the Middle East and Ngandong shows that although variable, the early modern sample has a greater number of the supposedly Ngandong–Australian regional features than the Ngandong sample itself. The results obtained clearly show that the Australian pattern could be derived from the early modern sample from Africa and the Middle East, and that actually the latter group of fossils form a better ancestral model population than the Ngandong hominids.

Two main conclusion can be derived from the examination of the incidence of the East Asian and Australian continuity traits in late Pleistocene fossils:

(1) East Asian Upper Pleistocene fossils do not show a consistent regional pattern in relation to the proposed East Asian features, and the Australian fossil specimens show only a limited regional pattern, largely related to a robust supraorbital and facial morphology.
(2) Although there is variation between individual specimens, the modern fossils from Africa and the Middle East show 15 of the 19 Australian continuity traits, while the Ngandong hominids only show 7 of the 19 features, either consistently or variably. Therefore, the early modern African and Middle Eastern sample represents a better ancestral model population for Australians than the archaic Ngandong hominids, based on the very features that have been proposed to reflect Australasian morphological continuity through time.

The evidence for continuity IV: Phylogenetic significance of the 'regional continuity traits'

None of the morphological features claimed by the Multiregional Model is unique to East Asian or Australian crania, and only a few have their most pronounced expression in these groups. These observations notwithstanding, it has been argued that what characterises the Ngandong–Australian evolutionary lineage is the combined expression of the continuity traits (Habgood, 1989; Wolpoff, 1992a). The combined occurrence of the morphological features measured as grades is indeed very high in Australian skulls. However, this could result from either a strong phylogenetic pattern as argued by Habgood (1989) and Wolpoff (1992a) or through the presence of functional constraints that affect the incidence and expression of more than one of these features simultaneously, and therefore, underlie their combined incidence. This is especially important in the case of the features under study, since these traits mainly represent robusticity, such as supraorbital and occipital tori, and their functional association with aspects of cranial size, shape and the masticatory apparatus have been intensely studied (Demes, 1987; Demes & Creel, 1988; Endo, 1970; Hrdlička, 1910; Hylander *et al.*, 1991; Rak, 1987).

The independence of expression of the regional 'continuity traits'

The aetiology of epigenetic traits and the fact that genetic and environmental factors tend to act through the same developmental pathways (Cheverud, 1988) makes an assessment of the heritability of the morphological features used by the Multiregional Model difficult. In order to establish whether the degree of expression of these features as measured in the grading system developed in this study can be related to measurements of cranial size, shape and muscular development, multivariate discriminant analyses were employed (Chapter 6).

The results of these analyses consistently show a strong predicting relationship between the regional features and other cranio-metrical variables. The relationships obtained for the grade features were strongest in the case of the supraorbital torus (98.5%), but all features could be predicted with at least 90% confidence. For all the traits studied, a set of between 14 and 23 variables was identified as jointly predicting their degree of expression. In the case of narial margins, in which the scoring represents categories rather than degree of development, the underlying relationships are relatively difficult to interpret. However, for the other nine features examined, although the number of variables involved meant that the relationships obtained and their anatomical interpretation were not simple, morphological patterns could be established.

Firstly, pronounced supraorbital ridges are found in skulls with a large dentition, large anterior temporal musculature, large skulls with a laterally retracted upper face and a backward position of minimum frontal breadth. In agreement with previous studies on the aetiology of the supraorbital torus, it is proposed that this feature is closely associated with the dimensions of the masticatory apparatus, although its expression is also related to specific cranial dimensions, like upper facial width and retraction. This feature is constrained by anatomical parameters, although it may be affected by the pressures posed on the facial architecture by a strong masticatory apparatus. Furthermore, it was found that the degree of development of the supraorbital ridges is associated with the development of six other features (depth of infraglabellar notch, zygomaxillary tuberosity, zygomatic trigone, sagittal keeling, lower narial margins and occipital tori), and that in the case of the first three it was the most important discriminating variable. These three features also have other discriminating variables in common, most notably molar size, indicating a strong relationship among the facial superstructures and between these and dental dimensions. The different association between each of these features and specific cranial dimensions could account for both the fact that not all crania with projecting supraorbital ridges present the other facial reinforcements, and the differences in regional distribution in these four features.

Secondly, there is a strong relationship between the development of a sagittal keel and large molars, large temporal muscles that are spatially close across the vault, and long frontal bone surfaces. Accordingly, maximum degree of sagittal keeling is observed in Australian crania, with large teeth and narrow vaults that result in spatially close temporal muscles, and among Inuit crania, with their extraordinarily large temporal muscles. These results support other studies (Hrdlička, 1910; Washburn, 1947) that suggest that sagittal keeling is mainly the result of the bilateral pressure of large temporal muscles during development.

The results of the discriminant analyses reveal that the 10 features studied are not independent of each other; they also show a strong correlation to general cranial dimensions. Variation in the predicting variables, mainly dental size, temporal muscle size and certain cranial metrical parameters affect the morphology of the regional features studied. This effect can be interpreted as partially originating from functional constraints during development, while the association to specific cranial dimensions reveals population differences in the anatomical conditions necessary for their expression. Therefore, the formation of the tori and tubercles studied, their presence and position, may result from a specific genetic basis for them, while the degree of expression may be explained as the result of interdependent osseous growth and adaptive processes, and in some cases, developmental patterns of stress.

Function and phylogeny

Anthropologists disagree as to which features should be used for establishing population relationships. Therefore, it is important to consider the theoretical basis for the use of the 'regional continuity traits' as phylogenetic features by the proponents of the Multiregional Model. Wolpoff (1992a) established the two parameters he considers necessary for a morphological feature to be a valid phylogenetic trait: (1) it must reveal a specific adaptation for functions that can be performed in alternative ways (he considers that, if only one anatomical response to a specific function is possible, evidence for it across taxa may reflect its independent evolution due to similar selective pressures; on the other hand, if more than one anatomical response is possible, evidence for any one response across taxa is supportive of a phylogenetic relationship); and (2) the features should appear in combination, since the whole set or pattern characterises the ancestor–descendent relationship.

The first of these concepts does not take into account that although the biological relevance, or adaptive meaning, of morphological variables is the ultimate source of information for understanding the biology of past populations, most studies are restricted to using a number of variables to which, through their relationship to other variables and functional interpretations, only inferred adaptive values can be attributed. Comparisons between taxa may give consistent results, but cause and effect of the observed variable functional relationships are difficult to differentiate. If the comparisons are made without taking phylogenetic polarity, i.e. character-states, into account, we could attribute an adaptive interpretation to variable correlations without knowing the evolutionary origin of these relationships. Correlated morphological variables may have evolved to perform a function, and may be thus interpreted as homologous structures across taxa; or they may be the result of a third variable, independent of the structures under study; or still they may be maintained without performing the function for which they originally evolved (Gould & Vrba, 1982). In any of these cases, these adaptive complexes may represent either plesiomorphies or synapomorphies. Therefore, the recognition of morphological features as the only possible functional response to an inferred adaptation does not confirm their phylogenetic value. Furthermore, Wolpoff's (1992a) first point does not take into account that later Pleistocene hominid taxa are so close phylogenetically and morphologically, that genetic constraints, i.e. the amount of genetic variance available for selection to operate, are likely to have caused the same functional response to arise independently in different populations, whether different responses were possible or not. This is especially the case for tori and tubercles with their role as either compensatory structures to masticatory forces (Demes, 1987; Endo, 1970; Hrdlička, 1910; Rak, 1983, 1985, 1987; Russell, 1985; Trinkaus, 1987; Washburn, 1947) or spatial arrangements (Biegert, 1963; Hylander *et al.*, 1991; Moss & Young, 1960; Picq & Hylander, 1989; Ravosa, 1991a,b; Shea, 1986), especially in similarly sized hominids in whom these reinforcements would tend to optimise resistance effects in similar ways.

The second of Wolpoff's (1992a) parameters is the point also made by Habgood (1989), which proposes that the 'regional continuity traits' may not support morphological continuity in Australasia individually due to their plesiomorphic character, but that their combined occurrence in both Javanese *Homo erectus* and modern Australians is proof of continuity in that area. However, this position can be rejected on two grounds. Firstly, the morphological features used as evidence for evolutionary continuity are partly inter-dependent, and therefore cannot each be considered independent phylogenetic evidence; secondly, the degree of inter-trait correlation among some of these traits and between them and the dimensions of the skull and the masticatory apparatus, suggests that their combined occurrence results from functional constraints, rather than the independent inheritance of a number of characteristics. Therefore, without knowing the nature of the relationship between variables, their combined occurrence in a skull does not carry phylogenetic weight.

Functional adaptations usually involve a complex of features, and in ideal conditions, only the main underlying biological variable(s), for which a character-state could be attributed, should be used as data points for reconstructing phylogenetic relationships. These situations are rare, and although most workers are faced with unknown trait relationships (or the consistency of known relationships across taxa) the concept of trait polarity and independence of expression cannot be ignored. Therefore, due to inter-trait correlations and the plastic response to function of the features used by the Multiregional Model that would increase the chance of independent evolution across late Pleistocene hominid taxa, their use as independent phylogenetic markers is incorrect. Furthermore, because some of these features, like sagittal keeling, may be affected by masticatory forces during an individual's life, the underlying determining variables should be studied in terms of regionality and character polarity instead.

The Multiregional Model: A review

The Multiregional Model of modern human origins predicts that two sets of cranial traits will occur exclusively or with a higher incidence in

Australia and East Asia, reflecting morphological continuity from *Homo erectus* populations in these regions. As such, it derives an important part of the observed recent regional cranial diversity from differences already present in the regional archaic populations. According to this model, fossil *H. sapiens* in each area of the world will strongly resemble modern regional inhabitants. The results of this study strongly reject these predictions.

Firstly, in terms of the regionality of the features studied it was found that:

(1) no trait shows an exclusive regional incidence;
(2) a number of traits did not show a regional pattern;
(3) the majority of the features studied have a more pronounced expression or occur at a higher frequency in other regions than those proposed by the model;
(4) the combined occurrence of a number of features in Australian crania is related to common underlying functional and anatomical constraints rather than to the joint inheritance of a number of independent traits.

Secondly, in terms of morphological stability, the lack of special similarities between the Afalou and Taforalt fossils and the African sample, and between the Fueguian-Patagonian remains and recent East Asians, suggests a pattern of differentiation and morphological specialisation rather than continuous gene flow maintaining broad regional patterns. This was further tested by examining the proportion of East Asian and Australian 'regional continuity traits' that are consistently present in these regions' modern fossil record. It was shown that these features do not form an East Asian lineage even in the late Upper Pleistocene, while four features related to the facial and supraorbital morphology do seem to be consistently present in fossil and recent specimens in Australia. However these features that characterise Australian crania on a regional and temporal scale are not shared by this population and the Ngandong hominids.

Furthermore, this study found that the early modern African and Middle Eastern fossil sample represents a better ancestral model for all modern humans, not only because of those modern features shared with all recent crania and commonly highlighted by proponents of the Single Origin Model, but in the very features proposed by the Multiregional Model as reflecting an exclusive morphological continuity process from Javanese *H. erectus* to Australian aborigines.

The only possible conclusion from these results is that the morphological basis of the Multiregional Model of modern human origins is incorrect. There is no morphological continuity from archaic to modern people in Asia and Australia that may be identified through the list of features suggested. We have already seen that the bulk of the chronological and genetic data indicate a single origin of all modern humans in Africa. By refuting the one piece of evidence that stood against a single origin – regional morphological continuity – this work provides further basis for accepting a recent and single origin for all modern humans.

Modern human cranial diversity

The results and discussions presented in this section show that the morphological basis of the Multiregional Model is incorrect, and therefore, an alternative interpretation of the origins of the observed modern regional diversity in human cranial morphology is necessary. An alternative interpretation consistent with the fossil and genetic evidence will need to derive all modern human morphologies from a recent and single ancestral source. However, one of the differences between the Multiregional and the Single Origin models is that the former provides an explanation for the origins of diversity in modern cranial form, while the latter only focuses on the evolution of modern humans from an archaic ancestor, not exploring the evolutionary processes which created modern cranial diversity. Therefore, an alternative model for the Multiregional hypothesis that seeks to explain the origins of modern diversity as well as of a modern morphology would have to develop the evolutionary concepts behind the Single Origin Model into a more comprehensive hypothesis encompassing events in the Upper Pleistocene. The following section of this book develops some ideas towards such an alternative hypothesis. Assuming a single origin for all modern humans in Africa, the remaining chapters focus on the evolutionary patterns, processes and mechanisms that gave rise to the development of regional morphological diversity from a single source.

III The evolution of modern human cranial diversity from a single ancestral source

The previous section of this book examined what has been for a long time the orthodox view of modern human origins. Proponents of this view explain past and present diversity, at least partly, from the genetic input from archaic hominids into regional modern populations. As such, even some researchers who would accept the earlier appearance of modern humans in Africa assume that certain features that differentiate early moderns from recent populations reflect some degree of archaic morphological continuity in the form of interbreeding. This section examines these points by addressing the issue of how modern human cranial diversity could be derived from a single ancestral source. We have seen that the best explanation for the presence of a number of morphological features in East Asian and Australian crania, features especially chosen by those who believe that multiregional evolution took place, is not an archaic ancestry, but rather the differential retention of certain traits from a single ancestor, like the early modern humans in Africa and the Middle East. This section therefore, takes the parsimonious view that no significant archaic—modern interbreeding occurred, and that spatial and temporal differences among modern groups are the result of the process of population differentiation itself.

In order to investigate the processes that gave rise to modern cranial diversity, three aspects will be addressed. Firstly, the anatomical unity of all modern skulls will be examined in Chapter 8. This chapter reviews the main morphological parameters differentiating recent modern regional populations and the definition of a modern skull within the perspective of a single origin of modern humans. Secondly, Chapter 9 examines how the spatial and temporal global patterns of differentiation provide information on the processes involved by establishing the anachrony of change and its effect on the diversification process, and proposes cranial gracilisation as the common trend in the different evolutionary pathways that characterise the history of each population. Thirdly, Chapter 10, examines the role of geography in the biological process of diversification, and proposes a model of multiple dispersals as the evolutionary framework for interpreting the evolution of modern diversity. Chapter 11 integrates some of the main events in the history of modern humans within the framework of multiple dispersals, proposing a hypothesis for the evolution of modern human cranial diversity.

8 *Cranial variation in* Homo sapiens

Throughout this book it has been stressed that a single African origin of modern *Homo sapiens* is the interpretation most compatible with the chronological fossil evidence. The earliest modern fossils in the world are found in Africa, tens of thousands of years before other regions of the world. Recent demographic and genetic (nuclear and mtDNA) evidence support an African origin with later derivation of all other modern populations. The archaeological record of certain areas strongly indicates that replacement of archaic populations by immigrating early modern people occurred. A single origin followed by subsequent regional differentiation is the evolutionary pattern of all but possibly one mammalian taxa with large geographical ranges (Groves, 1992). And finally, studies on the uniqueness of the morphological features used as evidence for regional continuity, either among archaic hominids (Groves, 1989b; Habgood, 1989) or modern populations (Lahr, 1992, 1994, present study), clearly show that this particular set of traits does not reflect regional evolutionary lineages through the Pleistocene period. All these different lines of evidence say the same thing. A single origin model is the best evolutionary explanation for the origins of modern humans in the late Pleistocene.

There are three models of the origin of *H. sapiens* that are based on a single and recent African ancestral source:

(1) The Out of Africa Model (Stringer, 1992a; Stringer & Andrews, 1988; Stringer *et al.*, 1984) proposes that the origin of modern humans in Africa was followed by geographic expansion and replacement of the archaic populations of the world. According to this model, interbreeding between early moderns and archaics would have been minimal and irrelevant in terms of the modern gene pool.

(2) The Afro-European Sapiens Hypothesis or African Hybridisation and Replacement Model (Bräuer, 1982, 1984a,b, 1989, 1992a,b), proposes that the first modern humans appeared in Africa, where they are represented by the fossils from Klasies River Mouth, Omo1 and Border Cave, followed by geographical expansion with hybridisation with regional archaic populations. For this model, the latter process of gene

flow between moderns and archaics accounts for the Neanderthal-reminiscent features found in early European Upper Palaeolithic fossils.

(3) The Assimilation Hypothesis (Smith, 1985, 1992; Smith *et al.*, 1989; Smith & Trinkaus, 1992) proposes that from an African source of modern humans, gene flow into other regions of the world accelerated the evolutionary process while maintaining regional patterns of differences. According to this model, hybridisation between moderns and archaics would have been more extensive than that accepted by either of the other models, so that, instead of moderns with archaic features, transitional populations are to be found outside Africa. Furthermore, the proponents of this model do not consider that population movements from an African source were either extensive or necessary, the pattern of gene flow being compatible with a process that has been called 'geographically extended phyletic speciation' (Van Valen, 1986).

However, the last two of these models require archaic genetic contribution for explaining early modern diversity, and this archaic–modern gene flow cannot be confirmed. The archaeological and chronological evidence related to the appearance of modern humans in Europe show a pattern of population movement and replacement (Allsworth-Jones, 1993; Klein, 1992; Mellars, 1989, 1992). Furthermore, the absence of transitional forms throughout the world argues against gene flow as the main process of geographical expansion of the modern morphology. The fossil evidence for continuity in Central Europe can be questioned on the basis of the fragmentary nature of some very important remains, making an objective morphological assessment difficult, while the Asian evidence for continuity has been dealt with extensively in this study. Furthermore, as mentioned in Chapter 2, the recent studies by Waddle (1994) show that, statistically, the best model to explain the morphology of the early European modern fossils is not through interbreeding with Neanderthals, but as descendants from early African or Middle Eastern modern humans. These points do not support the notion of extensive archaic–modern gene flow (the Assimilation Hypothesis) or less substantial interbreeding within regions (the Afro-European Hypothesis), which are unsatisfactory explanations for the origins of modern diversity.

If the genetic evidence of a single recent origin of modern humans (Cavalli-Sforza *et al.*, 1988) with minimal archaic genetic contribution (Cann *et al.*, 1987; Harpending *et al.*, 1993; Stoneking & Cann, 1989; Vigilant *et al.*, 1991; Wilson & Cann, 1992) is taken into account, the best evolutionary interpretation of the data is given by the Out of Africa Model.

However, two criticisms of this hypothesis can be made on morphological and theoretical grounds: (1) that it derives all modern humans from an already fully modern ancestor (giving grounds to arguments that it is not possible to derive certain primitive features observed in some recent populations from this source), and (2) that a process of single origin, migration and replacement is far too simplistic to account for the different spatial and temporal patterns in modern cranial diversity. This chapter will deal with the first of these points.

Modern morphology

As mentioned above, one of the main criticisms of the Out of Africa Model is that several modern fossils fall outside the limits of recent cranial morphological or 'modern' variation: Kow Swamp (Thorne, 1977), Skhūl 5, Qafzeh 6, Mladeč 1 and 5, Cro-Magnon 3, Fish Hoek, Keilor (Howells, 1989; Kidder *et al.*, 1992), Skhul IV and IX (Corruccini, 1992), Border Cave (van Vark *et al.*, 1989) and Klasies River Mouth (Caspari & Wolpoff, 1990; Rightmire & Deacon, 1991; Wolpoff & Caspari, 1990). Furthermore, the teeth from the sites of Mumba Rock Shelter, Equus Cave and Die Kelders Cave are also larger than in recent crania (Bräuer & Mehlman, 1988; Grine & Klein, 1985; Grine *et al.*, 1991). In order to be able to assess whether these various fossils are indeed modern or not, it becomes necessary to review what are the characteristics of a modern skull.

Day & Stringer (1982) and Stringer *et al.* (1984) have tried to define this common cranial morphology and produced a list of features that characterise modern *Homo sapiens* in relation to other late Pleistocene populations:

I: vault relatively short and high (BBH/GOL $>$ 0.70; VRR/GOL $>$ 0.64).
II: high frontal bone (FRA $<$ 134°).
III: parietal bone long and well-curved in mid-sagittal plane (PAA $<$ 138°).
IV: parietal arch long and high, inferiorly narrow, superiorly broad (ASTBR/ASB $>$ 1.19).
V: occipital bone long, narrow and not markedly projecting (OCA $>$ 113°).
VI: small dentition, particularly anterior teeth.
VII: weak supraorbital torus, divided into medial and lateral portions.
VIII: true external occipital protuberance.
IX: canine fossa.
X: well-developed mental eminence.

These authors consider that at least three of the above metrical features have to be present for a skull to be described as modern. Table 8.1 shows

Table 8.1. *Mean values of the six metrical features considered to characterise modern humans (Day & Stringer, 1982)*

	BBH/GOL	VRR/GOL	FRA	PAA	ASTBR/ASB	OCA
Region	x̄ (N)	x̄ (N)	x̄ (N)	x̄ (N)	x̄ (N)	x̄ (N)
EU	0.71 (24)	0.66 (24)	128.9 (25)	132.4 (25)	1.21 (25)	118.2 (24)
SSA	0.73 (24)	0.67 (25)	125.3 (26)	134.4 (26)	1.23 (26)	122.1 (26)
SEA	0.77 (25)	0.71 (26)	129.5 (26)	131.7 (26)	1.26 (26)	122.2 (25)
EA	0.75 (41)	0.69 (42)	129.3 (42)	132.9 (42)	1.24 (42)	120.7 (42)
AUS	0.71 (38)	0.65 (36)	130.8 (38)	132.4 (38)	1.25 (38)	116.7 (37)
F/P	0.73 (28)	0.67 (28)	135.7 (29)	133.5 (29)	1.24 (28)	116.3 (29)
AF	0.74 (23)	0.67 (33)	132.2 (32)	132.3 (37)	1.27 (34)	118.8 (33)
TAF	0.74 (9)	0.66 (19)	132.2 (19)	134.0 (20)	1.28 (20)	121.5 (17)
NAT	0.74 (8)	0.68 (15)	127.0 (24)	133.1 (25)	1.27 (21)	122.1 (18)
TOT	0.73 (220)	0.67 (248)	130.2 (258)	133.2 (268)	1.25 (260)	119.6 (251)

BBH/GOL, VRR/GOL: vault shape in terms of height relative to length; FRA: frontal angle; ASTBR/ASB: proportions of parietal arch; PAA: parietal angle; OCA: occipital angle.

the values for the above metrical features in the populations studied here. These results show that for most variables, the mean values of all nine populations fall within the limits established by Day & Stringer (1982). The exception is frontal angle, for which the Fueguians and Patagonians have on average a flatter frontal bone than is specified as 'modern' by Day & Stringer (1982). Furthermore, most of the crania present a supraorbital torus broken into medial and lateral portions and they also show an external occipital protuberance. Only in terms of the distinctiveness of a canine fossa, was it observed that an important percentage (10%) of skulls lacked a clear definition of this trait.

The use of mean values, however, ignores the ranges of variation in these variables. Accordingly, Wolpoff (1986a) found that in a sample of Australians and Tasmanians, 34% and 43% fell outside the limits for modern frontal and occipital angles respectively. Furthermore, in a study of the Kow Swamp and Coobool Crossing crania, Wolpoff (1992a) found that 29% (9/31) failed to present at least three features within the 'modern' limits of Day & Stringer (1982), concluding that no global definition of *H. sapiens* is possible because of regional continuity. In the populations studied here, although the mean values are all but one within the 'modern' limits for all six metrical variables, a certain percentage of cases fall outside these limits (Table 8.2). Furthermore, these percentages vary according to population (Figures 8.1a,b,c). However, these results do not show the pattern observed by Wolpoff (1986a) of higher percentage of flat frontals

Table 8.2. *Percentage of cases in each region that fall outside the 'modern limits' for the six metrical variables established by Day & Stringer (1982)*

Variable	ALL (%)	EU (%)	SSA (%)	SEA (%)	EA (%)	AUS (%)	F/P (%)	AF (%)	TAF (%)	NAT (%)
BBH/GOL <0.70	1.4	37.5	8.3	4.0	4.9	23.7	3.6	8.7	–	12.5
VRR/GOL<0.64	3.7	16.7	12.0	–	–	41.7	10.7	9.1	15.8	20.0
FRA>134°	19.4	8.0	7.7	11.5	7.2	17.2	55.2	37.5	31.6	–
PAA>138°	13.1	8.0	19.2	7.7	9.5	18.4	10.4	10.8	20.0	16.0
ASTBR/ASB <1.19	15.0	24.0	23.1	11.5	16.7	15.8	14.3	14.7	5.0	4.8
OCA<113°	14.4	25.0	3.9	4.0	9.5	29.7	20.7	18.2	5.9	–
$\bar{x}$	14.5	19.9	16.4	6.5	8.0	24.4	19.2	16.5	13.1	8.9

BBH/GOL, VRR/GOL: vault shape in terms of height relative to length; FRA: frontal angle; ASTBR/ASB: proportions of parietal arch; PAA: parietal angle OCA: occipital angle.

among Australian crania (the Fueguian-Patagonians, Afalou and Taforalt have more cases of frontal bones flatter than 134° than the Australian sample), although this population does show the largest number of skulls with an occipital angle smaller than 113°. The Europeans show the greatest number of cases with low vaults (as measured from basion), but if height is taken from porion, the greatest number of cases outside the 'modern range' is found in Australians. The Taforalt and sub-Saharan Africans show slightly more cases of a parietal angle greater than 138°, while it is the Europeans who show the greatest number of skulls outside the limit specified for the relative size of the anterior and posterior portions of the parietal bone. The Southeast Asians, with their gracile and rounded crania, are the group best defined by the parameters of Day & Stringer (1982).

The distribution of crania according to the parameters set by Day & Stringer (1982) reiterate the importance of both central tendencies and variances in interpreting the samples studied, and contrary to Wolpoff (1992a), do not support the claim that the definition of a modern skull was developed on the basis of European cranial variation. The different proportion of individuals from each population that fit a 'modern' morphology as defined by Day & Stringer (1982) suggests that in order to encompass all of modern human variation such a definition has to be broadened. This conclusion is similar to that reached by Brown (1990) in his assessment of the applicability of this definition to the material from Coobool Creek, Swanport and Murray Valley. Therefore, I will use the information from recent and sub-recent modern crania to establish the range of variation among modern populations and the levels of distinctiveness observed.

Regional morphologies

Part I of this book examined the degree of regionality of the 'continuity traits' by means of comparing their incidence in modern crania from five broadly defined regional samples (Europe, sub-Saharan Africa, Southeast Asia, East Asia and Australia), and four narrowly defined groups (a sample of Fueguian and Patagonian crania, and three Epi-Palaeolithic groups from the Natufian sites of the Middle East and the sites of Afalou and Taforalt in North Africa). The main results of these analyses show that the majority of the regional distributions proposed by the Multiregional Model were not confirmed. However, of the 31 traits or trait expressions examined, 26 show a strong, if not exclusive, regional distribution. Nine of these features have a strong regional expression in Australian crania, eight in the Fueguian-Patagonian sample and five in the Mediterranean fossil groups, showing that these groups are different from most other recent populations in certain aspects of cranial morphology. The incidence of the same regional features was then examined in some late Pleistocene archaic and modern fossil specimens, and found that these characteristics, rather than unique to some specific recent modern populations, were part of early modern human cranial variability, suggesting that among recent groups the degree of differentiation or retention of ancestral traits varied markedly.

Many studies have dealt with the problem of population distances, as expressed in metrical and non-metrical cranial traits (Brace *et al*, 1989; Brothwell & Krzabowski, 1974; Howells, 1973, 1976; Katayama, 1988, 1990; Knip, 1971; Ossenberg, 1986; Pietrusewsky, 1973, 1976, 1983; Tagaya & Katayama, 1988; Turner II, 1987, 1989; Wright, 1992), usually involving extensive multivariate analyses. Some of these have specifically concentrated on Australian groups in order to disclose the relationships of this very distinct modern population (Brown, 1967; Brown, 1989; Fenner, 1939; Giles 1976; Habgood, 1986; Kellock & Parsons, 1970; Pietrusewsky, 1979, 1984; Turner II, 1990a; Webb, 1989). However, the aim of this chapter is not to establish the distances between the populations being

---

Figure 8.1 Frequencies of the values of the six traits used by Day & Stringer (1982) to define a modern cranial form in the nine recent cranial samples studied. The line across all graphs represents the 'cut-off' point proposed by Day & Stringer (1982); below each population's graph are the percentage of cases that fall within the modern parameters in each population.
(a) Vault shape (BBH/GOL), (VRR/GOL).
(b) Frontal angle (FRA), parietal angle (PAA).
(c) Parietal proportions (ASTBR/ASB), occipital angle (OCA).

(a)
Frequencies of values of vault shape (BBH/GOL)

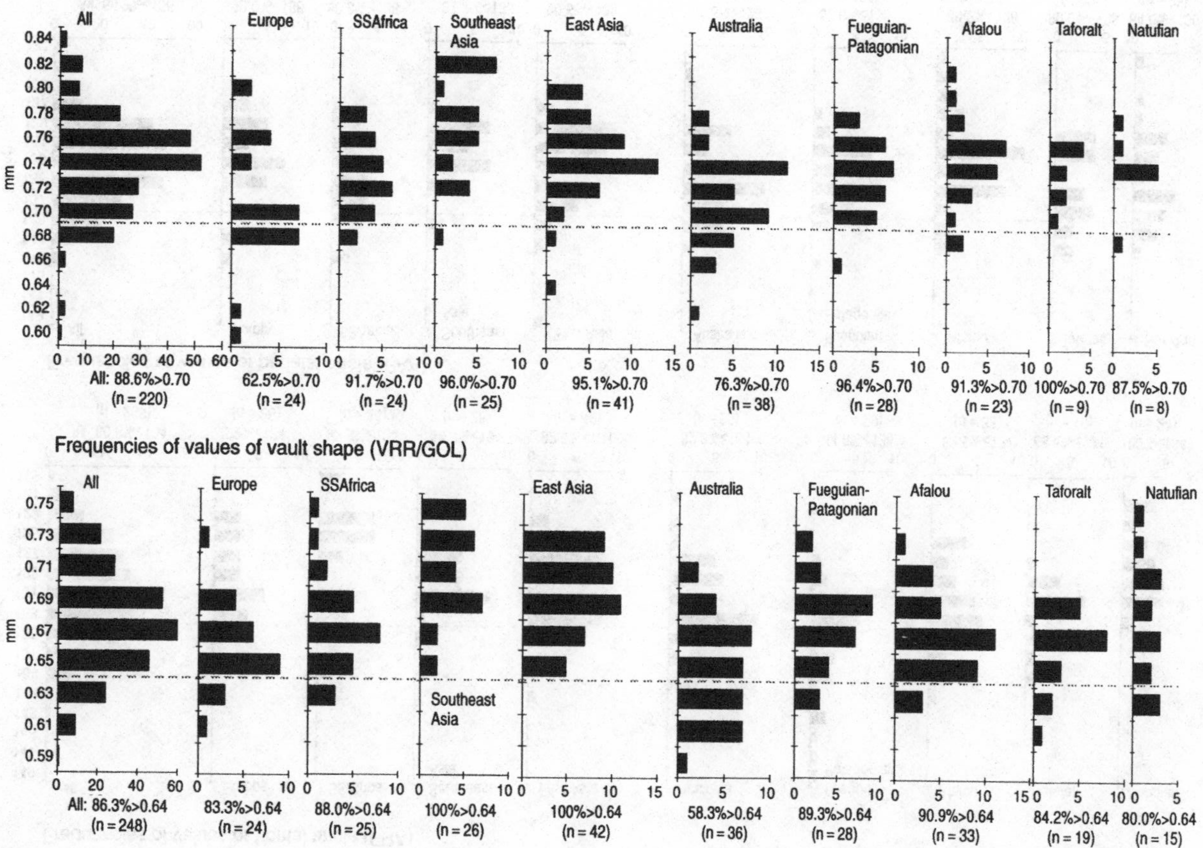
All
Europe
SSAfrica
Southeast Asia
East Asia
Australia
Fueguian-Patagonian
Afalou
Taforalt
Natufian
mm
All: 88.6%>0.70 (n = 220)
62.5%>0.70 (n = 24)
91.7%>0.70 (n = 24)
96.0%>0.70 (n = 25)
95.1%>0.70 (n = 41)
76.3%>0.70 (n = 38)
96.4%>0.70 (n = 28)
91.3%>0.70 (n = 23)
100%>0.70 (n = 9)
87.5%>0.70 (n = 8)
Frequencies of values of vault shape (VRR/GOL)
All: 86.3%>0.64 (n = 248)
83.3%>0.64 (n = 24)
88.0%>0.64 (n = 25)
100%>0.64 (n = 26)
100%>0.64 (n = 42)
58.3%>0.64 (n = 36)
89.3%>0.64 (n = 28)
90.9%>0.64 (n = 33)
84.2%>0.64 (n = 19)
80.0%>0.64 (n = 15)

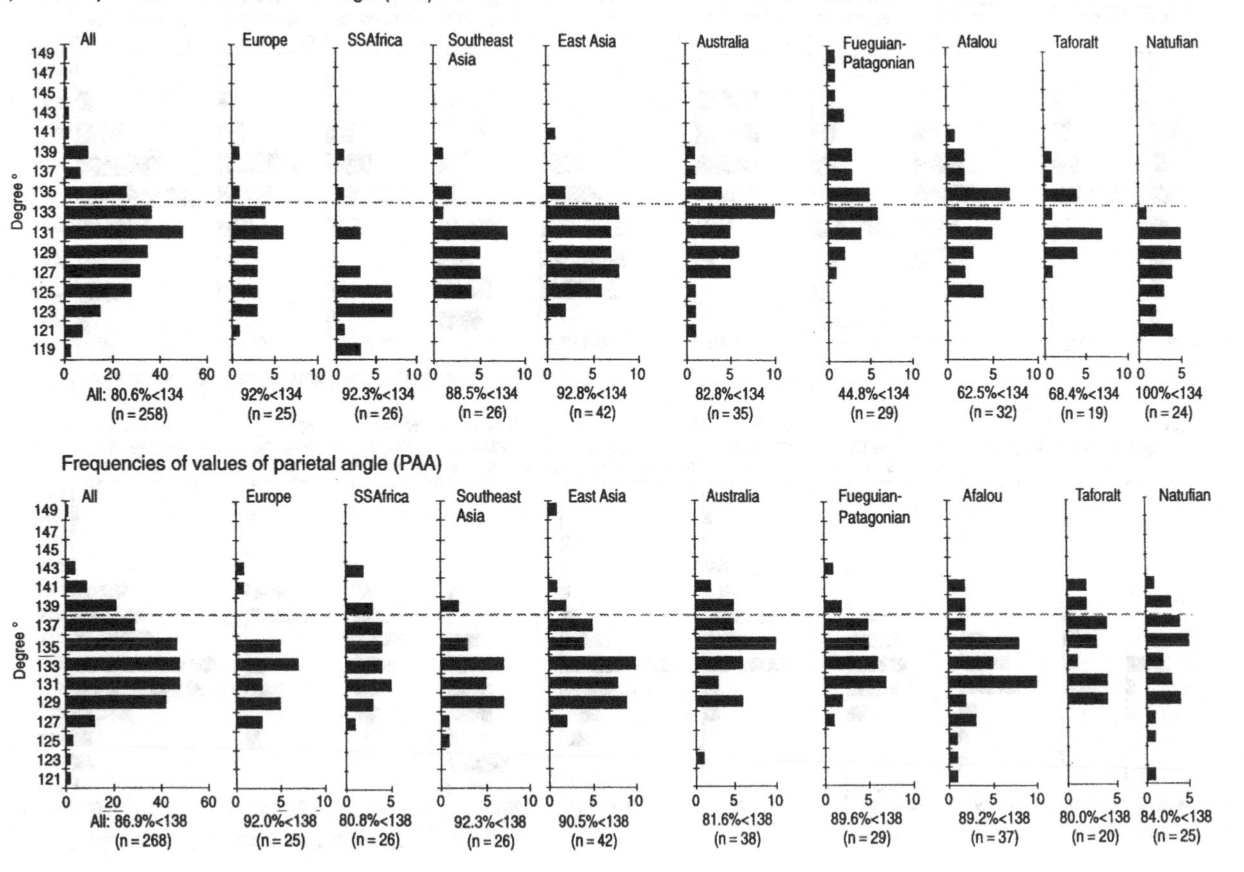
(b)
Frequencies of values of frontal angle (FRA)
Degree °
All
Europe
SSAfrica
Southeast Asia
East Asia
Australia
Fueguian-Patagonian
Afalou
Taforalt
Natufian
149
147
145
143
141
139
137
135
133
131
129
127
125
123
121
119
All: 80.6%<134 (n = 258)
92%<134 (n = 25)
92.3%<134 (n = 26)
88.5%<134 (n = 26)
92.8%<134 (n = 42)
82.8%<134 (n = 35)
44.8%<134 (n = 29)
62.5%<134 (n = 32)
68.4%<134 (n = 19)
100%<134 (n = 24)
Frequencies of values of parietal angle (PAA)
Degree °
All
Europe
SSAfrica
Southeast Asia
East Asia
Australia
Fueguian-Patagonian
Afalou
Taforalt
Natufian
All: 86.9%<138 (n = 268)
92.0%<138 (n = 25)
80.8%<138 (n = 26)
92.3%<138 (n = 26)
90.5%<138 (n = 42)
81.6%<138 (n = 38)
89.6%<138 (n = 29)
89.2%<138 (n = 37)
80.0%<138 (n = 20)
84.0%<138 (n = 25)

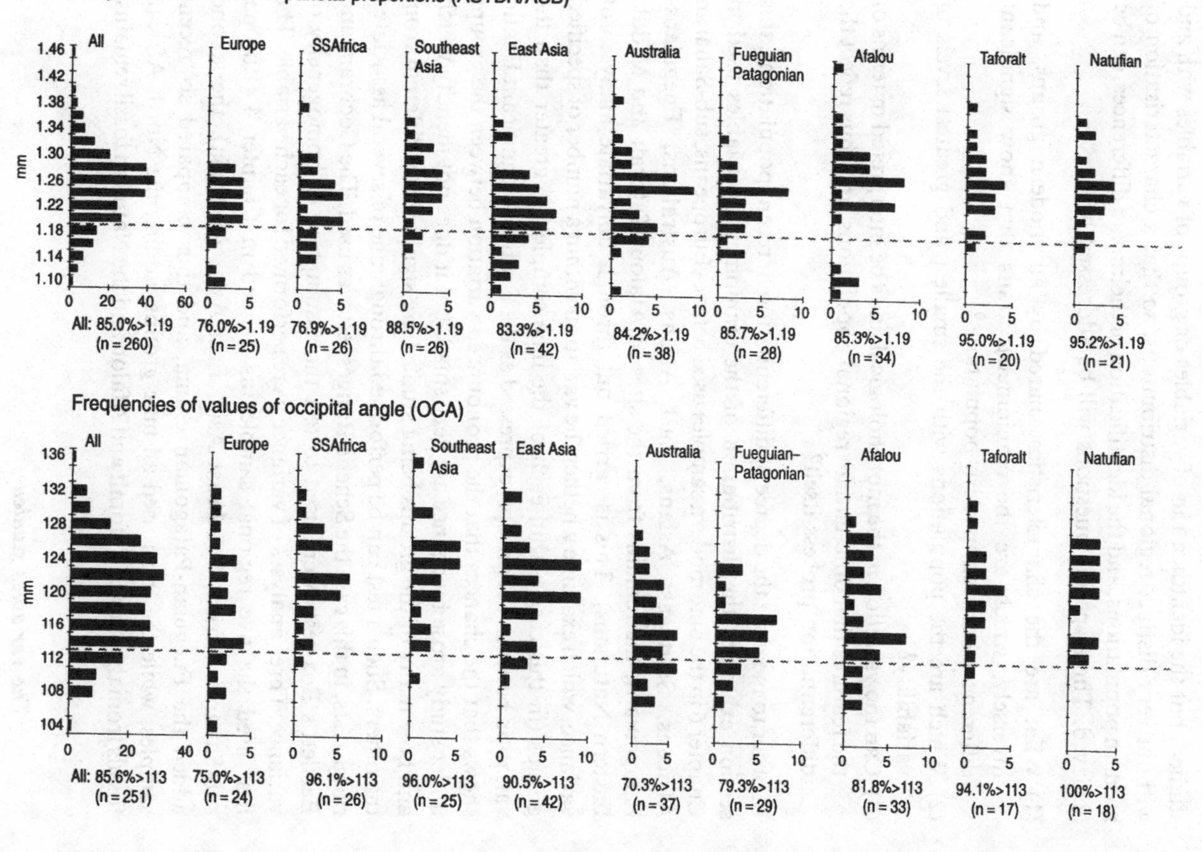
(c)
Frequencies of values of parietal proportions (ASTBR/ASB)
All
Europe
SSAfrica
Southeast Asia
East Asia
Australia
Fueguian Patagonian
Afalou
Taforalt
Natufian
mm
All: 85.0%>1.19 (n = 260)
76.0%>1.19 (n = 25)
76.9%>1.19 (n = 26)
88.5%>1.19 (n = 26)
83.3%>1.19 (n = 42)
84.2%>1.19 (n = 38)
85.7%>1.19 (n = 28)
85.3%>1.19 (n = 34)
95.0%>1.19 (n = 20)
95.2%>1.19 (n = 21)
Frequencies of values of occipital angle (OCA)
All
Europe
SSAfrica
Southeast Asia
East Asia
Australia
Fueguian–Patagonian
Afalou
Taforalt
Natufian
mm
All: 85.6%>113 (n = 251)
75.0%>113 (n = 24)
96.1%>113 (n = 26)
96.0%>113 (n = 25)
90.5%>113 (n = 42)
70.3%>113 (n = 37)
79.3%>113 (n = 29)
81.8%>113 (n = 33)
94.1%>113 (n = 17)
100%>113 (n = 18)

studied, but the identification of variables or groups of variables with and without very distinct regional distributions, so that a characterisation of modern populations, and the identification of their main differences, can be achieved. Three specific questions will be addressed here:

(1) What are the characteristics shared by all modern groups, and inversely, which are the craniometric traits that show significant differences between modern populations?
(2) Which are the populations with the smallest and greatest levels of variability?
(3) Can the variation in the morphological units be interpreted in terms of the features that differentiate regional populations, and thus reflect the differentiation process itself?

In order to measure the degree of differentiation in recent populations, this section examines the distribution of the craniometric variables used in Chapter 6 in the nine modern samples used before (Europeans, sub-Saharan Africans, Southeast Asians, East Asians, Australians, Fueguians-Patagonians, the remains from the sites of Afalou, Taforalt and Middle Eastern Natufians). This is carried out through univariate analyses of variance, which examine whether the variation within a number of specified groups (in this case within each of the nine samples) is greater than the variation between the groups, expressed as the F ratio. A significant F ratio shows that the chances that the pronounced variation between the groups under study arose by chance are very small, but it does not identify which and how many of the groups being studied are significantly different from the others. Such a test can be performed through one of several associated range tests, in this case the Scheffe Ranges test was used. The Cochrans and Bartlett's Box tests were used to test the assumption of homogeneity of variance. Three analyses of variance were performed for each variable. The first used the five regional samples as defined in Chapter 4 (Europe, sub-Saharan Africa, Southeast Asia, East Asia, Australia); the second added the Fueguian-Patagonian crania, and thus compared six recent samples; while the third used all nine groups, adding the North African fossils from the sites of Taforalt and Afalou and the Natufian fossil remains.

The variables studied

The metrical variables (97 in all) were grouped into 10 categories reflecting facial, cranial and individual bone dimensions (a description of the variables used can be found in Appendix II, Chapter 6 and Table 8.1):

(1) Main cranial dimensions:
length: GOL, NOL
height: BBH, VRR
breadth: XPB, ASB, AUB, EAMEAM, WCB, ZYB
distance face/cranium: BNL, NAR, FMR, EKR, ZOR, ZMR, SSR, BPL, PRR, AVR

(2) Facial dimensions:
height: NPH, NLH, OBH
breadth: FMB, EKB, JUB, ZMB, NLB, WIOB, WNB, NFRB
protrusion: NAS, SSS
alveolar: MAB, PALL, PALD, FP-PR, FPA
malar: XML, IML, WMH, ZMZM, LMLM
prognathism: GI, PROG1, PROG2, PROG3.

(3) Basicranium:
dimensions: FOL, FOB, BAH, HOP, WBAB
depth: D1, D2

(4) Mastoids: MDH, MDB

(5) Lateral cranial dimensions; PTE, PTEBR, PTEAST, ASTLAM, PTELAM, ASTBR

(6) Frontal bone:
dimensions: FRC, FARCH, XFB, WFB, WFBZFS, WFBORB, SOS, GLS
angle: FRS, FRF, FRA

(7) Parietal bone:
dimensions: PAC, PARCH
angle: PAS, PAF, PAA

(8) Occipital bone:
dimensions: OCC, OARCH
angle: OCS, OCF, OCA

(9) Temporal muscle:
size: STLHCH, STLLCH
proximity: SILLPA, SILLOC
temporal fossa: TFL, TFB

(10) Molar dimensions:
buccolingual: BLM1, BLM2
mesiodistal: MDM1, MDM2

Cranial differentiation in recent populations

Features shared by all recent populations

The results of the analyses of variance show that only a third (28/93, 30.1%) of the variables have a homogeneous distribution in the five regions studied. This shows that there is a high degree of population differentiation in cranial dimensions. The variables that are not significantly different in the five regional samples are cranial height (BBH), various measures of the distance of the upper face to the cranium (BNL, EKR, ZMR), orbital height (OBH), measurements of upper facial breadth, with the exception of bifrontal breadth (EKB, JUB), maximum malar length (XML) and inter-malar length (LMLM), distance of the posterior alveolar margin to the Frankfort Plane (FPAL), most aspects of basicranial dimensions (EAMEAM, FOL, FOB, BAH, HOP, D1), mastoid height and breadth (MDH, MDB), length of the frontal bone (FRC, FARCH), most parietal dimensions (PTEBR, PAC, PARCH, PAS, PAF, PAA), minimum nasal breadth and naso-frontal breadth (WNB, NFRB).

However, the homogeneity observed in the above mentioned cranial dimensions is much reduced when either a specialised cranial sample like the Fueguian-Patagonians or chronologically earlier populations are included in the comparisons. Once the Fueguian-Patagonian sample is included, there is an increase in significance in the F ratio of 26 variables, similarly when the Mediterranean Epi-Palaeolithic groups are included. This increase in the scale of the population differences observed indicates how much the variation between modern populations is affected by which populations are being sampled. Clearly, a specific population like the Fueguian-Patagonians is highly differentiated in relation to samples defined on a regional scale, while the *circum*-Mediterranean Epi-Palaeolithic fossils, representing in conjunction a regional sample, indicate important differences in the character of morphological variation through time. Only four cranial dimensions have a homogeneous distribution in all nine modern samples, namely parietal angle (PAA) and certain basicranial measurements (FOL, FOB, D1).

Considering each of the five modern populations individually, the following generalisations are possible:

(1) The Australians show the largest number of significant differences – 30/93 (32.3%) variables significantly different from the other four populations (VRR, XPB, AUB, ZOR, SSR, BPL, PRR, AVR, FMB, NLB, NAS, MAB, PALL, GI, PROG1, PROG2, PROG3, IML, WMH, XFB, WFBZFS, WFBORB, GLS, PTE, OCA, SILLPA, BLM1, MDM1, BLM2, MDM2), 43.3% of which they share with sub-Saharan Africans.

(2) The sub-Saharan Africans are significantly different from other populations in 22/93 (23.7%) measurements (VRR, XPB, AUB, WCB, ZYB, BPL, PRR, AVR, NLH, NAS, PALL, GI, PROG1, PROG2, PRG3, STLHCH, STLLCH, SILLOC, XFB, FRS, FRF, FRA), 59.1% of which they share with Australians.

(3) The Europeans are significantly different in 10/93 (10.7%) variables (ASB, D2, FP-PR, ZM-ZM, ASTLAM, PTELAM, ASTBR, SILLOC, TFL, GLS).

(4) The East Asians are significantly different in 7/93 (7.5%) measurements (PTEAST, NPH, ZMB, SSS, FP-PR, ZM-ZM, SOS).

(5) The Southeast Asians have the least (6/93, 6.4%) significant differences (GOL, NOL, NAR, OARCH, OCS, OCF).

Therefore, in a comparison of broad regional populations, all groups provide a degree of variation not present in the others, but the variability observed in recent cranial morphology is significantly increased if Australians or sub-Saharan crania are included. The Australian and African samples share a pattern of differentiation in a number of cranial dimensions. These similarities between the two groups are largely concentrated in two anatomical complexes. In terms of main cranial dimensions, the two groups share a relatively low vertex height and a narrow vault (VRR, XPB, AUB, XFB), and in terms of facial dimensions they share a long palate, and a sagitally and coronally prognathic face (BPL, PRR, AVR, PALL, GI, PROG1, PROG2, PROG3, NAS). It is interesting to note that like the Australians, the sub-Saharan Africans have a significantly longer palate, but contrary to the Australians, both the palate breadth and dental dimensions are much reduced in the African crania. Australians and sub-Saharan Africans are very different from each other in other dimensions, such as the configuration of the frontal bone (GLS, FRS, FRF, FRA) and measurements related to the size of the temporal muscle (ZYB, STLHCH, STLLCH, SILLOC). This pattern of similarities and differences in different portions of the skull between Australians and Africans explains why these two groups appear either closely or distantly related in multivariate distance analyses, depending on which variables are used.

As mentioned before, the inclusion of the Fueguian-Patagonian and Mediterranean Epi-Palaeolithic samples largely increases the number of significantly different values observed. These groups are so differentiated in relation to the regional samples that some of the previously detected differences between the regional groups lose statistical significance. This has the effect of making the recent regional samples appear more similar to each other. Although true in relative terms, this fact does not reflect a close similarity between recent regional populations (what we have seen above only exists for about a third of cranial dimensions) but rather the level of specialisation of the South American and fossil groups. The Fueguian-Patagonian sample is significantly different from other populations in 38.7% of the cranial measurements taken (36/93), the Afalou in 47.3% (44/93), the Taforalt in 36.6% (34/93) and the Natufians in only 14.0% (13/93). The latter population is on average, more similar to the recent regional samples than the fossils of Afalou and Taforalt. There are two possible reasons for this to be so. On the one hand, the Natufian peoples are the precursors of the Middle Eastern Neolithic farmers (Bar-Yosef & Valla,

1991), and they seem to be relatively smaller and more gracile than their North African contemporaries. On the other hand, the Natufian sample used here is a composite from several sites, therefore both geographically and temporally more varied than the crania from the sites of Afalou and Taforalt. In that sense, the definition of the Natufian group is closer to that of the regional samples than to the North African fossils.

Levels of variation within modern populations

In order to examine how populations differ in the variability they present, the coefficients of variation of the 93 variables for each of the nine samples was calculated. Table 8.3 shows, for each variable, the populations with the smallest and largest coefficients of variation among either the five regional samples (first column) or among all nine samples studied (second column).

Considering the five regional samples first, it is clear that the most homogeneous population, characterised by showing the smallest variation in the largest number of cranial dimensions, are the Australians (34/93, 36.5%), followed by the East Asians (25/93, 26.9%) and the Europeans (20/93, 21.5%). The most heterogeneous population, characterised by showing the greatest variation in the largest number of cranial dimensions, are the sub-Saharan Africans (36/93, 38.7%), followed by the Southeast Asians (20/93, 21.5%). Because of the narrow definition of the other four recent samples studied, especially the Fueguian-Patagonian population, we would expect to find smaller levels of within-group variation. However, none of these four groups shows a degree of homogeneity comparable to the Australian crania, although the Fueguian-Patagonian group has a relatively large number of measurements for which they vary the least (26/93, 28.0%). These are mainly concentrated in cranial size. The Epi-Palaeolithic fossil samples show between 10 and 15 minimum values of the coefficient of variation.

Levels of variation and morphological differentiation

An indication of the degree and pattern of differentiation in each group may be derived from the integration of the data on the distribution of significant differences among the various populations studied with the levels of variation in different morphological units of the skull. These characterisations are summarised below.

Of the variables in which the Europeans differ significantly from other populations, they show minimum within-group variation in the proximity of left and right temporal muscles across the occipital (SILLOC) and in temporal fossa length (TFL). This population has the shortest temporal fossae and the muscles set wide apart across the vault. They are particularly homogeneous in this differentiation, as they also are in terms of main cranial dimensions and a small posterior dentition. They can be further characterised by having wide skulls, with a low position of maximum parietal breadth, a flexed basicranium (the only cross-population differentiation in basicranial dimensions) and large lateral dimensions.

In relation to other recent populations, the sub-Saharan Africans have significantly shorter temporal muscles (STLHCH), one of the few homogeneous features of the group. They are also significantly different in several main cranial dimensions, but for a number of these (VRR, WCB, ZYB, BPL, AVR) and frontal angle (FRA), they show the greatest within-group variance of the five regional samples. We may conclude that many of these morphological differentiations do not extend across the range of sub-Saharan groups. In spite of this great variability, they have on average low vertex radii and narrow skulls, reflected in low values of maximum breadth, biauricular breadth, minimum cranial breadth and bizygomatic breadth (which accentuate the enlarged appearance of the parietals and parietal bosses in this group), prognathic alveolar regions (reflected both in the distance palate-porion/basion and in the relative differences among the sagittal facial radii), retracted upper faces, small nasal heights, wide interorbital breadths, long and deep palates and rounded frontal bones.

The Southeast Asians also show high within-group variation, expressed in their biauricular, biasterionic and bimeatal breadths, some lateral dimensions, facial height and prognathism, glabellar projection and posterior dental size. They have short skulls, with the face tucked well under the cranium, and short and rounded occipitals. They do not show any facial specialisations.

Most of the East Asian characterisations are based on the face – pronounced total and subnasal height, facial breadth, coronal facial flatness and lack of supraorbital projection. The great facial height, not accompanied by an increase in palate-hormion height, is associated with the curvature of the posterior alveolar plane of the maxilla. They are very homogeneous in their subspinale subtense values (SSS), while although they are significantly different from other populations in their bizygorbitale breadth (ZMZM) and supraorbital projection (SOS), there is great variation in these values within the population. They are relatively homogeneous in cranial length, breadth, prognathism, basicranial dimensions and posterior dental size.

Table 8.3. *Populations that show the maximum and minimum within-group variance, expressed in terms of the coefficient of variation. In the first column a comparison of the five regional samples; in the second column all nine samples included*

	Min/Max S^2			Min/Max S^2			Min/Max S^2	
Main cranial:			*Facial:*			*Basicranial:*		
GOL:	*EA/SSA*	*F+P/SSA*	NPH:	*AUS/SSA*	*AUS/TAF*	FOL:	*AUS/EU*	*AF/EU*
NOL:	*EA/SSA*	*F+P/NAT*	NLH:	*AUS/SEA*	–	FOB:	*EA/AUS*	*AF/TAF*
BBH:	*AUS/EU*	–	OBH:	*AUS/SSA*	*AUS/TAF*	BAH:	*AUS/SSA*	*TAF/NAT*
VRR:	*AUS/SSA*	*NAT/SSA*	FMB:	*AUS/SSA*	*F+P/SSA*	WBAB:	*EA/EU*	–
XPB:	*EA/EU*	*F+P/EU*	EKB:	*AUS/SSA*	*F+P/NAT*	D1:	*SSA/AUS*	–
ASB:	*EA/SEA*	*NAT/SEA*	JUB:	*AUS/SSA*	*F+P/SSA*	D2:	*EA/AUS*	*EA/TAF*
AUB:	*AUS/SEA*	*NAT/SEA*	ZMB:	*EA/SEA*	*F+P/SEA*	HOP:	*AUS/EU*	*AUS/TAF*
EAMEAM:	*EA/SEA*	*NAT/AF*	NLB:	*AUS/SEA*	*AUS/NAT*	*Occipital:*		
WCB:	*AUS/SSA*	*F+P/SSA*	WIOB:	*EU/SSA*	*AF/SSA*	OCC:	*AUS/EA*	*AUS/F+P*
ZYB:	*EU/SSA*	*F+P/SSA*	WNB:	*EU/SSA*	*F+P/SSA*	OARCH:	*AUS/SSA*	*TAF/NAT*
BNL:	*AUS/SSA*	*F+P/AF*	NFRB:	*AUS/SSA*	*AF/SSA*	OCS:	*AUS/EU*	*F+P/EU*
NAR:	*EU/SSA*	*F+P/NAT*	NAS:	*AUS/EU*	*AUS/NAT*	OCF:	*AUS/SSA*	*NAT/SSA*
FMR:	*EU/SSA*	*EU/TAF*	SSS:	*EA/EU*	*F+P/EU*	OCA:	*SSA/EU*	*F+P/EU*
EKR:	*EU/AUS*	*EU/TAF*	MAB:	*AUS/SSA*	*NAT/SSA*	*Temporal muscle:*		
ZOR:	*EU/SSA*	*EU/TAF*	PALL:	*EU/SEA*	*NAT/SEA*	STLHCH:	*SSA/EA*	*SSA/TAF*
ZMR:	*EU/SSA*	*EU/NAT*	PALD:	*EA/SEA*	*EA/AF*	STLLCH:	*EU/SEA*	*F+P/TAF*
SSR:	*EA/SSA*	*F+P/TAF*	FP–PR:	*SEA/SSA*	*TAF/AF*	SILLPA:	*SSA/AUS*	*NAT/F+P*
BPL:	*EU/SSA*	*F+P/NAT*	FP–AL:	*EU/EA*	*EU/NAT*	SILLOC:	*EU/EA*	*AF/EA*
PRR:	*EU/SEA*	–	GI:	*EA/SEA*	*EA/NAT*	TFL:	*EU/SSA*	*F+P/SSA*
AVR:	*AUS/SSA*	*F+P/NAT*	PROG1:	*AUS/SEA*	*AUS/AF*	TFB:	*EA/SSA*	*F+P/SSA*
			PROG2:	*EA/SEA*	*EA/AF*			
			PROG3:	*AUS/EU*	*AUS/NAT*			

Table 8.3 *(cont.)*.

Variable		
Lateral dimensions:		
MDH:	*AUS/SEA*	*NAT/SEA*
MDB:	*EU/SSA*	*AF/SSA*
PTE:	*SEA/EU*	*TAF/EU*
PTEBR:	*EA/EU*	*F+P/TAF*
PTEAST:	*EU/SEA*	*EU/NAT*
ASTLAM:	*AUS/SSA*	*TAF/SSA*
PTELAM:	*EA/SSA*	–
ASTBR:	*AUS/SSA*	*F+P/SSA*
Parietal dimensions:		
PAC:	*EA/EU*	*AF/EU*
PARCH:	*AUS/EU*	*AF/TAF*
PAS:	*AUS/EA*	–
PAF:	*AUS/EA*	*AUS/F+P*
PAA:	*SEA/EA*	*SEA/NAT*

Variable		
Malar:		
XML:	*AUS/SSA*	*AF/SSA*
IML:	*SEA/SSA*	*AF/SSA*
WMH:	*AUS/SEA*	*AF/TAF*
ZM–ZM:	*SSA/EA*	*AF/NAT*
LMLM:	*EA/SSA*	*F+P/SSA*
Dental:		
BLM1:	*EA/AUS*	*NAT/F+P*
MDM1:	*EA/EU*	*TAF/EU*
BLM2:	*EA/SEA*	*AF/SEA*
MDM2:	*SSA/SEA*	*TAF/SEA*

Variable		
Frontal:		
FRC:	*SEA/SSA*	*F+P/TAF*
FARCH:	*EU/SEA*	*AF/TAF*
XFB:	*SEA/EU*	*TAF/EU*
WFB:	*EU/AUS*	–
WFBZFS:	*EA/AUS*	–
WFBORB:	*EA/AUS*	*AF/AUS*
SOS:	*EU/EA*	*F+P/EA*
GLS:	*AUS/SEA*	*TAF/SEA*
FRS:	*SEA/EA*	*NAT/F+P*
FRF:	*EA/AUS*	*NAT/F+P*
FRA:	*SEA/SSA*	*TAF/F+P*

See Appendix II for explanation of variables.

The Australians appear both as the most differentiated and the most homogeneous, the homogeneity highlighting the differentiation. Australian crania are significantly different from the other populations in a large number of cranial dimensions in which they show the least variation (VRR, AUB, AVR, FMB, NLB, NAS, MAB, PROG1, PROG3, WMH, GLS). They do show however, maximum within-group variation in four important variables. These are values for which the Australians are, on average, widely different from the other regional samples, namely the position of minimum frontal breadth (WFBZFS, WFBORB), the proximity of temporal muscles across the vault (SILLPA – not related to absolutely larger muscles, but to the effect of large muscles on a narrow skull) and dental size (BLM1). This pattern probably belies an on-going process of reduction in dental and muscle size (to which the position of minimum frontal breadth may be related), to a point where, at a population level, we observe both very large and very small values. They are characterised by having low vertices and narrow skulls, and share with sub-Saharan Africans a narrow basicranium, a prognathic alveolar region, which in the case of Australians also extends to the subnasal area, long palates and retracted upper faces. However, they also show some facial differentiation not present in the African group, like wide bifrontal, nasal and palatal breadths, and long and vertically short malars, lacking the wide interorbital widths seen in the Africans. We will see in Chapter 9 that these facial differences between the Australians and sub-Saharan Africans can be closely linked to the different levels of facial robusticity in the two groups.

The high level of morphological differentiation observed in the North African fossils and the Fueguian-Patagonian sample is mostly related to size. These three populations are much larger and more robust than most recent crania. Table 8.4 shows how much larger or smaller in percentage terms these three groups are than the sub-Saharan African average value, taken here as reference, for all those measurements in which the South American and/or North African fossils are significantly different from other populations. These groups are larger than the recent reference value in most of the dimensions examined. If only measurements that reflect size in Table 8.4 are considered (i.e. ignoring subtenses, angles and distances from specific features – SSS, PALD, SILLPA, SILLOC, FRS, OCS, GLS, WFBZFS, WFBORB and FRA), the Fueguian–Patagonian sample, the Afalou and the Taforalt crania are on average 6.3%, 8.4% and 8.6%, respectively, larger than the sub-Saharan African population. The exceptions are a few variables for which the fossil and Fueguian-Patagonian dimensions are comparable or slightly smaller than the sub-Saharan African mean (PRR, FARCH, WFB, FRS), and in five variables in which the South American and

Table 8.4. *Size comparison between the Fueguian-Patagonian (F-P) and north African Epi-Palaeolithic fossils with the sub-Saharan (SSA) regional sample as reference. The variables presented are those for which the Afalou (AF), Taforalt (TAF) and South American samples (F-P) are significantly different from the regional samples*

Variables	AF (% SSA)	TAF (% SSA)	F-P (% SSA)	Variables	AF (% SSA)	TAF (% SSA)	F-P (% SSA)
BBH	+7.2	+6.8	+4.5	MDH	+15.0	+14.6	+3.3
ASB	+5.9	+4.2	+4.6	MDB	+18.9	+15.0	+14.1
EAMEAM	+8.6	+7.3	+9.6	PTEBR	+7.5	+4.6	+1.9
ZYB	+10.0	+11.4	+13.0	PTEAST	+6.4	+8.8	+11.3
BNL	+6.2	+8.9	+5.2	PTELAM	+3.4	+5.1	+1.9
FMR	+8.6	+7.9	+7.6	ASTLAM	+5.9	+5.3	+0.9
EKR	+7.6	+9.9	+6.3	ASTBR	+9.2	+8.0	+5.1
ZOR	+4.7	+8.2	+3.4	FARCH	+4.1	−1.6	+0.8
ZMR	+5.6	+13.8	+5.0	XFB	+9.2	+6.5	+4.7
PRR	−1.6	+0.8	+2.1	WFB	+3.6	+0.5	−1.4
NPH	+1.7	−2.7	+14.3	WFBZFS	+16.7	+33.9	+18.3
NLH	+9.0	+7.8	+14.7	WFBORB	+20.9	+29.1	+14.5
OBH	−7.7	−7.4	+9.8	GLS	+195.0	+168.4	+184.2
FMB	+5.3	+5.0	+3.5	FRS	−8.9	−13.5	−17.4
EKB	+5.4	+3.5	+3.1	FRF	+21.4	+11.9	+18.4
JUB	+8.6	+9.1	+6.6	FRA	+5.5	+5.5	+8.3
ZMB	+9.2	+4.2	+7.8	PARCH	+6.3	+7.3	+0.7
SSS	−4.1	−19.1	+1.2	OCC	+3.6	+8.0	+5.1
NLB	+6.7	+5.2	−5.6	OCS	+9.3	+8.5	+17.8
WIOB	+12.7	+11.3	−12.7	STLHCH	+15.9	+16.3	+16.0
WNB	+31.3	+33.7	+1.2	STLLCH	+10.4	+12.6	+18.4
NFRB	+34.0	+42.4	+9.4	SILLPA	−3.7	−11.2	−15.8
MAB	+8.0	+8.0	+2.5	SILLOC	0.0	−4.1	−24.6
PALL	−2.1	+1.1	+1.4	TFL	+0.5	+5.4	+11.6
PALD	−23.6	+5.0	−2.1	TFB	−1.3	+16.6	+19.2
FPPR	+10.9	+7.1	+16.8	BLM1	+6.1	+7.0	+1.7
FPAL	+10.1	+8.5	+6.9	BLM2	+9.2	+5.9	−0.8
XML	+8.4	+14.1	+6.1				
IML	+6.3	+16.1	+4.4				
WMH	+15.7	+13.1	+14.0				
ZMZM	+6.3	−7.1	+2.0				
LMLM	+12.1	+13.0	+10.0				

See Appendix II for explanation of variables.

North African crania do not share the same pattern of difference to the African sample. Among the latter are the two measures of facial height (NPH, OBH) in which the North African fossils have very low values and the Fueguian-Patagonians very large; these groups also differ in the level of subnasal projection (SSS) showing the pronounced alveolar flatness of the

North African fossils, nasal breadth (NLB), which is negatively related to the sub-Saharan African mean in the South American material, and BLM2, also smaller in the latter group.

Population differences in the context of defining a modern cranial form

The above results on the character and degree of differences in cranial morphology in recent and sub-recent populations show that there has been a large amount of differentiation in the evolution of modern populations. This differentiation has taken different pathways according to the different demographic and genetic histories of each group. This is implied not only by the differences in morphology that characterise each group, but also on the paucity of features common to all modern skulls. The samples that represent narrowly defined populations (Fueguian-Patagonian and the Epi-Palaeolithic fossils) have a very strong influence on the levels of differentiation observed. This is particularly interesting once it is considered that the size of most human populations in the past must have been comparatively small, with the well-known possibilities for accelerated genetic processes favouring differentiation acting on them. This has two main implications for our interpretation of past morphologies. On the one hand, it tells us that we have probably sampled only a small part of the variation encompassed by all past and present modern humans. On the other, it shows that when sampling narrowly defined groups we may be observing particular patterns that not only do not generalise with other modern groups, but that may never occur again in later times. Therefore, it is consistent with these observations on the Fueguian–Patagonian and Epi-Palaeolithic crania that some modern fossil specimens fall outside the expected ranges of variation referred to as 'modern'.

Another aspect of cranial morphology that the Fueguian-Patagonian, Afalou and Taforalt samples have in common is their large size. As seen in Table 8.4, they are approximately 10% larger than recent crania in most cranial dimensions. This is consistent with the observations by Kidder *et al.* (1992), who found that early Middle Eastern modern humans and early European Upper Palaeolithic fossils are between 10 and 15% larger than more recent European skulls. Table 8.5 shows similar calculations in relation to the sub-Saharan African sample of the fossil specimens used in Chapter 5. It is clear that greater size was a characteristic of earlier modern populations, retained in a recent group like the Fueguians. If the variation related to size of the skull is taken out, so that the remaining variation

Table 8.5. *Size comparison (in terms of percentages) between the fossil crania used in Chapter 5 with the sub-Saharan regional sample as reference. In bold those values at least 10% larger than the recent sub-Saharan African mean; in parentheses values below the African mean*

	Fossil crania																								
	Ngan	Ndu	Dj Ir	LH18	Omo I	Flor	Q9	Q6	SK5	SK4	NEG	Oha	EGev	Pred	UC101	UC103	Liu	Min. I	Min IV	Tep	L.St6	WaI	Wa II	Kei	Kan
GOL	12.1	0.9	**10.4**	**13.2**	**16.6**	**12.7**	8.7	8.2	6.5	**14.3**	(1.3)	1.5	3.7	**13.2**	**14.3**	4.3	6.5	(0.7)	(1.3)	0.0	4.3	**14.3**	-	**11.0**	5.4
BBH	(6.1)	-	-	-	-	-	-	-	(7.6)	-	(1.5)	1.5	(4.6)	1.5	3.0	**10.7**	4.7	0.8	-	5.3	3.8	3.8	-	8.4	9.9
XPB	7.3	0.0	**13.4**	2.1	-	**13.4**	5.8	8.1	6.6	9.6	2.8	7.3	**11.1**	**10.4**	7.3	(4.6)	8.1	**10.4**	0.0	8.1	(2.4)	12.6	-	8.1	6.6
ASB	**19.5**	4.4	**15.7**	8.2	**11.0**	-	4.4	9.1	9.1	-	4.4	2.5	8.2	3.5	**14.8**	1.6	0.6	2.5	-	**11.9**	5.4	3.5	-	**12.9**	**11.0**
AUB	**24.1**	9.4	**24.1**	6.8	**13.7**	-	7.6	**12.8**	**12.8**	**12.8**	5.0	8.5	7.6	**20.6**	**18.0**	7.6	7.6	8.5	-	9.3	5.9	**21.5**	-	**16.3**	**12.0**
NAR	**19.6**	-	**12.1**	**26.1**	**30.4**	-	-	4.5	(1.9)	-	(6.2)	0.2	5.2	**19.6**	**15.3**	9.9	8.8	(1.9)	-	(1.9)	0.0	9.9	-	**18.5**	7.7
FMR	**24.0**	-	**10.8**	-	**47.7**	-	0.0	**13.5**	**13.5**	-	(3.7)	8.2	(7.6)	**24.0**	**18.7**	**14.8**	**16.1**	**13.5**	-	0.3	9.5	**18.7**	-	**18.7**	6.9
NPH	-	**12.6**	**18.7**	-	-	**21.8**	**12.6**	2.0	**18.7**	-	(4.1)	3.5	-	**17.2**	**14.1**	5.0	(1.0)	(4.1)	(11.7)	-	(8.7)	6.5	-	9.6	**11.1**
NLH	-	**17.9**	**15.8**	-	-	**26.3**	**13.7**	(5.3)	-	-	(1.0)	9.5	-	**15.8**	**26.3**	**15.8**	3.1	(1.0)	(3.1)	5.3	(3.1)	5.3	-	9.5	**17.9**
NLB	-	1.1	**23.6**	8.6	-	**16.1**	**23.6**	**19.8**	-	**16.1**	1.1	(6.4)	-	(2.6)	**19.8**	1.1	(2.6)	(10.1)	-	(10.1)	4.9	**19.8**	**12.3**	1.1	**16.1**
FMB	1.4	(0.6)	**15.3**	-	**16.3**	**23.3**	7.3	**14.3**	**13.3**	-	(0.6)	6.4	(1.6)	9.3	8.3	7.3	0.4	2.4	(4.6)	1.4	-	**12.3**	-	7.3	7.3
MAB	-	-	**17.0**	**10.7**	-	-	9.2	**10.8**	9.2	**13.9**	(4.8)	4.5	-	6.1	7.6	4.5	3.0	4.5	-	(3.3)	-	9.2	**24.8**	**12.3**	**24.8**
WMH	-	-	**22.2**	-	**22.2**	9.2	0.4	4.8	4.8	**17.9**	(25.8)	0.4	(12.7)	4.8	**22.3**	(3.9)	0.4	(12.7)	(25.8)	**17.9**	4.8	**13.5**	-	**17.9**	**17.9**
FOL	**23.2**	3.6	-	-	-	-	-	-	**17.6**	-	(4.8)	0.8	-	**17.6**	9.2	**14.8**	0.8	6.4	-	-	3.6	**14.8**	-	**14.8**	9.2
XFB	6.2	-	8.0	1.8	-	**21.2**	4.4	**12.4**	1.8	-	9.7	9.7	7.1	**15.9**	7.1	8.8	8.8	(0.9)	(5.3)	8.8	0.0	8.0	-	**14.1**	4.4
WFB	8.2	(3.5)	**12.0**	4.8	**16.2**	**24.5**	8.9	**13.1**	4.8	**11.1**	(2.5)	8.9	3.7	7.9	6.8	**12.0**	0.6	(7.7)	(14.9)	2.7	(0.4)	0.6	5.8	4.8	(1.4)
FRC	4.3	-	(2.9)	4.3	**16.9**	(1.1)	5.2	7.9	(2.9)	7.9	2.5	(1.1)	**11.5**	7.9	2.5	(3.8)	6.1	(9.2)	(11.9)	4.3	1.6	1.6	-	0.7	6.1
PAC	(10.2)	-	6.1	5.3	5.3	-	**16.8**	(9.0)	8.8	7.0	(1.9)	0.0	(1.9)	5.3	**10.6**	7.9	6.1	(1.0)	(9.9)	(6.3)	(1.9)	6.1	-	8.8	7.9
OCC	(13.2)	(12.4)	-	-	4.6	-	3.5	1.4	(3.9)	-	(0.7)	5.6	-	5.6	0.3	1.4	(0.7)	(0.7)	(1.8)	**22.7**	**11.0**	**15.3**	-	**13.1**	6.7

See Appendix II for explanation of variables.

Ngan: Ngandong; Ndu: Ndutu; Dj Ir: Djebel Irhoud; LH18: Ngaloba; Flor: Florisbad; Q9: Qafzeh 9; Q6: Qafzeh 6; SK5: Skhūl 5; SK4: Skhūl4; NEG: Nahal Ein Gev; Oha: Ohalo; EGev: Ein Gev; Pred: Predmost; Liu: Liujiang; Min I and IV: Minatogawa I and IV; Tep: Tepexpán; LSta6: Lagoa Santa 6; WaI and II: Wajak I and II; Kei: Keilor; Kan: Kanalda reflects only shape, as done by W.W. Howells (1989), fossil and recent modern humans share a cranial morphology that differentiates them from all other archaic crania. This conclusion has an important implication. If all early and later moderns are to be considered fundamentally modern on the basis that they share other important derived features with recent people (Aiello, 1992), such as a fully modern bipedalism (Day *et al.*, 1991; Kennedy, 1992; Rak, 1990, 1992), the limits of what is described as modern have to be wider and comprise some anatomical features that would currently be considered archaic, like a robust morphology and varying degrees of overall cranial size.

If all modern humans have a single common African ancestor, two conditions have to be met: (1) the ancestral populations should be undifferentiated in relation to recent populations (i.e. it will not present African or any other groups' characteristics), and (2) it should not be derived in terms of later moderns (i.e. it should include within its range of variation certain features that are currently considered 'archaic', like large dentitions, large size and cranial tori and ridges). Studies of the African late Middle to early Upper Pleistocene fossils have shown that the first of these conditions is met – these remains do not present a special resemblance to recent African populations (Ambergen & Schaafsma, 1984; Rightmire, 1981; van Vark, 1986), a fact that, contrary to the position taken here, is considered by proponents of the Multiregional Model as discrediting the Out of Africa hypothesis (Wolpoff, 1989a,b). However, although these fossils can be considered undifferentiated in relation to recent crania, in order to satisfy the second condition mentioned above, the early undifferentiated modern human could not have looked 'fully anatomically modern' in the sense that this description is used today.

Among the African archaic *Homo sapiens* sample, three temporal and morphological groups have been recognised (Bräuer, 1984b, Rightmire, 1984):

(1) An early group represented by Kabwe, Ndutu and Bodo, which has some archaic traits not present in any modern skull, and shares its derived features, like brain enlargement, with other archaics outside Africa, these traits becoming plesiomorphous for later Pleistocene hominids in general.
(2) A late Middle Pleistocene group represented by Laetoli 18, Florisbad and Djebel Irhoud, characterised by a very variable morphology and generally considered as an archaic–modern transitional sample. This group shares with modern crania a smaller, flatter face and more rounded cranium, while still presenting large teeth and strong superstructures.

(3) An early Upper Pleistocene group represented by the early modern fossils of Klasies River Mouth, Omo 1 and Border Cave, which show more gracile crania, without pronounced cranial superstructures (although some of the remains from Klasies River Mouth present varying degrees of robusticity and large sexual dimorphism, Caspari & Wolpoff, 1990; Smith, 1992; Wolpoff, 1989a; Wolpoff & Caspari, 1990;).

The first group is clearly early, and its morphological characteristics align it with other archaic hominids of the Middle Pleistocene. However, the definition of the transitional and modern samples is less clear. The transitional sample does indeed show morphological differences, a mixture of archaic and modern features, that distinguishes it from the modern remains, although the chronological differences are not so clear. Both groups are composed of early and late fossil specimens, but there is an important overlap between the transitional late fossils and the early modern ones. The transitional sample covers from possibly 190 ka (Djebel Irhoud) to about 130–120 ka (Omo 2, Ngaloba). The early modern sample covers from 130–120 ka (Omo 1, Klasies River Mouth) to about 80 ka (Border Cave). Therefore, we may conclude that morphological arguments played an important role in defining these units.

This grouping of fossils has been very useful for the confirmation of the presence of a modern morphology among late Middle to early Upper Pleistocene African hominids, in the sense that by restricting the composition of the third sample only to modern specimens, excluding those of similar age that show archaic features, the 'modernity' of the sample is stressed. However, the existence of early moderns in Africa and the Middle East is not being questioned here. If all late Middle to early Upper Pleistocene African hominids are treated as a unit, separating instead this group from very early or very late specimens (like Djebel Irhoud and Border Cave), the result is a variable ancestral population in which individuals with both archaic and modern features are found. This is consistent with the levels of variability expected in the ancestral population. Accordingly, a specimen like Border Cave, which has already undergone a degree of differentiation and gracilisation in relation to earlier fossils, would be excluded from the ancestral population from which all fossil and recent crania are derived. It is among a late Middle to early Upper Pleistocene heterogeneous African sample, including fossils such as Ngaloba, Omo 1 and 2 and Klasies River Mouth, that some of the specialisations towards a modern morphology are found, together with a variable distribution of primitive traits relating to cranial size, facial prognathism and large dentitions, that makes this group a likely ancestral source of all later modern peoples.

Temporal patterns in the evolution of modern humans

If several lines of evidence are considered, chronological, morphological, archaeological and genetic, an explanation of observed modern cranial variation should be sought in the context of a single common origin of all modern peoples in Africa. The robust early modern and Australian crania have a number of features that depart from what is generally considered a modern morphology. These features have been traditionally interpreted as resulting from gene flow from archaic populations. However, the different morphologies observed in past and present modern humans can be interpreted as reflecting spatial and temporal differences in the ranges of variation, rather than resulting from hybridisation with archaic populations. If this interpretation is accepted, it follows that the definition of a modern morphology has to encompass these archaic features present in modern populations to varying degrees. In an evolutionary perspective, the broadening of the parameters used to define modern *Homo sapiens* implies that the ancestral population must have had high levels of variability which included the primitive features observed in these later robust groups. Such a variable ancestral population may be represented by the African fossils of Ngaloba, Omo 1 and 2 and Klasies River Mouth. The following chapters investigate the processes that could have given rise to cranial diversity from this variable ancestral source throughout the Upper Pleistocene.

9 *Morphological differentiation from a single ancestral source*

I have argued that there are two important aspects of the Out of Africa Model that can be criticised. The first of these is the ancestral source itself. Proponents of this model suggest that the early modern fossils of Omo 1, Klasies River Mouth and Border Cave may be taken to represent the ancestral population of all modern humans. However, some of these specimens are relatively derived in morphology in the sense that they have already lost a number of primitive features that can be observed in later material from some parts of the world, and it was argued that the ancestral population should represent a more variable group, including fossils such as Ngaloba, Florisbad and Omo 2, without the comparatively late and derived Border Cave remains.

The second criticism of the Out of Africa Model relates to the subsequent process of geographical expansion. Once a single origin of all modern humans is assumed, we are faced with the need to explain the development of the different regional morphologies. As it currently stands, the Out of Africa Model does not explore the possible processes that developed such differences beyond stating that the single origin is followed by geographical expansion and differentiation. The analysis of the regional morphologies in the previous chapter shows that recent populations differ not only in their morphological characterisation, but also on the pattern of such characterisation and their level of distinctiveness. Furthermore, some group characterisations can be made at a regional level, while others are much more restricted, either spatially or temporally. It is clear from these data that the evolutionary processes that shaped the history of modern populations in the Upper Pleistocene varied from group to group and place to place, according to ecological, historical and demographical characteristics. It follows that the process of modern human differentiation from a single ancestral source was a complex one.

In order to investigate the evolution of modern human diversity I will focus on two points – the pattern in the temporal and spatial distribution of morphologies, and the evolutionary process by which this pattern could have been developed. Therefore, this chapter will centre on examining morphological patterns in the modern human fossil record, while

Chapter 10 will examine geography as an important part of the evolutionary process that created such patterns.

The patterns of morphological differentiation

In order to disclose the patterns of morphological differentiation in modern human populations through time it is necessary to examine the fossil record. However, the human fossil record throughout the Upper Pleistocene is very irregular in quality and quantity. Figure 9.1 shows the approximate distribution of important human fossils during the Upper Pleistocene in the different regions of the world. It is clear that there is virtually nothing for the early part of the period, not even in Africa where we know modern humans were present from an early date. This makes the task of disclosing the patterns very difficult, for the first fossil specimens available for each region are already differentiated from each other and it is not always clear whether they represent the first modern occupants of the area. There is no solution to this problem until more fossils are found. This chapter will mainly focus on the later part of the Upper Pleistocene, attempting to identify continuity vs. discontinuity and anachrony vs. synchrony in cross-regional morphological comparisons, so that general and specific patterns may be found and an attempt made at outlining the specific regional histories.

Temporal and spatial trends

In order to obtain a measure of spatial and temporal differentiation during the Upper Pleistocene, morphological change through time was assessed by calculating the percentage difference from a referential skull in a number of fossil specimens and recent populations. The fossil crania examined are those used in previous chapters, with the addition of a number of specimens and groups measured by W.W. Howells, who kindly provided the data for this study (Table 9.1). The reference skull chosen was Skhūl 5, on the basis that this early modern specimen is virtually complete (although it is missing a relatively large portion of the subnasal maxillary area, almost all the measurements being examined could be obtained in the non-reconstructed preserved portions of this specimen). This choice does not imply that Skhūl 5 represents the ultimate ancestor of all moderns, it is only used as a modern temporal reference point around 100 ka ago. Figures 9.2 to 9.4 show the distribution of change in a number of features through time and space in the late Pleistocene. normalised on the dimensions of Skhūl 5.

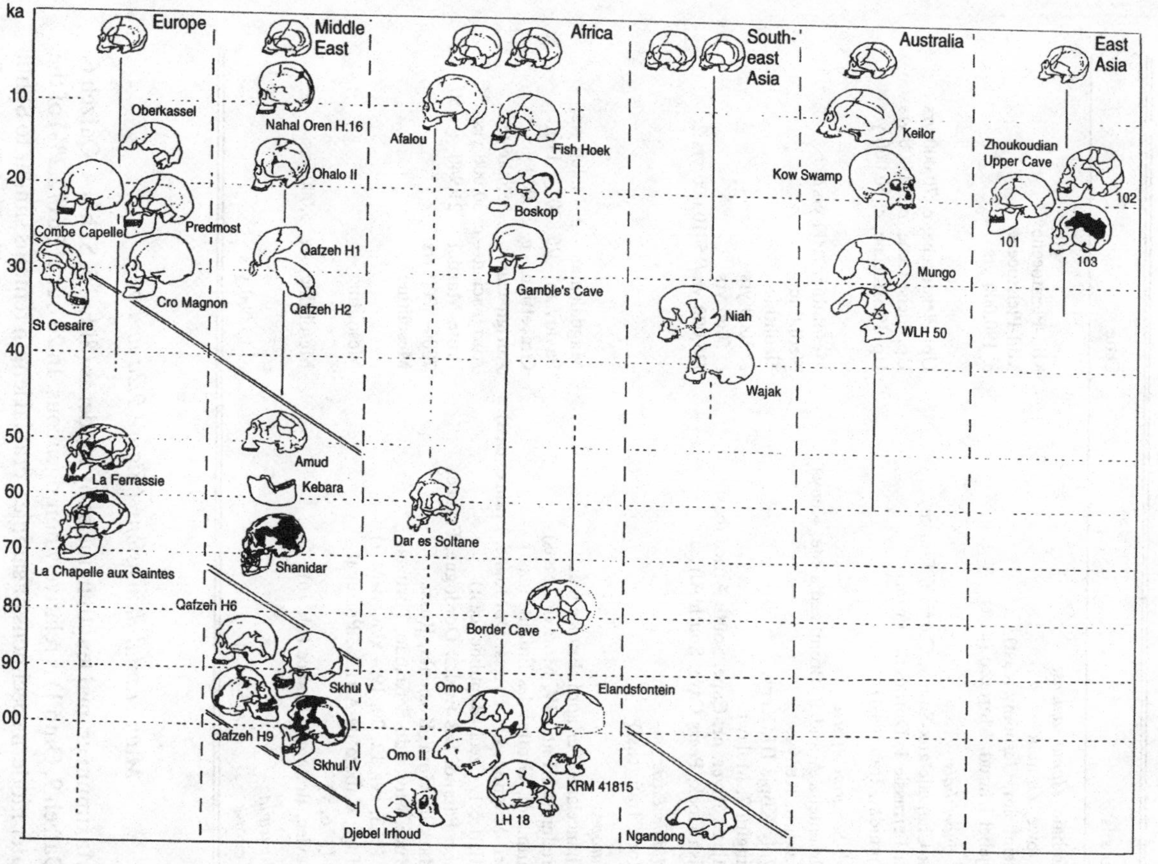

Figure 9.1 Temporal distribution of main fossil modern human crania in various regions of the world, showing the pronounced asynchrony in the appearance of modern humans throughout the world.

Table 9.1. *Additional fossils included, (data from W.W. Howells)*

Fossils	Date
'archaic' *Homo sapiens:*	
Kabwe, Zambia	Mid-Pleistocene
Steinheim, Germany (cast)	Mid-Pleistocene
Djebel Irhoud, Morocco (cast)	*c.* 190,000 yrs
Neanderthals:	
La Chapelle-aux-Saints, Corrèze, France	Up. Pleistocene, *c.* 70,000 yrs
La Ferrassie 1, Dordogne, France	Up. Pleistocene, *c.* 70,000 yrs
Shanidar, Iran (cast)	Up. Pleistocene, *c.* 70,000 yrs
sub-Saharan Africa:	
Elmenteita A and B, Bromhead's site, Kenya	Mesolithic, 7000 yrs?
Nakuru, Kenya	Neolithic
Willy Kopje III, Kenya	Neolithic
Fingira 2, Malawi	*c.* 3000 yrs
Mjates River, nr. Cape, South Africa (cast)	*c.* 5000 yrs
Fish Hoek, Peers Cave, South Africa	?35,000 yrs/15–10,000 yrs ?
Middle East:	
Hotu, Iran (cast)	
Europe:	
Chancelade, Perigord, France (cast)	Magdalenian – *c.* 14,000 yrs
Predmost III and IV, Moravia (casts)	Pavlovian – 25,820 ± 170
Grimaldi, Grottes des Enfants (cast)	Gravettian – *c.* 25,000 yrs
Cro-Magnon, Les Eyzies, Dordogne, France (cast)	Aurignacian – *c.* 35,000 yrs
Mladeč 1, Czech Republic (cast)	Aurignacian – *c.* 30,000 yrs
Abri Pataud, Les Eyzies, Dordogne, France	Proto-Magdal. – 21,940 ± 250
Markina Gora, Kostenki, Russia	7000–8000 yrs
Muge, Moita do Sebastião, Portugal (1, 3, xvi, 12, 20, 19 – xvii, vi-M 7)	Mesolithic
Muge, Cabeça de Arruda, Portugal (3, 6, v)	Mesolithic
Teviec, Bretagne, France (11, 16)	Mesolithic, – *c.* 5,000 yrs.
Australia:	
Keilor	*c.* 13,000 yrs

Main cranial dimensions (Figure 9.2a and b)

Maximum cranial length in the early moderns (Skhūl 5, Skhūl 4, Qafzeh 6, Qafzeh 9, Omo 1) is relatively homogeneous. In 26 out of 36 (72.2%) of the later modern specimens or samples, cranial length remains similar to Skhul 5 or smaller. The specimens that are longer than Skhūl 5 are early Europeans (Predmost III, Cro-Magnon, Mladeč, Grimaldi), some Africans (Elmenteita, Willy Kopje, Fish Hoek), Keilor, Wajak I and UC101. There is a clear temporal trend for reduction in cranial length in the later part of the Pleistocene, but this reduction/trend seems to apply to different regions at different times. Cranial height (BBH) is not available for any other early modern but Skhul 5, so that the early levels of variability remain unknown. All later modern crania have a greater cranial height than this specimen. The differences are least (within the 5% range) in the specimens of Ein Gev, Fingira and Mjates River. However, the small size and gracility of these specimens suggests that rather than a retained condition, these small values of basi-bregmatic height are a consequence of an overall pronounced reduction in size in these specimens.

In terms of the distance of the face from the vault (BNL, NAR, FMR), we observe that the archaic specimens (Neanderthals and Kabwe, and to a lesser degree, Ngandong) have much greater values of basion–nasion length (around 25% greater) than Skhūl 5. Most modern skulls also show large distances from basion to nasion, but the differences with Skhūl 5 are much less pronounced (maximum of 15% greater). A small number of crania show values smaller than Skhūl 5, namely Grimaldi, Ein Gev, Nahal Ein Gev, Fish Hoek, Mjates River and Tepexpán. Therefore, there is no clear trend throughout the period, although it is clear that in all modern human crania the distance of the face to the vault is smaller than in archaic fossils like Kabwe, Neanderthals or Ngandong. There is great variability in upper facial radii in early modern specimens, ranging from values smaller than Skhūl 5 to more than 30% larger in Omo 1. In terms of facial retraction, however, the difference between nasion and fronto-malar radii show relatively homogeneous values in all cases. On the other hand, the fossils of Kabwe, Djebel Irhoud and Neanderthals consistently show large values of NAR (+14.3%, +14.3%, +22% respectively) and small of FMR (−9.3%, −2.3%, = Sk.5 respectively), i.e. mid-facial protrusion associated with lateral facial retraction (about 15–25% greater difference between NAR and FMR than in Skhul 5). The Ngandong hominids show a much smaller retraction of the zygomatico-frontal process in relation to the mid-face. Among later moderns, a majority of crania show values of NAR comparable to Skhūl 5 (within ± 5%). The exceptions are Predmost (+22%) and Cro-Magnon (+25.3%) and Teviec (+11%) in Europe, Afalou and Taforalt (+8.1%, +7.9%), Keilor (+20.9%) and Kanalda (+9.9%) in Australia, UC101 (+17.6%) and UC103 (+12.1%) in east Asia, Fueguians-Patagonians (+7.8%), Wajak (+12.1%) and Willy Kopje (+15.4%). However, these specimens also have larger values of FMR, so that the shape in terms of lateral facial retraction is not changed, only the dimensions of the measurements. Actually, all but six of the 36 later modern fossils or samples have a similar degree of upper facial retraction, approximately 10–15% greater than in Skhūl 5.

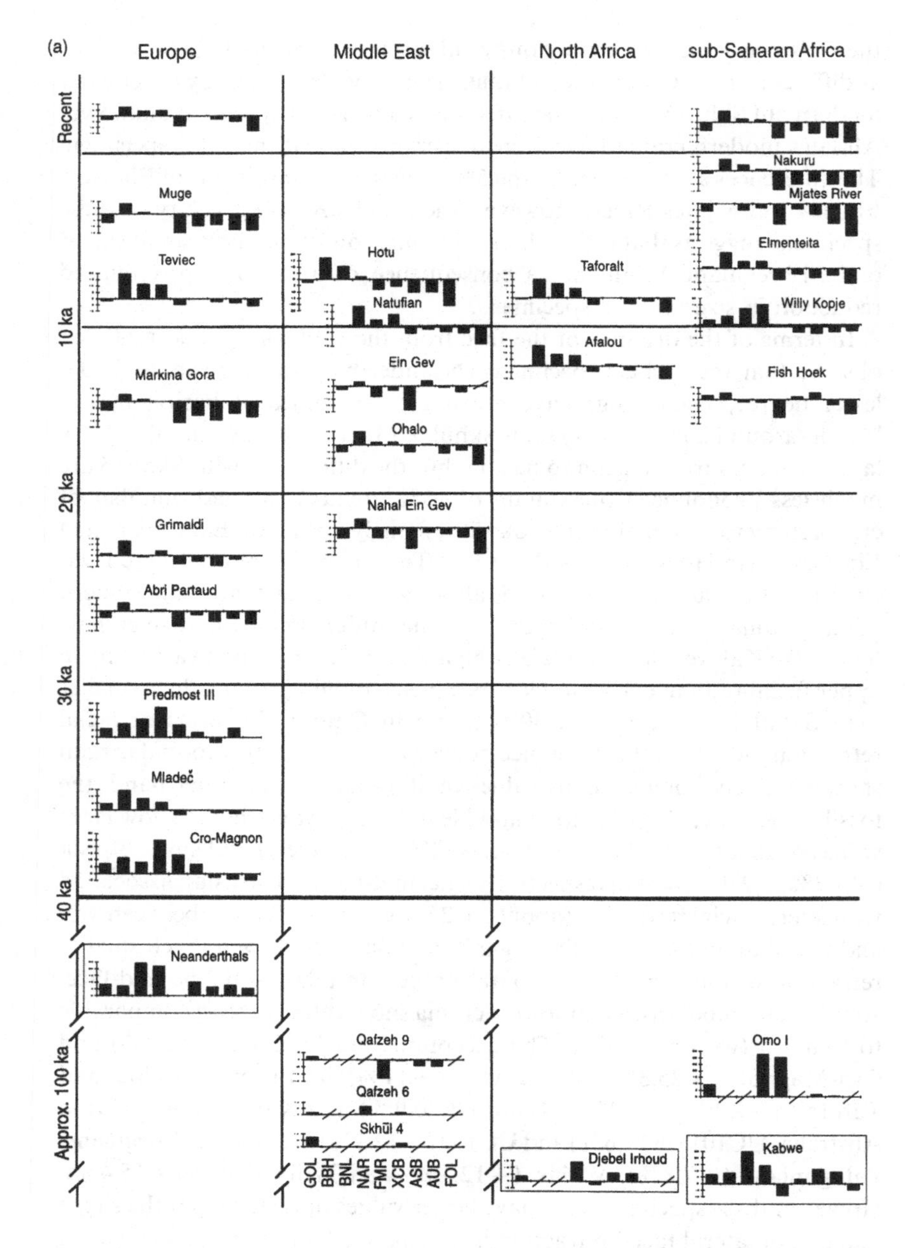

Figure 9.2 (a,b) Temporal variation in main metrical cranial dimensions (GOL, BBH, BNL, NAR, FMR, XCB, ASB, AUB, FOL) of some fossil modern human crania in various regions of the world. Values expressed in terms of percentage larger or smaller than Skhūl 5, taken here as reference.

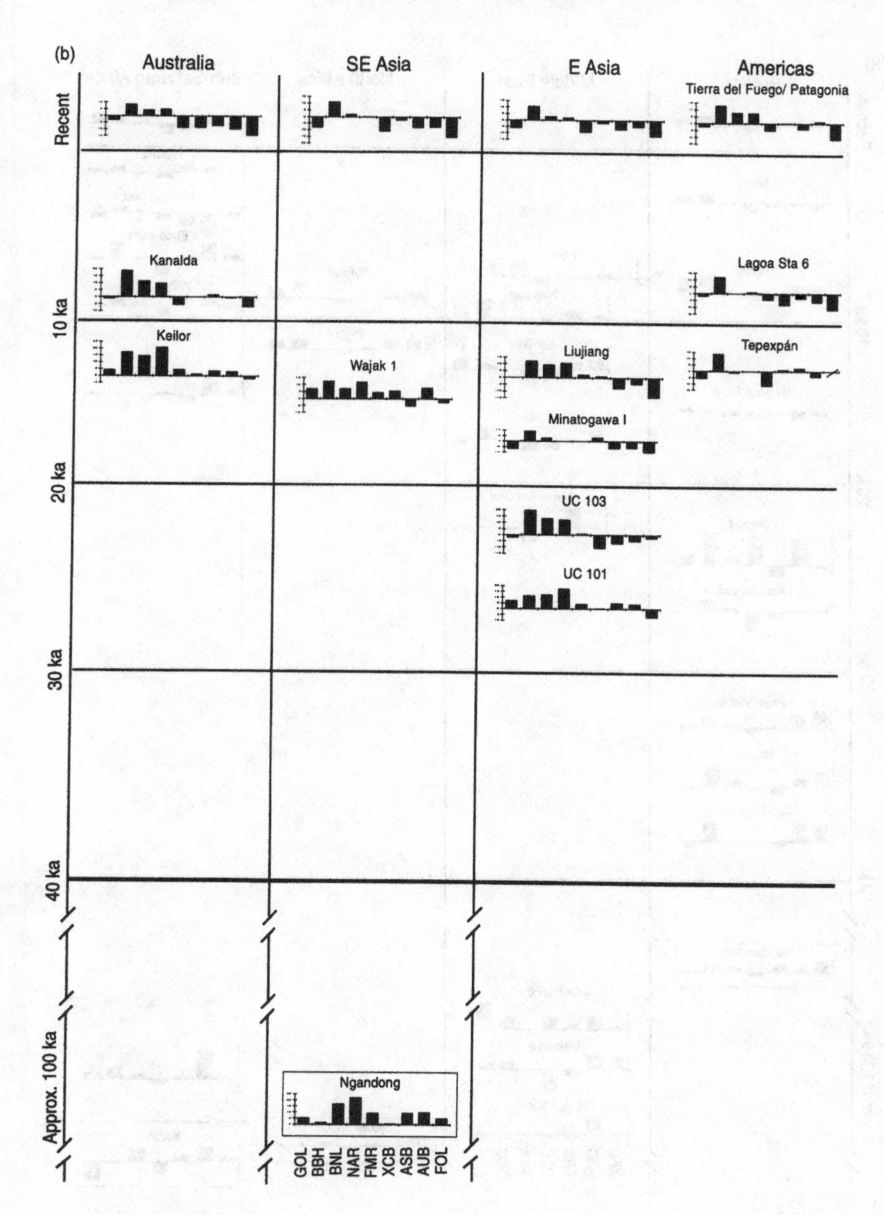
(b)
Australia
SE Asia
E Asia
Americas
Tierra del Fuego/ Patagonia
Recent
Kanalda
Lagoa Sta 6
10 ka
Keilor
Liujiang
Tepexpán
Wajak 1
Minatogawa I
20 ka
UC 103
UC 101
30 ka
40 ka
Approx. 100 ka
Ngandong
GOL
BBH
BNL
NAR
FMR
XCB
ASB
AUB
FOL

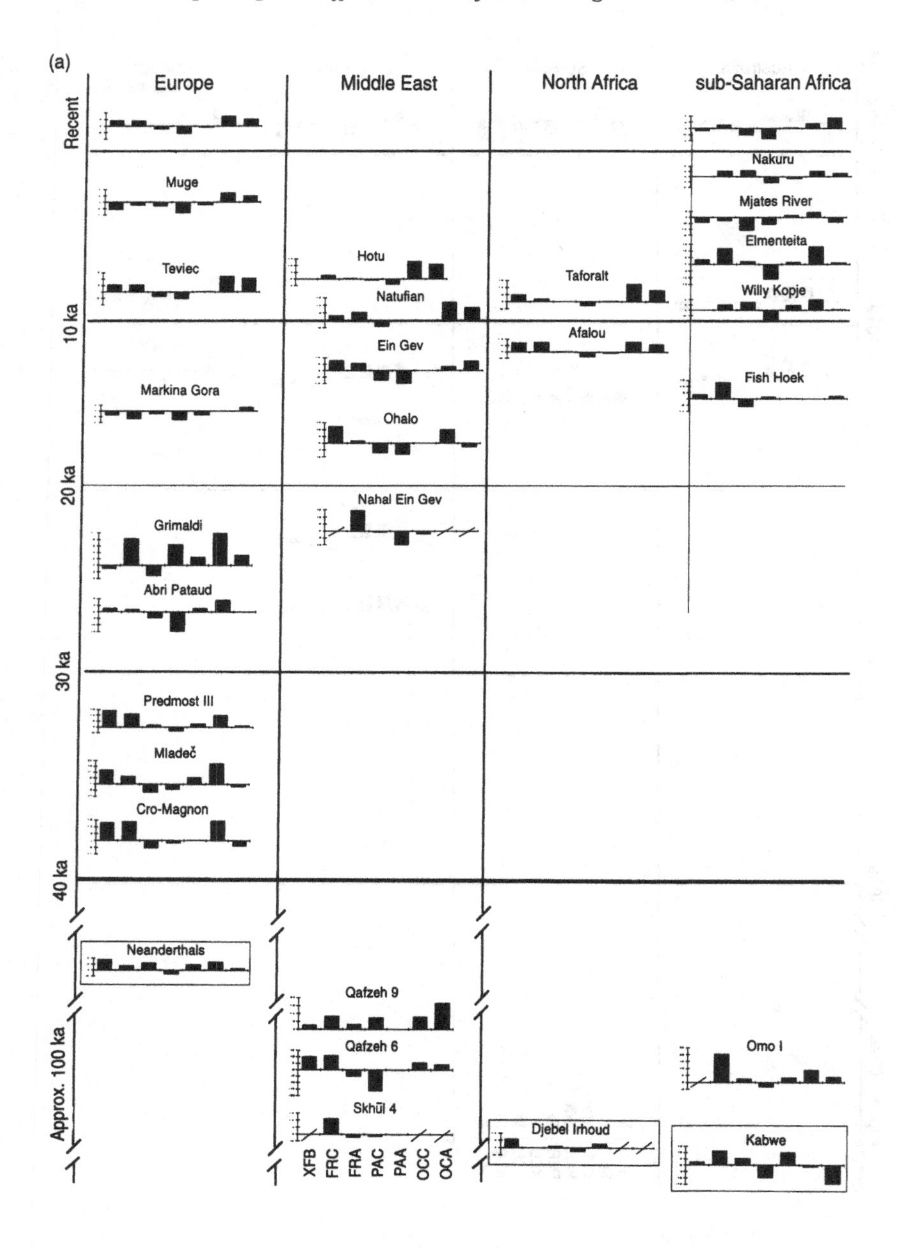

Figure 9.3 (a,b) Temporal variation in frontal, parietal and occipital chords and angles (XFB, FRC, FRA, PAC, PAA, OCC, OCA) of some fossil modern human crania in various regions of the world. Values expressed in terms of percentage larger or smaller than Skhūl 5, taken here as reference.

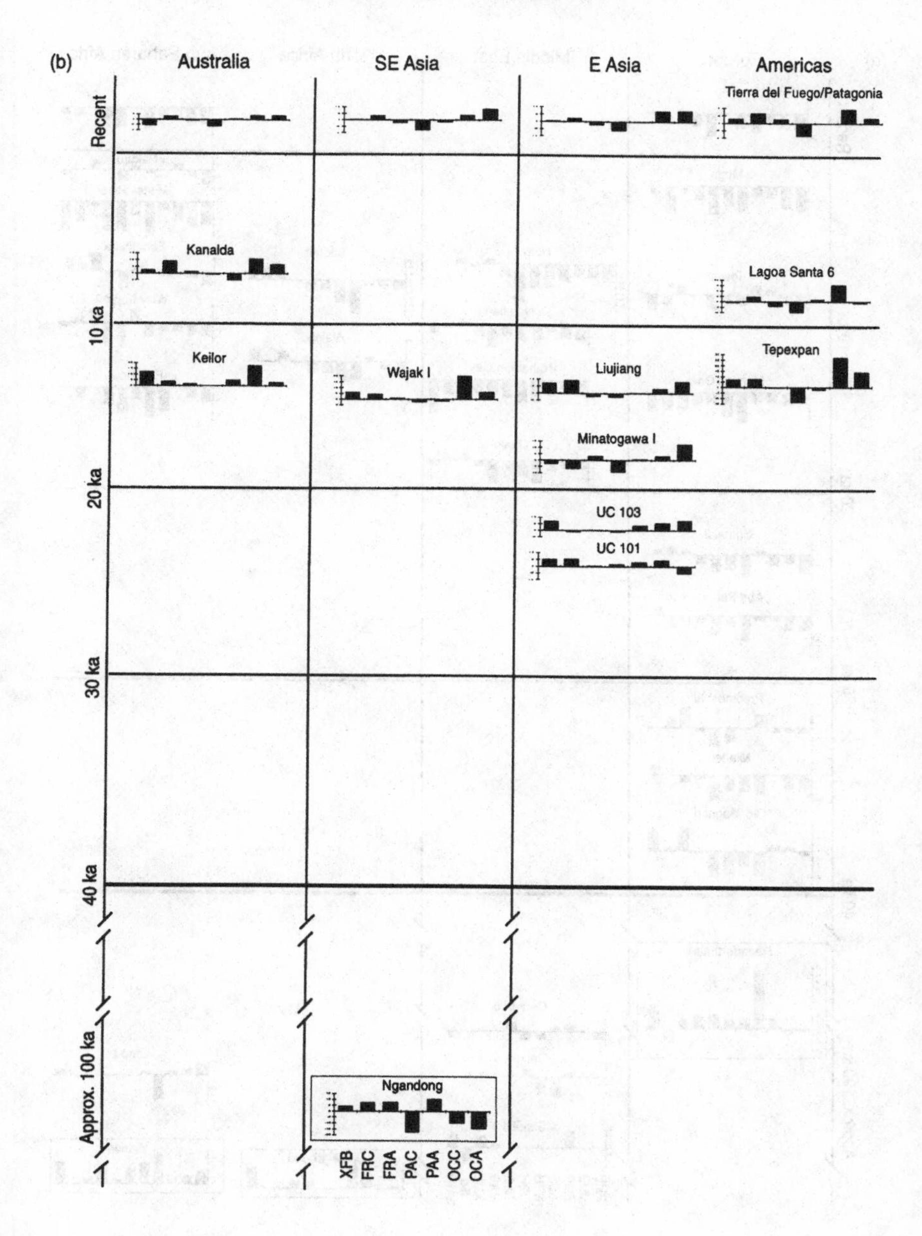
(b)
Australia
SE Asia
E Asia
Americas
Recent
10 ka
20 ka
30 ka
40 ka
Approx. 100 ka
Tierra del Fuego/Patagonia
Kanalda
Lagoa Santa 6
Keilor
Wajak I
Liujiang
Tepexpan
Minatogawa I
UC 103
UC 101
Ngandong
XFB
FRC
FRA
PAC
PAA
OCC
OCA

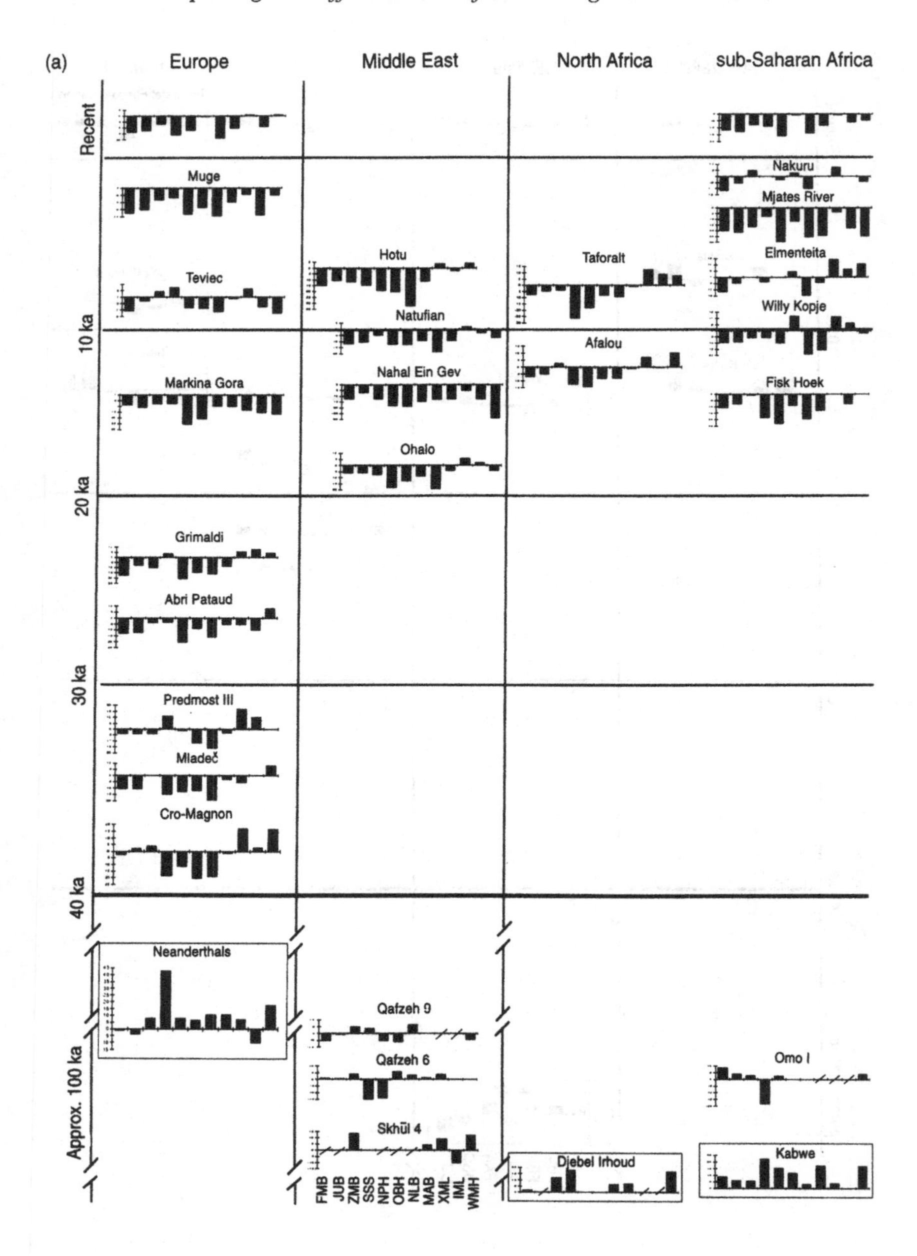

Figure 9.4 (a,b) Temporal variation in main facial dimensions (FMB, JUB, ZMB, SSS, NPH, OBH, NLB, MAB, XML, IML, WMH) of some fossil modern human crania in various regions of the world. Values expressed in terms of percentage larger or smaller than Skhūl 5, taken here as reference.

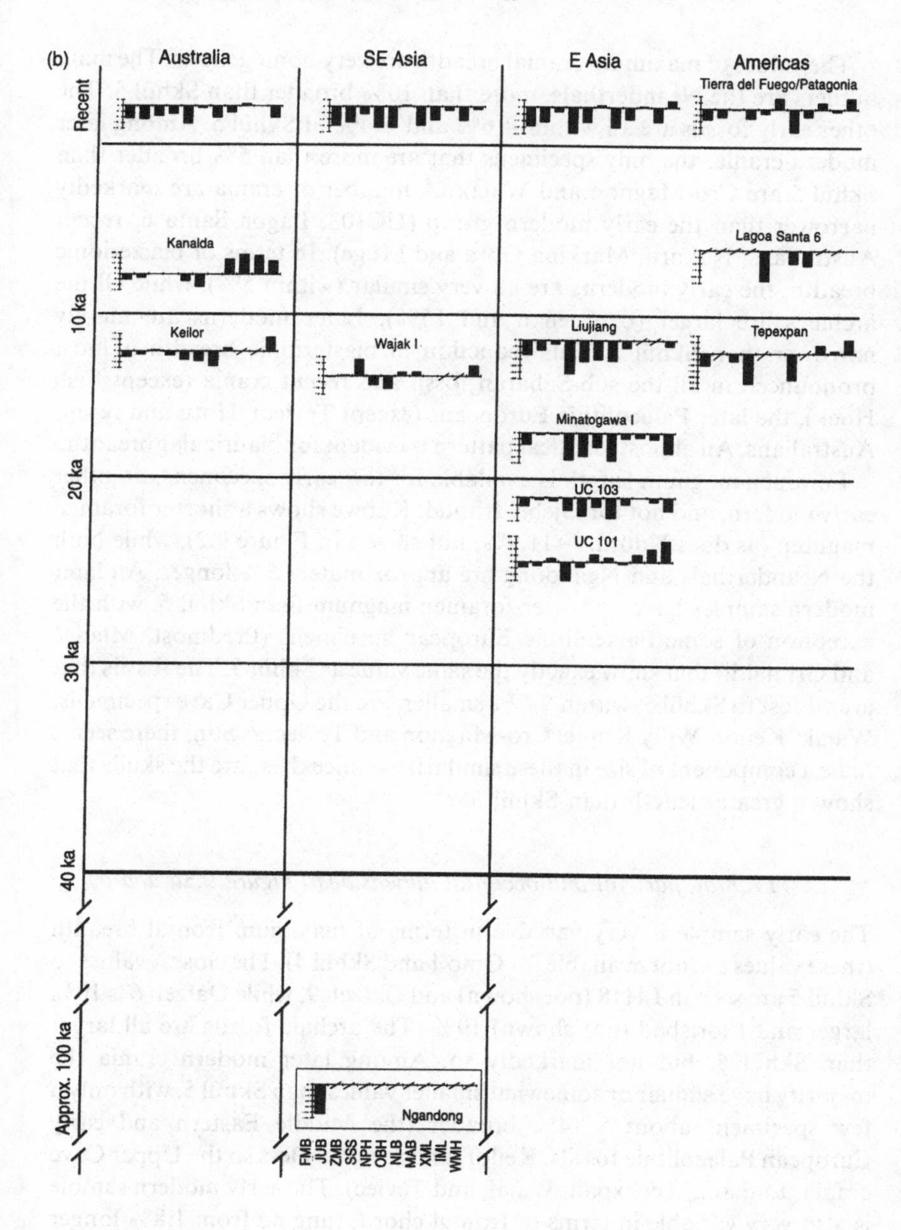
(b)
Australia
SE Asia
E Asia
Americas
Tierra del Fuego/Patagonia
Recent
Kanalda
Lagoa Santa 6
10 ka
Keilor
Wajak I
Liujiang
Tepexpan
Minatogawa I
20 ka
UC 103
UC 101
30 ka
40 ka
Approx. 100 ka
Ngandong
FMB
JUB
ZMB
SSS
NPH
OBH
NLB
MAB
XML
IML
WMH

The values of maximum cranial breadth are very homogenous. The main outliers are the Neanderthals, more than 10% broader than Skhūl 5. The other early fossils are all within +6% and −4% of Skhūl 5. Among later modern crania, the only specimens that are more than 5% broader than Skhūl 5 are Cro-Magnon and Wajak. A number of crania are markedly narrower than the early modern group (UC103, Lagoa Santa 6, recent Australians, Nakuru, Markina Gora and Muge). In terms of biasterionic breadth, the early moderns are all very similar (within 5%), while all the archaics are larger (between 6 and 15%). Later moderns are mainly narrower than Skhūl 5. This reduction in biasterionic breadth is most pronounced in all the sub-Saharan fossil and recent crania (except Fish Hoek), the later Palaeolithic Europeans (except Teviec), Hotu and recent Australians. An almost identical pattern is evident for biauricular breadth.

Foramen magnum length is available for few early specimens, no other early modern, and not for Djebel Irhoud. Kabwe shows a shorter foramen magnum (as does Ndutu: −11.9%, not shown in Figure 9.2), while both the Neanderthals and Ngandong are approximately 5% longer. All later modern samples have a shorter foramen magnum than Skhūl 5, with the exception of some Palaeolithic European specimens (Predmost, Mladeč and Grimaldi) that show exactly the same value as Skhūl 5. The fossils that are closest to Skhūl 5, within 5–7% smaller, are the Upper Cave specimens, Wajak, Keilor, Willy Kopje, Cro-Magnon and Teviec. Again, there seems to be a component of size in these similarities, since these are the skulls that show a greater length than Skhūl 5.

Frontal, parietal and occipital dimensions (Figure 9.3a and b)

The early sample is very variable in terms of maximum frontal breadth (these values are not available for Omo 1 and Skhūl 4). The closest values to Skhūl 5 are seen in LH18 (not shown) and Qafzeh 9, while Qafzeh 6 is 10% larger and Florisbad (not shown) 19%. The archaic fossils are all larger than Skhūl 5, but not markedly so. Among later modern crania the majority have similar or somewhat smaller values than Skhūl 5, with only a few specimens about 8–14% broader (the Middle Eastern and early European Palaeolithic fossils, Keilor, and relatively less so the Upper Cave crania, Liujiang, Tepexpán, Wajak and Teviec). The early modern sample is also very variable in terms of frontal chord, ranging from 1.8% longer than Skhūl 5 (Florisbad) to 20.4% longer (Omo 1). The great majority of later moderns show very similar values to Skhūl 5, with the exception of a few markedly larger frontals (10–15%) seen in Ein Gev, Predmost and Cro-Magnon, and relatively less so in Tepexpán, Kanalda, Afalou and

Grimaldi. In terms of frontal angle, the values are much less variable than expected. Among the archaic group, Djebel Irhoud is almost identical to Skhūl 5, while Kabwe, Neanderthals and Ngandong are respectively 5.2%, 6% and 7.9% flatter. The early modern sample shows a wider range of variation, from 5.1% more rounded frontals in Qafzeh 6 to 7.4% flatter in LH18. The majority of later crania show more rounded frontals than Skhūl 5, some markedly so (−10.7% in Matjes River). The exceptions are Minatogawa I (3.5% flatter), Fueguians-Patagonians (2.7%), Elmenteita (2.2%), Predmost (2%), Keilor (1.7%), Kanalda (1.4%), Wajak (1.1%), Hotu (0.7%), UC101 (0.4%), UC103 (0.4%), but the values are generally all very close. The trend for more rounded frontals is very clear in all sub-Saharan African, and most Middle Eastern and European crania.

In terms of parietal chord, the archaic and early modern samples show great variability. The average parietal length in Ngandong is 17.5% shorter than in Skhūl 5, but this difference is almost matched by Qafzeh 6. On the other hand, the parietal of Qafzeh 9 is 7.4% longer than Skhūl 5. A majority of later crania (21/36) have parietals that are very much shorter. A group of fossils, however, show values either slightly larger (UC101, Fish Hoek, Grimaldi) or only slightly smaller than Skhūl 5 (Mladeč, Cro-Magnon, Predmost, Afalou, Taforat, Kanalda, Keilor, Wajak, Liujiang, UC103, Hotu, Natufians). All the archaic hominids have flatter parietal angles than Skhūl 5. This is most pronounced in Ngandong and Kabwe (10.3% and 10.1% flatter respectively). Among the early modern group, values of parietal angle range from virtually identical to Skhūl 5 in Skhūl 4, Qafzeh 6 and 9, to 6.5% flatter in LH18. Later crania are all very similar to the Skhūl and Qafzeh sample, only two skulls (Keilor and Mladeč) have a parietal slightly more than 5% flatter. These results are consistent with those of the previous chapter which identify parietal angle to be one of the few shared features of all modern humans.

The Ngandong and African archaic samples show the smallest values of occipital chord of all the crania (−9.7% in Ngandong, −8.9% in Ndutu and −1.1% in Kabwe). Neanderthal occipitals are longer (+7%). The early modern values range from 5.5% to 8.9% longer (Qafzeh 6 and Omo 1 respectively). The later specimens all have longer occipital bones than Skhūl 5, varying in value from 28% longer in Tepexpán, 20% in Wajak, 17.8% in Mladeč and Keilor, 15.5% in Lagoa Sta. 6 and Fingira, 14.4% in Cro-Magnon, down to equal size to Skhūl 5 in Fish Hoek and Markina Gora. As expected, Ngandong and the African archaic fossils have a pronouncedly more angled occipital than all later crania. The Neanderthals and the early modern fossils of Omo 1 and Qafzeh 6 have similar values of occipital angle to Skhūl 5 (+1.7%, +3.9% and +4.2% respectively), while Qafzeh 9 has a much more rounded occipital bone (+16.6%). Later crania vary from those that are slightly more angled (UC101, Cro-Magnon, Mladeč, Ohalo, Lagoa Santa 6 and Mjates River), to those that have similar or slightly greater values than Skhūl 5 (Predmost, Abri Pataud, Grimaldi, Markina Gora, Elmenteita, Willy Kopje, Fingira, Fish Hoek, Nakuru, Fueguian-Patagonians, Keilor, Australians), to all those that have occipital angles more than 5% greater than Skhūl 5.

Facial dimensions (Figure 9.4a and b)

The temporal distribution of values of fronto-malar breadth is very interesting. Among the archaics, the Ngandong hominids have on average a much smaller value than Skhūl 5, as does Ndutu (−10.5% and −12.3% respectively). Kabwe shows greater upper facial breadth, as does Florisbad. Neanderthals, Djebel Irhoud, Omo 1 and Qafzeh 6 all have very similar values around the reference, while Qafzeh 9 is 5.3% smaller. All later modern crania have smaller values than Skhūl 5, most markedly so. The smallest values are those of Mjates River (−20.2%), Grimaldi and Muge (−17.5%). Only seven modern fossils have relatively broad upper faces (0–5% smaller than Skhūl 5) – Wajak I, Cro-Magnon, Predmost, UC101, UC103, Keilor, Kanalda and Ohalo. These are many of the 'large' crania identified before. In terms of bijugal breadth, the temporal distribution of values is very similar to fronto-malar breadth, with the difference that the early moderns are all very similar in this dimension. As it was the case for fronto-malar breadth, all later crania have smaller values than the early modern sample. The crania in which this difference is least pronounced (up to 5%) are the same as listed above (Wajak I, Cro-Magnon, Predmost, UC101, UC103, Keilor, Kanalda, Ohalo), with the addition of Taforalt, Teviec and Minatogawa I. Bimaxillary breadth follows a different pattern. The available archaic measurements (Kabwe, Djebel Irhoud and Neanderthals) are all larger than Skhūl 5 (+5.9%, +11.8% and +8.1% respectively), while the only early modern that is markedly different is Skhūl 4 (+13.7%). Among later crania, the majority are smaller than Skhūl 5, except for a group that show values slightly above or below the reference. Many of these are the same crania identified above, although there are some clear regional patterns as well. In eastern Asia, the Upper Cave specimens are similar to Skhūl 5 and this condition remains a characteristic of all subsequent East Asian crania (Liujiang, Minatogawa, recent East Asians and Fueguians-Patagonians). Large bimaxillary breadths are also seen in Keilor and Kanalda. In Europe, Cro-Magnon, Predmost, Mladeč and Abri Pataud have broad bimaxillary dimensions, and this early Upper Palaeolithic feature is observed in the later crania of Taforalt, Afalou and Teviec, the latter two groups intensifying the trend to show values above Skhūl 5. In sub-Saharan Africa there is a lot of variation, ranging from values above Skhūl 5 (Nakuru) to those very much smaller (Mjates River).

All the archaic crania have much greater subnasal prognathism, measured as subspinale subtense, than the early modern group. This is most pronounced in Neanderthals, which are 43.3% greater than Skhūl 5, followed by Kabwe (+22.2%) and Djebel Irhoud (+18.5%). The early moderns show a different pattern. Two specimens are similar or identical to Skhūl 5 (Qafzeh 9 and Skhūl 4), while two show a much more pronounced facial flatness (Qafzeh 6 and Florisbad, −14.8% and −18.5%). The great majority of later crania are clearly less prognathic than Skhūl 5. This is most pronounced in all Middle Eastern, North African, East Asian and Amerindian crania. In Southeast Asia, there is a marked reduction in prognathism from Wajak to recent, also seen to a smaller extent in Australia. The Upper Palaeolithic population of Europe shows a wide range of variation, some specimens like Cro-Magnon and Mladeč with relatively flat faces (as do recent Europeans), while others like Predmost and to a lesser degree Grimaldi and Teviec, are more prognathic than Skhūl 5. Similarly, sub-Saharan African remains vary from a prognathism identical to Skhūl 5 (Nakuru) to pronouncedly flatter faces (−25.9% in Fingira).

In terms of facial height, the largest face is found in Kabwe (+15.4%), followed by the Neanderthals (+8.1%). Djebel Irhoud and Florisbad have similar values to Skhūl 5, while Qafzeh 9 and specially Qafzeh 6, have much shorter faces (−5.1%, −14.1%). All later moderns have shorter faces, but the difference is least in UC101, Predmost III, Elmenteita, Nakuru, recent Fueguian-Patagonians, Keilor, Kanalda and Teviec. Orbital height follows a similar pattern. Kabwe has much taller orbits (+11.8%), as do Neanderthals (+6.8%) and Ndutu (+5.9%). Djebel Irhoud, Florisbad and LH18 are identical to Skhūl 5, while the orbits in Qafzeh 6 are taller (+5.9%) and in Qafzeh 9 shorter (−5.9%). Although not fully uniform, there is less variation in this feature among the early modern group than for upper facial height. Interestingly, the pattern among later moderns is very mixed. Middle Eastern and North African fossil crania all have short orbits. European Palaeolithic crania have short orbits, less so Mesolithic skulls, while recent crania approach the size of Skhūl 5. In the Asiatic populations, the fossils are variable (UC101 has comparatively tall orbits and all the other specimens have smaller values). Tepexpán and Lagoa Santa 6 are identical in orbital height to Skhūl 5, and both recent East Asians and Fueguian-Patagonians have values above Skhūl 5 (especially the Fueguian-Patagonian crania). This is consistent with the findings in Chapter 8 which show that current East Asians have particularly tall orbits in relation to other recent populations. Wajak, like UC101, has tall orbits, a feature observed in recent Southeast Asian crania. Keilor and Kanalda have comparatively shorter orbits than recent Australians. Finally, in sub-Saharan African crania a whole range of values are observed that do not conform to clear temporal trends, from orbits as tall as Kabwe in Willy Kopje, to very short ones in Fingira. There seems to have been a reduction in orbital height from the earliest moderns to middle Upper Pleistocene specimens, a trend that was then reversed in most populations in recent times.

There is a large degree of variation in nasal breadth among the archaic fossils. This portion of Skhūl 5 is reconstructed. All but three later crania have much smaller nasal breadths than the archaic (especially Neanderthals) and early modern fossils. The three exceptions are UC101, Wajak 1 and Kanalda.

Maxillary breadth is greater than Skhūl 5 in all the archaics, specially so in Kabwe (=17.1%). The early moderns show all comparable values, within 5% of Skhūl 5. With few exceptions, a general trend for reduction in maxillary breadth can be identified among later crania. This trend is clearly visible in all recent crania (less so in Australia), but varies among the fossils. It is clear in the Middle East and Europe, with Teviec (and Afalou and Taforalt) retaining the earlier condition. Wajak, Keilor, Kanalda and the Upper Cave remains also have values similar to the early moderns. As for other measurements, a whole range of values are observed among sub-Saharan crania, from identical to Skhūl 5 in Nakuru to almost 25% smaller in Mjates River. The only later crania with broader maxillas than Skhūl 5 are the two Australian fossils, Keilor and Kanalda (=2.9% and +14.3% respectively).

The three measures of malar size do not follow the same temporal and spatial patterns. Maximum malar length is available for few early specimens, and all of these show greater values than Skhūl 5 (+3.8% in Kabwe, +7.0% in Neanderthals, +9.4% in Skhūl 4 and +3.8% in Qafzeh 6). The variation among later crania is very large, from much greater values (+18.9% in Kanalda, and +17% in Predmost and Cro-Magnon), to much smaller malars (−13.2% in Markina Gora). This variation does not follow a clear temporal or spatial trend. In terms of inferior malar length, Neanderthals and Skhūl 4 are 10% smaller, while both Kabwe and Qafzeh 6 are identical to Skhūl 5. A slight temporal trend of reduction in inferior malar length is observed in all regions except Australia and sub-Saharan Africa, where there is much variation among sub-recent crania. The pattern of change in minimum malar height is clearer. Archaic hominids have tall malar bones (+16.7% in Kabwe and Djebel Irhoud, +17.9% in Neanderthals), while the early modern sample is again very variable (+16.7% in Omo 1, +12.5% in Skhūl 4, +4.2% in Florisbad, Qafzeh 6 identical to Skhūl 5, and −4.2% in Qafzeh 9). Taller malars than Skhūl 5 are observed in Wajak I, Keilor, Kanalda, UC101, Tepexpán, recent East Asians, Fueguian-Patagonians, Afalou, Taforalt, Elmenteita, Hotu, early Palaeolithic Europeans and Teviec, and Fish Hoek is identical to Skhūl 5. All recent populations, except East Asians and Fueguians-Patagonians, have much reduced malar heights.

The above descriptions disclose three important points. Firstly, that given the group of variables being considered, the archaic crania examined (Ngandong, Ndutu, Kabwe, Djebel Irhoud and Neanderthals) can be clearly separated from the modern skulls, including the early modern sample. Secondly, that the early modern group shows not only great variability, but that this variation is not consistent throughout cranial dimensions. And thirdly, that the main trend throughout the period is one of reduction in size of most cranial dimensions. These points will be discussed below.

Early modern variability

Chapter 5 examined the likelihood of either the Ngandong hominids or the early modern sample representing the ancestral condition for the Australian morphology, and it was found that most of the features usually considered typical of Javanese and Australian crania are observed within the large range of variation of the late Middle to early Upper Pleistocene modern humans. This high degree of variability in the ancestral group was then argued in Chapter 8 to be an important feature in explaining the diverse levels of variation observed among recent modern populations, which include the retention of a number of archaic features in some groups. On this basis, it was suggested that the ancestral early modern group be composed not only of those crania that are clearly 'modern', i.e. fully derived in relation to archaic *Homo sapiens*, but that it should include a number of specimens of possible similar age that show a greater incidence of archaic features. In this light, the African ancestral population should be represented by the fossils of Florisbad, LH 18, Omo 1, Omo 2 and Klasies River Mouth, possibly excluding the Border Cave remains. The early fossils of Skhul and Qafzeh in the Middle East could be either part of the range of this population, or an early dispersal.

The comparisons presented in Figures 9.2 to 9.4 are very important in confirming the nature of early modern cranial variation. It could have been argued that claims of ancestral variability in the early modern group are an artefact of the composition of the group, i.e. that the variability is being created by the inclusion of archaic crania within the sample. However, the results show that the pattern of variation is not consistent throughout the different variables studied (Figure 9.5), indicating that the level of variation is not being affected just by one or other specimen. Accordingly, these mixed morphological patterns are reflected in the debated 'modernity' of many of the remains. We have seen that the majority of researchers generally consider the particularly heterogeneous group of Florisbad, Omo 2 and LH18 archaic (Dreyer, 1936; Singer, 1958; Smith, 1985, 1992), although claims of modernity have also been made for some of these specimens (Howells, 1959, 1974). But even some of the more generally accepted early modern remains have been questioned. Wolpoff & Thorne (1991) claim that the single cheekbone fragment among the Klasies River Mouth remains falls outside the range of modern Africans, and it is even larger and more robust than African archaics. However, we have seen that the morphological range of any one spatial or temporal modern population cannot be taken to assess the modernity of a specific trait, especially recent Africans with their comparatively gracile skeleton. Accordingly, Smith *et al.* (1989) cite the zygomatic KRMS 16651 as typically modern. The remains of Klasies River Mouth are indeed very fragmentary and very variable, interpreted as high levels of sexual dimorphism within a single fully modern population (Bräuer, 1984b, 1989; Deacon & Shuurman, 1992; Rightmire, 1984, 1986, 1989). Kingdon (1993) argues that the high levels of polymorphism in the Klasies River Mouth population are consistent with a decline in the selective pressure that maintains levels of variation. These high levels of polymorphism, extended to a broader early modern human sample, are consistent with the observations of several researchers that there are no strictly clade features that characterise the transition from archaic to modern *Homo sapiens* (Habgood, 1989; Rightmire, 1986; Smith, 1992; Smith *et al.*, 1989). The early modern population was morphologically variable, with varying levels of size, robusticity and certain primitive features later lost by most modern populations.

General trends in modern human differentiation

The description of Figures 9.2 to 9.4 indicates that there is a reduction in most cranial dimensions from the early modern to recent populations. This is true of maximum cranial length (GOL), facial radii (NAR, FMR), biasterionic and biauricular breadth (ASB, AUB), foramen magnum length (FOL),

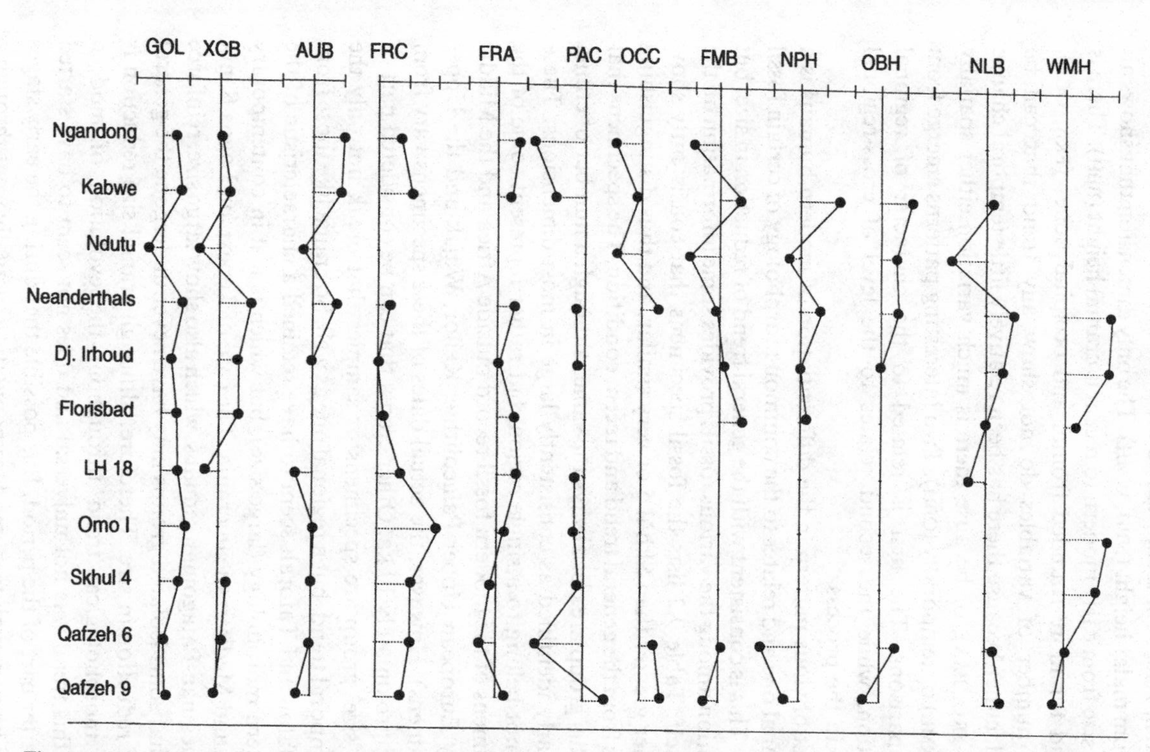

Figure 9.5 Variability in some archaic (Ngandong, Kabwe, Ndutu, Neanderthals, Djebel Irhoud) and some early modern (Florisbad, LH18, Omo 1, Skhūl 4, Qafzeh 6, Qafzeh 9) crania in relation to the values observed in Skhūl 5 for some main cranial dimensions, showing that the variation observed in the early modern sample cannot be attributed to specific specimens but to an overall high level of variability in various dimensions.

maximum frontal, fronto-malar, bijugal, bimaxillary, nasal and malar breadths (XFB, FMB, JUB, ZMB, NLB, MAB), parietal chord (PAC), and facial and minimum malar height (NPH, WMH). The only dimension that shows an increase in size from early modern to recent is cranial height (BBH). There is also a trend for more rounded frontal and occipital bones (FRA, OCA). Finally, a number of variables do not show any trend that can be generalised, either because there has been relatively little temporal change (BNL, XCB, FRC, PAA), or because there is much variation either spatially (OCC, SSS, XML) or temporally (OBH). Two interesting patterns emerge from these comparisons. The first is related to the character of regional differentiation, while the second relates to the level of cross-regional synchrony in the process.

The most obvious pattern in the differentiation of modern humans is a cross-regional one and relates to the common morphology of certain fossil specimens. This is consistent with the general trend of reduction in size, but its distribution among the various fossils provides some information on the process itself. Table 9.2 lists the fossil specimens that consistently show values either greater than Skhūl 5 or very similar, and thus distinguishing themselves from the general trend and recent condition. The specimens that make up this group are not always the same, although a number of crania can be clearly identified as consistently large in most dimensions. These crania do not belong to a single region, but rather represent some of the early specimens of the modern fossil record outside Africa and the Middle East (early European Upper Palaeolithic, Keilor, Wajak and the Upper Cave specimens). However, the actual dates of these specimens vary from 35 ka to approximately 10 ka. Other crania that can be considered part of this 'large size' group are specimens or samples that break not only the general temporal trend, but a regional one as well. So crania like those from Teviec, Afalou and Taforalt seem to have retained a characteristic of the early modern morphology (large size), but which is lost in contemporary European and Mediterranean crania. The case of Tepexpán, Lagoa Santa and the Fueguian-Patagonian remains, which also show large size, reinforce the view that some isolated groups may have never undergone the general process of reduction in size. Therefore, although cranial size reduction is the main morphological trend throughout the evolution of modern diversity, this trend was not universal and does not seem to have started until the latter part of the period. It is possible that during the early stages of modern human evolution, modern populations actually went through a period favouring large size, reflected in early specimens worldwide, to then undergo a process of size reduction. Cranial size reduction seems to have occurred at different moments and at different rates in various modern

Table 9.2. *Fossil specimens that show values similar to or above Skhūl 5 in some important cranio-metrical variables*

Trait	EU	ME/NA	SSA	AUS	SEA	EA/AM
GOL	Ml., CM, Pr, Grim		FH, WK, Elm	Kei	Wa	UC101
NAR	CM, Pr, Tev	Af, Taf	WK	Kei, Kan	Wa	UC101,103,F/P
XCB	CM				Wa	
FOL	Ml, CM, Pr, Grim, Tev		WK	Kei	Wa	UC101, 103
XFB	Ml, CM, Pr, Tev	Oh, NEG		Kei	Wa	UC101, 103, Liu, Tep
FRC	CM, Pr, Grim	EG, Af		Kan		Tep
FRA	Pr	Ho	Elm	Kei, Kan	Wa	UC101, 103, Min, Tep,F/P
PAC	Ml, CM, Pr	Ho, Nat, Af, Taf		Kei, Kan		UC103, Liu
PAA	Ml			Kei		
OCC	Ml, CM		Fin	Kei	Wa	Tep, LS6
OCA	Ml, CM	Oh	MR			UC101, LS6
FMB	CM, Pr	Oh		Kei, Kan	Wa	UC101, 103
JUB	CM, Pr, Tev	Oh, Taf		Kei, Kan	Wa	UC101, 103, Min
ZMB	Ml, CM, Pr, Tev	Af, Taf	FH, Elm, Nak	Kei, Kan		UC101, 103, Liu, Min EA, F/P
NPH	Pr, Tev		Elm, Nak	Kei, Kan		UC101, F/P
NLB				Kan	Wa	UC101
MAB	Ml, CM, Pr, Tev	Af, Taf	Nak	Kei, Kan	Wa	UC101
WMH	Ml, CM, Pr, AP Grim, Tev	Af, Taf, Ho	FH, Elm	Kei, Kan	Wa	UC101, Tep, LS6 EA, F/P
	EU: Europe	ME/NA: Middle East/North Africa	SSA: sub-Sah. Africa	Aus: Australia	SEA: Southeast Asia	EA: East Asia/Americas
	Ml: Mladec CM: Cro-Magnon Pr: Predmost Grim: Grimaldi AP: Abri Pataud Tev: Teviec	Af: Afalou Taf: Taforalt Ho: Hotu EG: Ein Gev Oh: Ohalo NEG: Nahal Ein Gev Nat: Natufians	FH: Fish Hoek WK: Willy Kopje Elm: Elmenteita Fin: Fingira Nak: Nakuru MR: Mjates River	Kei: Keilor Kan: Kanalda	Wa: Wajak	UC101 UC103 Liu: Liujiang Min: Minatogawa Tep: Tepexpán LS6: L. Santa 6 EA: East Asians F/P: Fue-Patagonian

populations, independent of the main processes of population differentiation. So we see that, while in Europe there is a difference in cranial size between the very early and very late Upper Palaeolithic crania, as already pointed out by Kidder *et al.* (1992), this morphological shift seems to have occurred much later in North Africa and Australia, or not at all in parts of the Americas, where some populations (the Onas and Tehuelches of Tierra del Fuego and Patagonia) with cranial size similar to that of early modern humans lived until recent times. Furthermore, while in Europe the change in size may be associated with population movements (Howells, 1959, 1993), in East Asia (where a decrease in size is evident from the earlier Upper Cave remains of specimens like Liujiang or Minatogawa), this morphological change seems to have occurred within a regional lineage, as evidenced by the regional trends of facial flatness in East Asia discussed in Chapter 5 and seen in Figure 9.4. Lastly, in certain groups within sub-Saharan Africa and Southeast Asia, the change towards a smaller cranial size seems to have been taken a step further with the evolution of pygmy populations.

Therefore, besides the main trend of size reduction, Figures 9.2 to 9.4 also disclose very specific differences within each region. These differences relate to the degree of continuity in regional morphologies and the levels of variation in the fossil record of each area, and show that each recent population has undergone quite distinct regional processes that shaped their cranial morphology as seen today. So how can we explain both the common trend and the evolution of such diversity?

Gracilisation as a common mechanism of change

This and previous chapters have shown that there are great differences among modern populations in relation to the level of morphological specialisation, robusticity and size. We have described above an evolutionary pattern of differentiation from an ancestral source that identifies the polarity of change in the regional modern morphologies through time and a main common trend of decrease in size. It was also shown that this evolutionary trend, although expressed in almost all recent populations, did not occur gradually or synchronically across the different regions of the world. Here I will briefly explore one mechanism that would account for both such a common process of anatomical change in different populations and the anachrony of its expression throughout the Upper Pleistocene.

The results of the analyses in Chapter 6 not only showed that the cranial superstructures are correlated in their expression, but identified a number of anatomical dimensions related to the expression of robusticity. Some of these dimensions, like the size of the masticatory apparatus, cranial size and upper facial breadth were important discriminating variables throughout most of the analyses performed. As discussed before, this close relationship between features of cranial robusticity and the masticatory apparatus gives support to the findings of other studies which suggest a localised masticatory stress factor in the development of facial superstructures (Demes & Creel, 1988; Endo, 1970; Russell, 1985). However, the relationship between the superstructures and main cranial dimensions indicates that there are other important components in the determination of each of the features of robusticity examined, relationships more consistent with spatial models for the role of the superstructures (Hylander *et al.*, 1991; Moss & Young, 1960). In terms of our discussion on modern human differentiation, it becomes important to examine whether, given the findings in Chapter 6, the general trend of decrease in size can be explained as the result of cranial gracilisation, and whether such a process occurred through functional or phylogenetic pathways. In order to do this, we first have to investigate the strength of the relationship between robusticity and size.

In order to examine the relationship between all the cranial superstructures taken as a single complex 'robusticity' (measured non-metrically) and cranio-dental dimensions (metrical variables) canonical correlation analyses were used. These analyses, carried out in collaboration with Richard Wright (Lahr & Wright, in press) attempt to correlate sets of data, and thus differ from the discriminant analyses performed in Chapter 6 by treating all the non-metrical features together. In this way, the results obtained reflect the degree of association between robusticity and cranial dimensions, identifying, through the variable loadings on each vector, which variables within each data-set are contributing to the correlation. However, because the two data-sets do not behave symmetrically, a measure of the contribution of each set of variables (non-metrical or metrical) to the canonical vectors is necessary. This contribution was measured as the redundancy, which is quantified as the proportion of variance in metrical dimensions explained by the robusticity vector and vice-versa. Two canonical correlations were performed. The first of these, on a complete set of 198 recent skulls, correlated nine features of robusticity in the non-metrical set (supraorbital torus, zygomatic trigone, depth of infraglabellar notch, zygomaxillary tuberosity, rounding of the infero-lateral margin of the orbit, profile of nasal saddle, sagittal keeling, occipital torus and occipital crest) with metrical dimensions of the skull. The second canonical correlation focused on the facial superstructures in order to examine whether the relationships within a functional unit are stronger than across the cranium. This analysis was performed on a complete set of 134 skulls, and it correlated six features of cranio-facial robusticity (supraorbital torus, zygomatic trigone, depth of infra-glabellar notch, zygomaxillary tuberosity, rounding of the infero-lateral margin of the orbit and sagittal keeling) with cranio-dental dimensions.

The results of both canonical correlations were highly significant (CCA 1: 0.85, $p<0.0001$; CCA 2: 0.90, $p<0.0001$), showing that there is a strong relationship between specific cranial dimensions and degree of cranial robusticity. The metrical variables with the highest loadings on the first factor (Figures 9.6a,b), i.e. those dimensions that contribute most to the correlation of robusticity and metrical variables, are maximum length (GOL, NOL), upper facial breadth (FMB, JUB, EKB, OBB), facial-basal length (BNL), palatal size and protrusion (MAB, BPL), occipital flexion (OCS), and in the case of the second canonical correlation analysis also palato-dental size (PALL BLM1, BLM2), glabellar protrusion (GLS) and distance between the position of minimum frontal breadth and the zygomatico-frontal suture (WFBZFS). Cranial breadth is negligibly correlated with the first factor in both analyses. The loadings of all variables on the first factor are mainly positive or close to zero, indicating that the relationship represents size, i.e. gracile and robust forms are not related to alternative shapes but to an increase in size in the above mentioned traits. Furthermore, the redundancy results show that the cranio-dental metrical dimensions explain a large proportion of the variance in robusticity, while the non-metrical features explain comparatively less. In other words, the main direction of influence is from the metrics to the non-metrics, so that robusticity increases as cranial size increases. The metrical set in the first analysis explains 24.9% of variance in robusticity, while the non-metrical set explains 9.8% of metrical variance. In the second analysis, the metrical set explains 37.2% of facial robusticity, while the non-metrical set explains 11.0% of the metrical variance. These results not only confirm those obtained in Chapter 6 for each feature independently, but the fact that in the first canonical correlation analysis both the features of facial and occipital robusticity have high loadings on the first factor corroborates the view that all the cranial superstructures are correlated at some level to form a single complex of overall cranial robusticity. However, the coefficient and redundancy values of the second canonical correlation are larger than those of the first, also showing that the relationships are strongest within functional units.

Figure 9.7a,b shows the distribution of crania according to robusticity and size along the first canonical factor, by which the crania at the upper right quadrants of Figure 9.7 represent robust and large specimens, while those at the lower left quadrants represent gracile and small ones. The

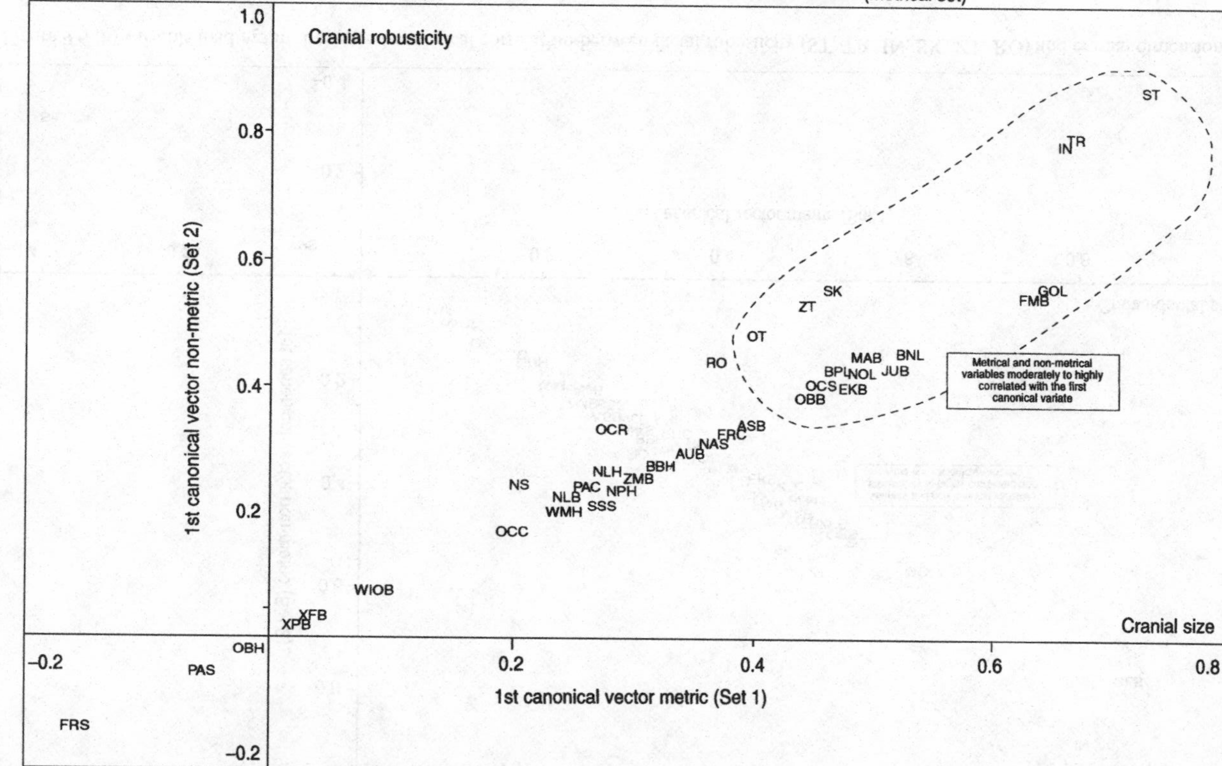

Figure 9.6 Canonical correlations between cranial dimensions and robusticity, expressed as cranial superstructures: Plot of variables along the metric and non-metric vectors of the first canonical variate.
(a) variable loadings in the first canonical correlation between overall robusticity (ST, TR, IN, SK, ZT, OT, RO, OCR, NS) and cranial dimensions.

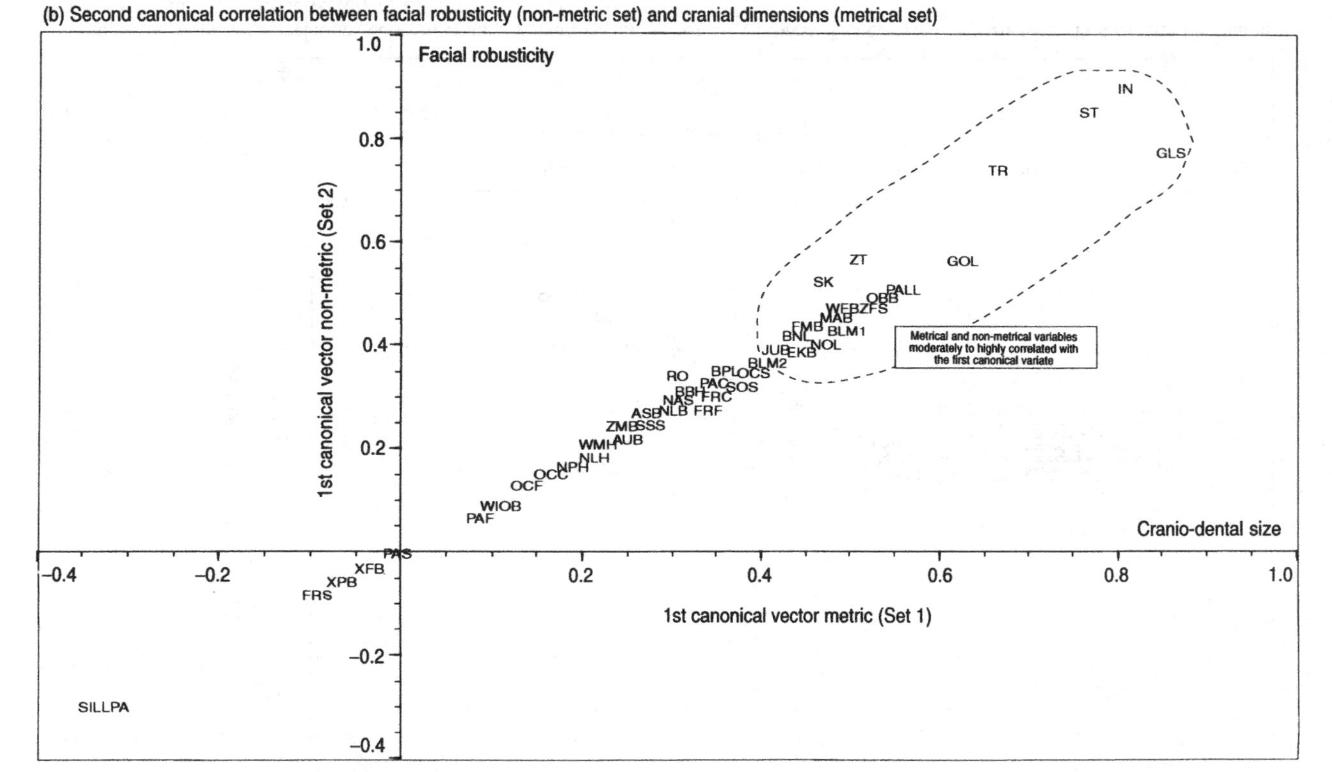

Figure 9.6 (b) variable loadings in the second canonical correlation between facial robusticity (ST, TR, IN, SK, ZT, RO) and cranial dimensions.

identification by population of the robust/large and gracile/small crania confirms that robust individuals are not equally found in all recent populations (Figure 9.8). Most populations are either robust or gracile. The 'robust' group of Figure 9.7a is composed of 15 Australians (65.2%), three Fueguian-Patagonians, two Afalou-Taforalt, one Ainu, one Maori and the Zkn UC101 specimen. The facially robust group contains the same specimens except the Maori and Zkn UC crania. Of the crania along the axis of factor 1 which are neither particularly robust or gracile, those towards the upper right quadrant are mainly Australian, Fueguian-Patagonian or Afalou-Taforalt. These are robust and large modern populations. The 'gracile' group of Figure 9.7a is composed of 13 sub-Saharan Africans, 13 eastern Asians, 11 Southeast Asians and seven Europeans, while the facially gracile group has a broader composition, including also Inuit, Indian, an Australian from the Northern Territory and the fossil of Nahal Ein Gev from Israel. These populations – the majority of recent populations – are relatively gracile. The fact that the majority of recent populations falls within small and gracile parameters is consistent with the size reduction trend discussed above. Only the recent Europeans have similar percentages of both morphological patterns, although none extreme. It is possible that the similar numbers of moderately robust and moderately gracile individuals among recent Europeans reflect the specific pattern of sexual dimorphism in this population.

An interesting aspect of the relationship between size and robusticity is that the 'size' complex does not include cranial breadth in its definition. The robust crania of Figure 9.7a have a mean cranial breadth of 137 mm, while the gracile ones are 136 mm, although the two groups differ in the amount of variation they contain (XPB robust: cv = 7.3; XPB gracile: cv = 4.4). In other words, the gracile group consists mainly of small and broad crania (and therefore, brachycephalic), while the robust group contains large crania that may be either broad or narrow. Clearly, the combination of cranial length and breadth not only creates size but also a given shape in terms of cranial index. In order to investigate the relationship between cranial size and shape in constraining the expression of the cranial superstructures a principal components (PC) analysis was performed. This analysis used the same 134 crania and variables of the second canonical correlation analysis. A plot of the first and second principal components confirms that the main structure in the data reflects size and robusticity as defined by the first canonical factor, accounting for 27.3% of variance in the first PC. The second PC (accounting for 11.4% of the variance) contrasts alternative cranial shapes – broad and gracile skulls against narrow and robust ones with large teeth (Figure 9.9a). The correlation of

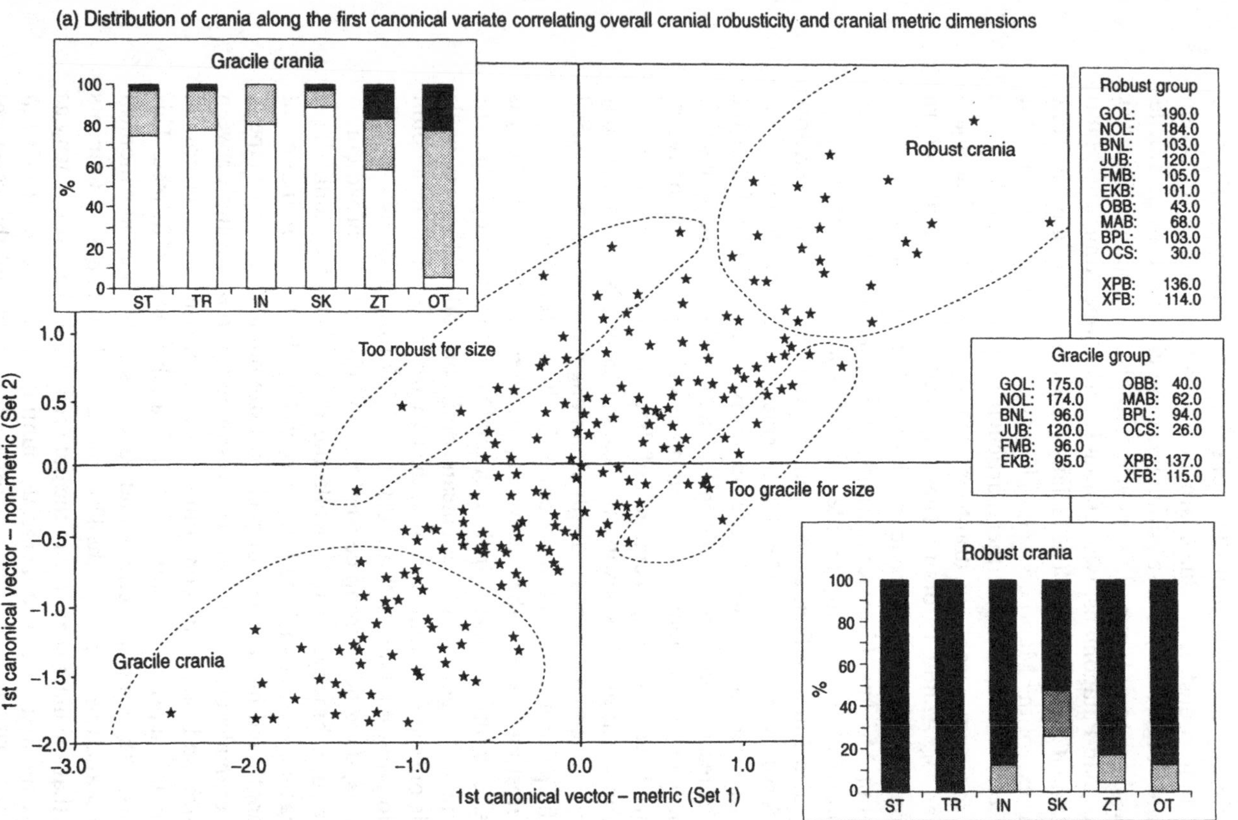

Figure 9.7 Plot of crania along the first canonical variate correlating (a) overall cranial robusticity or (b) facial robusticity with main cranial dimensions, separating robust crania on the upper right-hand quadrant from gracile ones on the lower-left quadrant. Histograms represent the expression of grades of development of cranial superstructures in two graphically defined 'robust' and 'gracile' groups, ranging from very pronounced (black) to absent (white); in boxes, mean values for the 'robust' and 'gracile' groups of the metrical variables moderately to strongly correlated with the metric vector, while the mean values of XPB and XFB illustrate the absence of correlation of these variables with the first canonical vector obtained.

(b) Distribution of crania along the first canonical variate correlating facial robusticity and cranial-dental metrical dimensions

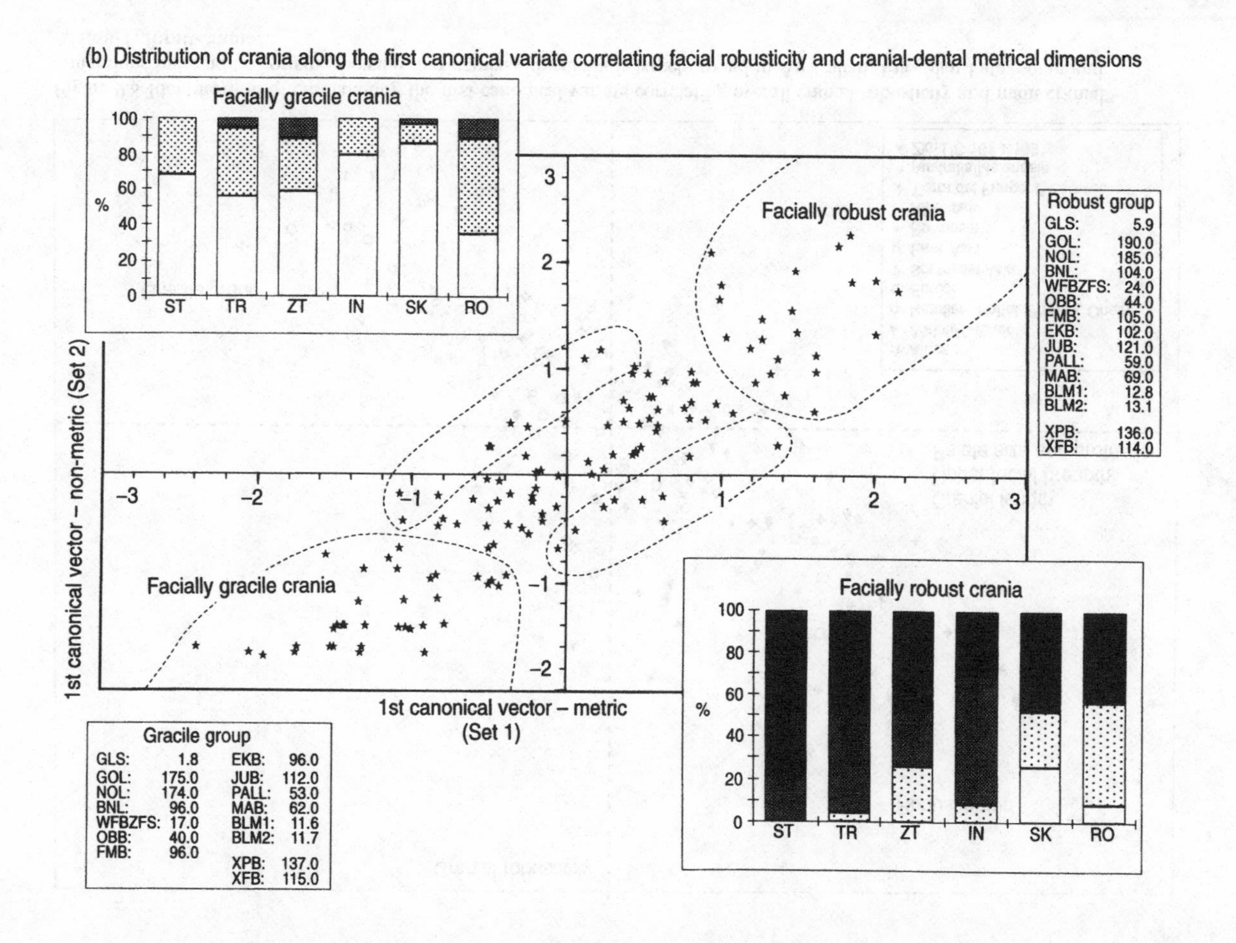

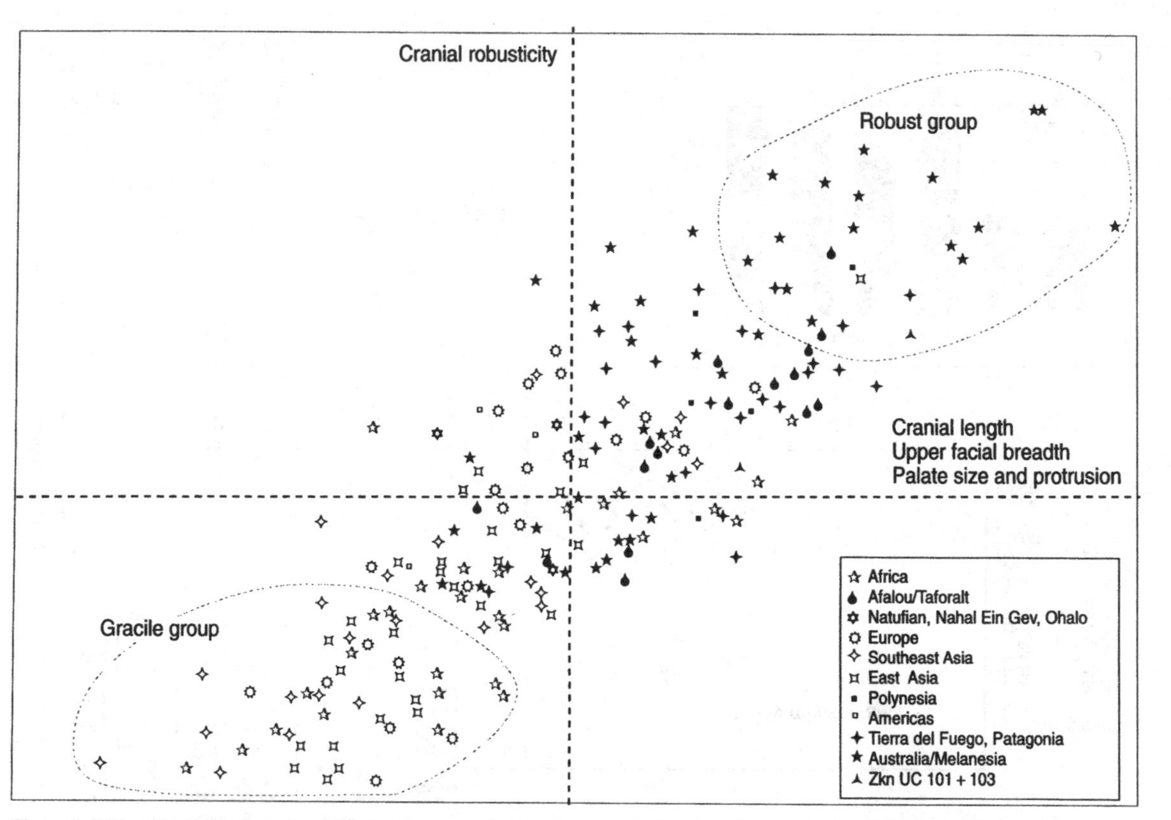

Figure 9.8 Identification of crania along the first canonical variate correlating overall cranial robusticity and main cranial dimensions by region of origin, showing that cranial robusticity is largely found in Australian, Fueguian-Patagonian and Afalou-Taforalt crania.

metrical and non-metrical variables along the first and second PCs (Figure 9.9a) creates a number of anatomical combinations, ranging from very broad and gracile to very large, narrow and very robust, which can be used to interpret the distribution of crania along the first two PCs (Figure 9.9b). The combination of cranial size and shape indicates that there are two anatomical circumstances that relate to the development of superstructures: the first and most common is in very long skulls with broad upper faces and large palates (independent of the breadth of the vault), and the second in very narrow skulls with large teeth (only moderately associated to cranial length and upper facial breadth). The skull with greatest robusticity is one that combines these two circumstances through very long and narrow vaults, with broad upper faces and large teeth (Figure 9.10).

What makes the relationship between the constraining parameters of size-shape and robusticity all the more remarkable, is its constancy throughout the evolution of modern diversity. We have seen in previous chapters that one of the most morphologically different modern populations are the aboriginal inhabitants of Australia. Cranially, this population is very distinct and part of this distinctiveness relates to its very large dentition and very pronounced robusticity. However, the large dentition and robusticity do not relate to the retention of early modern dimensions, since recent Australians have also undergone the size reduction trend of most modern populations, as is clear from Figures 9.2–9.4. Nevertheless, they do not depart from the relationship between large size and robusticity, as it is clear from Figures 9.7a and b. The means by which size reduction without gracilisation occurred in this population is clear when the crania in Figure 9.9b are identified by region of origin (Figure 9.11). While in most populations, size reduction occurred through reduction in cranial length and not breadth, resulting in a brachicephalic form, recent Australian crania seem to have reduced both in length and breadth, although not reducing upper facial breadth and in certain cases increasing palatal breadth (the fossils of Keilor and Kanalda are the only two cases among all the specimens shown in Figures 9.2–9.4 that show greater maxillary breadth than Skhūl 5). This morphological change seems to have not only maintained a relationship between the constraining parameters constant, but possibly intensified the expression of a number of traits related to the facial skeleton. It should be remembered that the features that seem to form an Australian lineage from the Upper Pleistocene (Chapter 5) are the surpraorbital ridges, zygomatic trigone, position of minimum frontal breadth and zygomaxillary tuberosity, all related to the robusticity of the facial and supraorbital skeleton. Furthermore, the extent of cranial breadth reduction in Australian specimens has been such that hyper-dolichocephalic crania are observed,

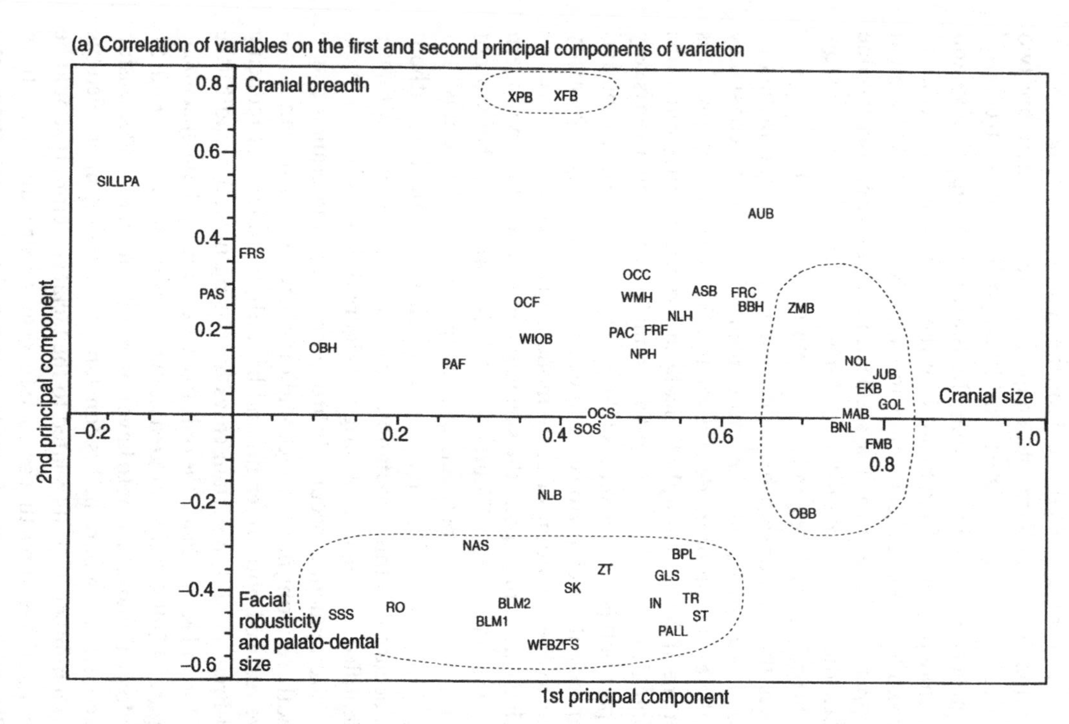

Figure 9.9 Plot of first and second principal components showing the correlation of cranial size on PC1 and cranial shape on PC2:
(a) Plot of variables along the first and second principal components of variation involving cranial dimensions and robusticity.

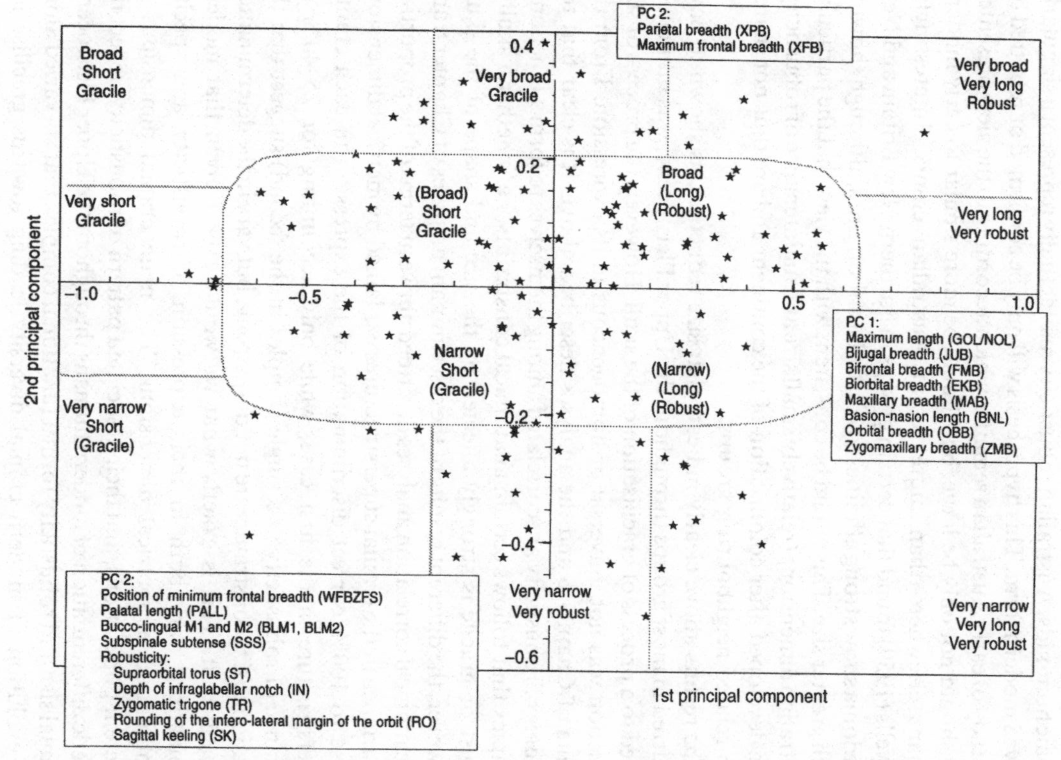

(b) Plot of crania along the first and second principal components of variation and possible anatomical interpreations of the combination of PC1 and PC2.

and these are clearly the most robust (bottom right-hand corner of Figures 9.10 and 9.11). Therefore, another interesting point emerging from these analyses is that instead of reflecting a primitive inheritance from Javanese archaic hominids, Australian robusticity may actually represent a singularly derived modern form. This hypothesis will only be confirmed or rebutted by the study of early Australian remains, which according to the view presented here should not only be large, but should be more similar to early modern humans elsewhere than more recent Australian crania. Interestingly, Howells (1959) found that certain groups in northern New Britain and New Caledonia show strong affinities to Australian aborigines, although showing much greater size. This would be consistent with the view that the particular Australian dimensions (relatively small size and maintenance of robusticity) were developed after colonisation of the continent and are thus not shared with related neighbouring groups.

The results above clearly indicate that size reduction, as shown to have occurred in most groups throughout the Upper Pleistocene, was intimately linked to a process of gracilisation of the skull. However, the process of size reduction was not universal or homogeneous in its expression. From the results of Chapter 6 and the analyses described above, it is clear that the features of robusticity are correlated among themselves to form a functional complex that follows certain anatomical constraints, i.e. whether a skull is robust or gracile is strongly affected by the size and shape of the skull. However, the different results of the discriminant analyses of Chapter 6 and the canonical correlation analyses described above in terms of the proportion of variance in the cranial superstructures explained by metrical dimensions (close to 100% correct discrimination of the expression in each cranial superstructure in the first case, while only accounting for 25–40% of variance in robusticity as a single complex in the second) suggest that the functional/size constraints are not only strong, but vary in the determination of each trait. This is exactly what we would expect given that modern populations vary both in their expression of robusticity and specific combination of metrical dimensions. The main cranial dimensions of different populations, and their associated pattern of robusticity, reflect the genetic differentiation of modern humans throughout the Upper Pleistocene. Cranial size and robusticity form a functional complex that co-varied along the evolution of modern cranial diversity leading towards gracilisation within each population lineage through varying dimensional changes in different populations.

It still remains to be discussed, however, whether the evolutionary unit of covariance formed by cranial size, shape and robusticity resulted from phylogenetic or functional processes. It has been argued that technological

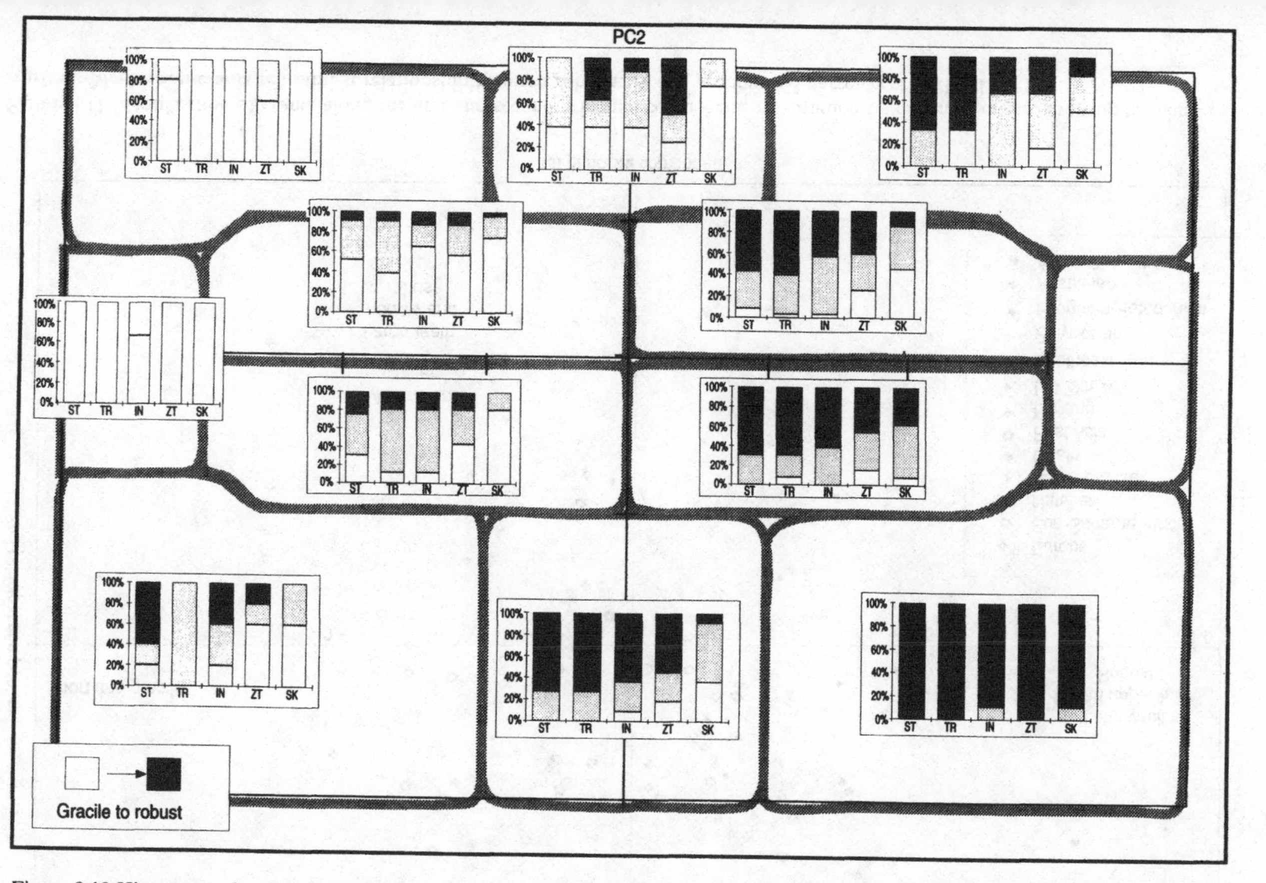

Figure 9.10 Histograms showing the expression of facial robusticity, in terms of grades of developement of the facial superstructures (ST, TR, IN, ZT, SK), in the 12 groups of crania defined graphically by the plot of the first and second principal components of variation, forming a gradient from gracile crania on the upper left quadrant (broad and short skulls) to very robust ones on the lower right quadrant (long and narrow skulls).

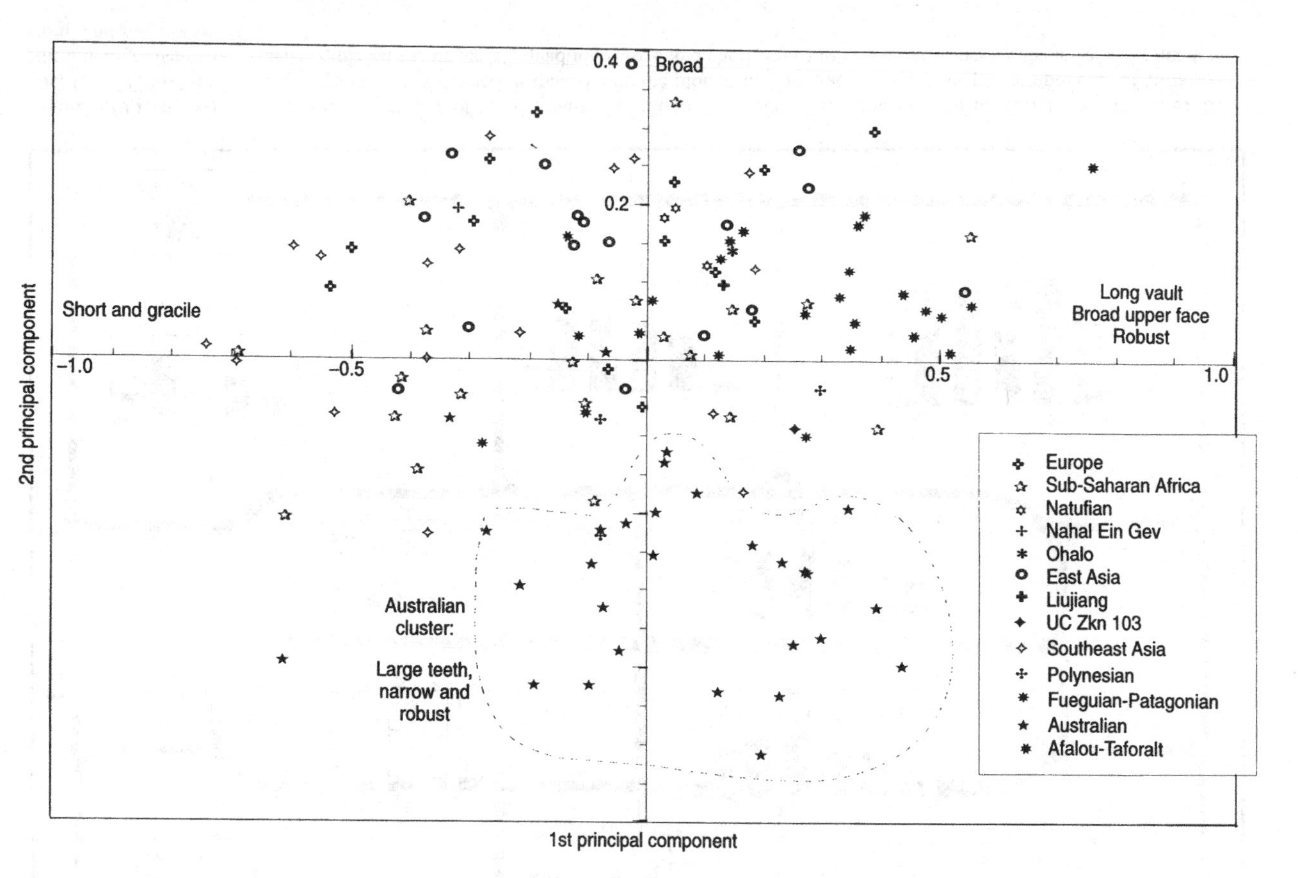

Figure 9.11 Identification of crania along the first and second principal components of variation by region of origin, showing the cluster of Australo-Melanesian crania reflecting a relationship between robusticity and pronounced narrowness of the skull.

developments released the selective pressure for a large masticatory apparatus (Brace, *et al.*, 1984) or for robusticity in general (Frayer, 1978, 1984). If that was the case, it is possible that in all those populations in which dental size and the size of the muscles of mastication (especially the size of the *temporalis* which is closely associated to cranial length) decreased, an entire functional complex was released. Thus robusticity and size would have been lost in parallel in all populations that underwent this process, its geographical distribution carrying no phylogenetic information. However, the association of cranial robusticity with technological developments is not clear. Among hunter–gatherers the whole range of modern human robusticity is observed, from the very gracile San/Khoi and Inuits, to the very robust Australians and Onas. Furthermore, as argued by Turner (1985b), the populations that load their teeth the most (Eskimos) do not show the greatest dental size or robusticity. Therefore, there is an incongruence between the mechanical sources of stress in the cranium and the expression of robusticity, suggesting that cranial gracilisation could not have been solely the result of a release in selective pressures. The alternative solution is that cranio-dental dimensions, with their effect on robusticity, have a strong genetic basis. This is consistent with the fact that, when treated multivariately, cranio-dental dimensions disclose genetic distances between populations (Brace, *et al.*, 1990, 1992; Howells, 1973, 1989; Pietrusewsky, 1992; Wright, 1992).

Therefore spatial variation in the size–shape–robusticity complex reflects specific adaptations and random genetic changes, and cranial size reduction and gracilisation may have been homoplasic in most populations. It can also be argued that some of the discontinuities observed in the modern fossil record (Kamminga & Wright, 1988; Stringer, 1992a; van Vark, 1990) reflect gene flow between modern groups, which together with a different morphology also introduced different levels of robusticity, affecting the rate of change within an area. The particular character and expression of gracilisation in each population reflect the evolutionary processes undergone by each group, and thus, the phylogenetic relationships between differentiating modern humans. In this light, it should be possible to explain the temporal distribution of morphologies in the Upper Pleistocene in terms of the specific history of each population.

Evolutionary mechanisms

By placing the last common ancestor in the African late Middle Pleistocene and considering the temporal and spatial differences in the morphology of modern humans throughout the Upper Pleistocene it is possible to conceive a much more complex course of events in the evolution of modern human cranial diversity than is currently suggested by a single origin and expansion out of Africa. This course of events, reflected in the patterns and trends described above, must have been defined by the specific evolutionary mechanisms acting on each population. The evolutionary mechanisms by which a biological population diversifies are well known, and relate to the genetic effects of selective pressures from particular environments and demographies. However, early modern human populations faced very different environments that also varied through time, so no single mechanism will be able to explain the evolution of diversity. Modern human diversification is intimately linked with the process of geographical expansion, and the character and extent of this expansion must have played a central role in defining the amount of gene flow between geographically separated groups and the chance effects of drift and founding. The next two chapters of this book examine the geographical processes that could account for the complex temporal and spatial distribution of cranial morphologies that created modern human diversity.

10 *Geographical differentiation from a single ancestral source*

The patterns and trends in modern human cranial morphology throughout the Upper Pleistocene show that it is possible to derive all modern morphologies from a single variable ancestral source in the late Middle to early Upper Pleistocene of Africa. However, from that single source modern human differentiation occurred at different rates and to different extents in different groups. Although anachronical, the processes that resulted in the differences between the various modern populations occurred within a common morphological trend of cranial size reduction and gracilisation. This chapter focuses on the evolutionary mechanisms that may have caused each population to diversify to the extent they did by disclosing the role of geographical expansions in the evolution of modern cranial diversity.

Geographical expansions and dispersals

Biogeographical comparisons indicate that evolutionary novelties tend to appear in small populations, which if successful, will expand geographically (Foley, 1989; Tchernov, 1992c). Groves (1992) discusses the effect of wide geographical expansions in the origins of diverse mammalian species. He finds that the mode of evolution of all but one of six mammalian families was through a single origin, expansion and replacement of other forms. Multiregional evolution may have occurred only in the case of the Javanese Rhino. Therefore, the pattern proposed for modern humans of a single origin and replacement (Bräuer, 1992a; Howells, 1976; Stringer, 1992a), and supported in this and previous studies (Lahr, 1992, 1994), is consistent with the evolutionary process acting on mammals in general. Groves (1992) also finds that the evolution of subspecies, i.e. regional variants of a species, has been widely anachronical, even within a single species, and unpatterned, i.e. there is no consistent relationship between level of differentiation and divergence time. These results are also fully consistent with the varying degrees and rates of change in modern human regional populations throughout the Upper Pleistocene discussed in Chapter 9.

The model of a single origin of all modern humans, followed by a process of geographical expansion that gave rise to regional populations, is therefore not only supported by several sources of data (morphological, chronological, genetic), but is also theoretically consistent in terms of mammalian speciation and sub-speciation events. Given the role of geography in the process of biological differentiation, the answer to many of the questions on the character, rate and timing of the evolution of modern human populations raised by the previous chapters will be found in the historical events that gave rise to these populations. As mentioned before, the modern human fossil record is very variable in completeness and distribution, so that the detailed reconstruction of regional histories can only be tentative. Particularly, the interpretation of the temporo-spatial changes in terms of *in situ* differentiations or gene flow, is hampered by widely differing contextual information for the fossils from different areas. Therefore, any attempt at reconstructing the evolution of modern cranial diversity has to do so within a precise theoretical framework that would provide a consistent interpretation of the data. This chapter develops a model based on the role of geographical dispersals as an evolutionary framework for interpreting the evolution of modern human diversity.

Evolutionary consequences of geographical expansions

The evolutionary consequences of geographical expansions have been recognised for a long time. The expansion of an animal's geographical range plays two major roles in the process of biological differentiation. Firstly, by exposing sub-groups of the expanding population to different environments, thus subjecting them to different selective pressures which will consequently create different adaptive responses and a level of polymorphism in the population. Secondly, by providing a degree of allopatry, with the possibility of fixation at a population level of specific environmental adaptations, any chance genetic changes and drift, as well as varying the level and rate of diffusion of evolutionary changes between the centre and edge of the geographical range. Therefore, several important biological consequences can result from geographical expansions – population extinction, speciation, reduction/increase in variability and admixture with more or less distantly related groups. Except speciation, all the biological consequences of geographical expansions concern the present discussion on the evolution of modern human diversity. Each of these consequences will have a different effect on the process of differentiation from an ancestral source.

The effect of geographical range expansions on the levels of genetic and morphological variation will vary according to three different circumstances. The first is the case of geographical range expansion without subsequent subdivisions. Such a situation would subject a single breeding population to diverse ecological and competitive pressures. This population would not be able to specialise towards maximising a single adaptive trend, a process that would favour genetic variability and polymorphisms to maintain high levels of adaptability. The second is the case of geographical range expansions with restricted gene flow between the extreme groups, giving way to a process of parapatric differentiation with resulting clines. And the third, is the case of geographical expansion followed by range break-up into small sub-units, increasing the chances of isolation of groups. The effects of the resulting allopatric conditions on small sub-populations are well-known, and may vary from extinction to the development of independent evolutionary trajectories. Geographically restricted populations will be able to specialise towards their specific environment, but specially through the effects of drift and founding, may lose variation and/or accelerate the rate of change. The rate at which these isolates will change depends largely on the size of the populations concerned.

The effects of genetic admixture, however, are very different. Expanding populations may come into contact with other groups, and the interaction with these will depend on the behavioural, morphological and genetic distances involved. Therefore, such interactions may result in either the displacement of the local population, or total admixture of local and immigrant groups, or on partial admixture, especially in uneven areas whereby pockets of the local population remain unmixed. In the case that admixture (whether total or partial) takes place, the direction of gene flow will also affect the outcome. If gene flow is unilateral, the original source of the immigrant population will not be affected while the local group will be subjected to the introduction of genetic novelties into their gene pool. The process of differentiation of the population receiving the gene flow will be halted, reduced or changed, depending on the intensity of admixture. If the direction of gene flow is bilateral, the result will be a homogenising effect, by which the distinctiveness in both gene pools involved would be diluted.

Finally, extinctions may also be an outcome of geographical expansions, the underlying factor being a combination of population size and environmental change through time. Although a dispersing group may not survive its attempt at establishing itself within a different environment, these wanderings are not real extinctions, as no differentiation from the source has occurred. However, if a population has expanded and established itself ecologically in a different area and then fails to survive a change in its environment (whether competitive or climatic), extinction has occurred.

At one time or another, all these processes have occurred within the evolution of modern humans. The time and place of these events varied with the nature, source and extent of the geographical dispersal involved. Recently, Tchernov (1992c) discussed the subject of geographical dispersals in detail, and set the concepts necessary for developing the framework of human dispersals in the Upper Pleistocene.

Defining dispersals

Tchernov (1992c) describes the temporal effects of geographical dispersals in the process of biological differentiation of species and defines a number of concepts related to geographical movements of populations. Dispersal is defined as equivalent to range expansion, 'a process whereby organisms are able to spread from their place of birth or phylogenetic origin to another locality' (Tchernov, 1992c:p.21), equal to range expansion. Therefore, a basic distinction is made between processes that imply an expansion of range and those that imply the establishment of new populations not in contact with the parental group. The former are dispersals, while the second may take the form of wanderings (regular unilateral movements beyond the species range), or immigrations, implying the establishment of a colonist population discontinuous from the parental source. Migrations are a special category of movement, for they imply the regular movement of a group beyond the range of the species, but from which most of the members of the group will return.

Dispersals may occur in three modes (Pielou, 1979): *jump dispersals* – across great distances in a short time, resulting in the successful establishment of colonists; *diffusions* – slower and gradual process across hospitable areas; *secular migrations* – very slow dispersals during which there is evolutionary change in the disperser groups. Clearly, these three types of dispersal will have very different evolutionary effects on population parameters.

A final concept needs to be defined. In the following section, the notion of *relic* populations will be used. In this context, 'relic' does not necessarily imply a biological condition but a phylogenetic one, although in some cases it may imply both. Therefore, by identifying a group as a 'relic', it is not implied that this group has remained unchanged or evolutionarily static since its divergence time, but rather that it represents a small population within the range of a larger group, the small population having diverged from an ancestral source earlier than the surrounding one. Furthermore, the ancestral source of the 'relic' group need not be the same as that of the surrounding population. Groves describes how true biological relic populations might exist 'surviving primitive species even in a highly speciose lineage; they are the unchanged stem species, and there is no logic in classifying them separately from the actual stem forms if such are known from the fossil record. Indeed, whether one of the daughter species differs significantly from its parent – i.e. has many, or any, autapomorphic features – might be one way of deducing the sort of speciation event occurred' (Groves, 1989a:29). As was shown in the previous chapter, all modern populations have differentiated from the common ancestor, and the stem population from which others differentiated disappeared as itself became differentiated. Australian aborigines have been often considered to represent a biological relic in the sense described by Groves. However, as it was argued in the previous chapters, this population has undergone some important morphological changes which differentiates it not only from all other modern populations, but from the ancestral forms as well.

Multiple dispersals in the history of modern humans

In collaboration with Robert Foley, the ideas on the temporal effects of geographical dispersals were developed further into an evolutionary framework for the origins of modern human diversity (Lahr & Foley, 1994). This framework is based on a single origin of modern humans, but it explores the concept of multiple dispersals through time as the mechanism giving way to the processes of adaptation, gene flow, drift, founding effect and extinction (discussed above). Such a mechanism of multiple dispersals would account for the different dates of modern human colonisation in different parts of the world and the character of modern morphological differentiation within large population groupings.

Although speciations did not occur in the evolution of modern human populations, the same theoretical principles are applicable to the differentiating processes of subspeciation, or to the appearance of morphs (Groves, 1989a; Lewontin, 1974), as was the case in modern human evolution. Accordingly, the principles of sympatric, parapatric and allopatric differentiation become important for understanding the mechanism of modern human geographical differentiation. Although geographic isolation or allopatry has been much stressed as a necessary precondition for differentiation (Eldredge & Gould, 1972; Mayr, 1942, 1963; Stanley, 1979), sympatric and parapatric events have been strongly defended (Bush 1969; ; Groves, 1989a; Maynard Smith, 1966; White, 1978). Although the outcome of any of these events is a differentiated morph, the mechanisms driving these processes are completely different. Allopatric differentiation results from a founder effect followed by the release of constraints on variation because of lower intraspecific competition, allowing for a period of marked variability and eventual fixation of a morph through genetic drift (Stanley, 1979). By definition, the new morph arises at the peripheries of a species range. On the other hand, *in situ* processes, through stasipatric mechanisms favouring homozygosity within a polymorphic species, or parapatric differentiation through the intensification of a hybrid barrier, a portion of the polymorphic species may result in a markedly differentiated morph (White, 1978). These processes of *in situ* differentiation are a reflection of the centrifugal speciation model of Brown (1957), which suggests that derived forms tend to appear at the centre of a species range, for there is more diversity in the centre than at the peripheries, and when the range breaks, a greater number of genetic combinations are possible at the centre and thus greater genetic change. Subsequent expansion of the new morph will establish it in relation to the peripheral unchanged populations. A similar evolutionary pattern of intra-species diversity was described by Thorne (1981) as the centre-and-edge hypothesis. The morphological results of these stasipatric or parapatric events are very different from those of founding effects. While the former will turn an existing polymorphic variant into a more discrete form, founding effects will usually result in the development of a new morph. In terms of the process of modern human differentiation, these mechanisms would have acted on different situations creating different patterns of variability and distinctiveness at the root of dispersal events.

The process of modern human differentiation starts with a single ancestral population that has suffered a genetic bottleneck and is at the phase of expanding itself demographically. The first dispersals from such ancestral source (x) may not have been geographically extensive, but they are at the core of modern human differentiation. If later dispersals, of a greater geographical extent, originated from groups that had already diversified from the ancestral source at an early stage, then the evolutionary processes acting on these early dispersed small populations (x1) would have determined the features shared by a large number of populations deriving from each of them (Figure 10.1). Temporary allopatric conditions must have been specially important in shaping early modern differentiation. However, even considering that Pleistocene human populations were significantly smaller, allopatric conditions must have been rare. Only in the case of colonisation of geographically marginal habitats by jump dispersals followed by range reduction, could such conditions of isolation have been achieved. These jump dispersals would have occurred at periods of very favourable conditions during which populations expanded rapidly. Such

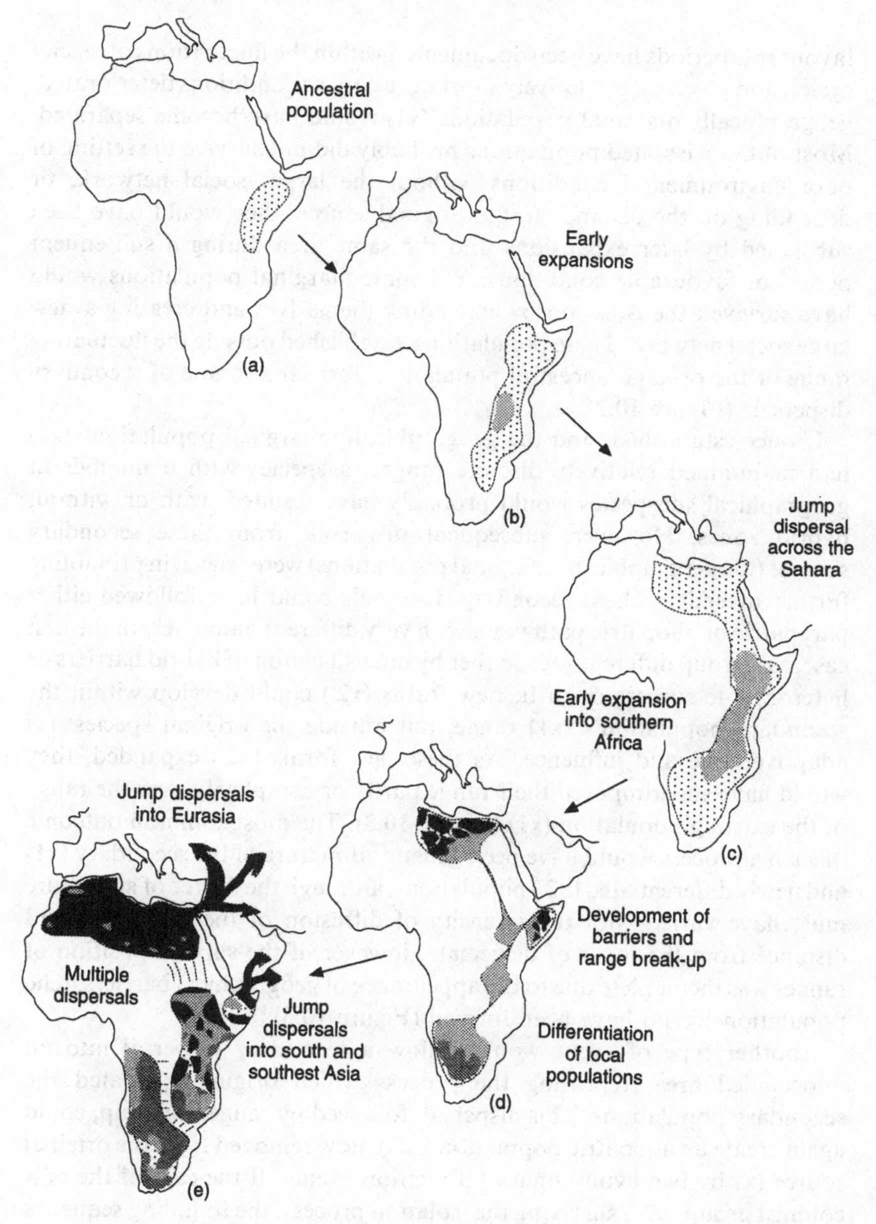

Figure 10.1 Modelled process of initial geographical differentiation of early modern population in Africa involving expansion of ancestral groups with central differentiation; early jump dispersals allowing the colonisation of northern and southern Africa; development of barriers and range break-up, resulting in partial population isolation and establishment of early differentiations; and subsequent population expansions (now from individual, somewhat differentiated populations) and jump dispersals.

favourable periods have been documented within the fluctuations of glacial cycles, and given their relatively short duration, as conditions deteriorated, geographically marginal populations (x1) would have become separated. Most of these isolated populations probably did not survive the setting of poor environmental conditions without the larger social network, or depending on the distance to the original source, they would have been subsumed by later expansions into the same area during a subsequent period of favourable conditions. Yet some marginal populations would have survived the isolation by expanding themselves and creating a new large social network. These populations, established outside the fluctuating range of the original ancestral population, form the source of secondary dispersals (Figure 10.2).

If once established, and the geographically marginal populations (x1) had maintained relatively discrete ranges, a species with a number of geographical subspecies would probably have resulted, with or without hybrid zones. However, subsequent dispersals from these secondary sources (the geographically marginal populations) were crucial in promoting further diversity. These secondary dispersals could have followed either parapatric or allopatric pathways, with very different outcomes. In the first case, as a group differentiated, either by intensification of hybrid barriers or heterozygote success or drift, new forms (x2′) could develop within the secondary population's (x1) range, but outside the original species' (x) adaptive zone and influence. As these new forms (x2′) expanded, they would have superimposed their range partly or completely over the range of the existing population (x1) (Figure 10.3). The most common outcome of such a process would have been genetic admixture of the secondary (x1) and newly differentiated (x2′) populations, although the degree of admixture must have varied with the intensity of diffusion of the new form and distance from the focus of dispersal. However, if the superimposition of ranges was incomplete due to the appearance of geographical barriers, relic populations could have been formed (Figure 10.3).

Another type of event would follow a secondary dispersal into an unoccupied area, repeating the process which originally created the secondary populations. This dispersal, followed by range break-up, could again create an allopatric population (x2″), now removed from the original source (x) by two evolutionary bifurcation events. If the case of the new colonist group (x2″) surviving the isolation process, the founding sequence of effects described before would take place. This process would also create a new form (x2′) from the secondary source (x1), but differently from the case of stasipatric or parapatric differentiation, the founding event would most probably result in a new morphological combination of traits.

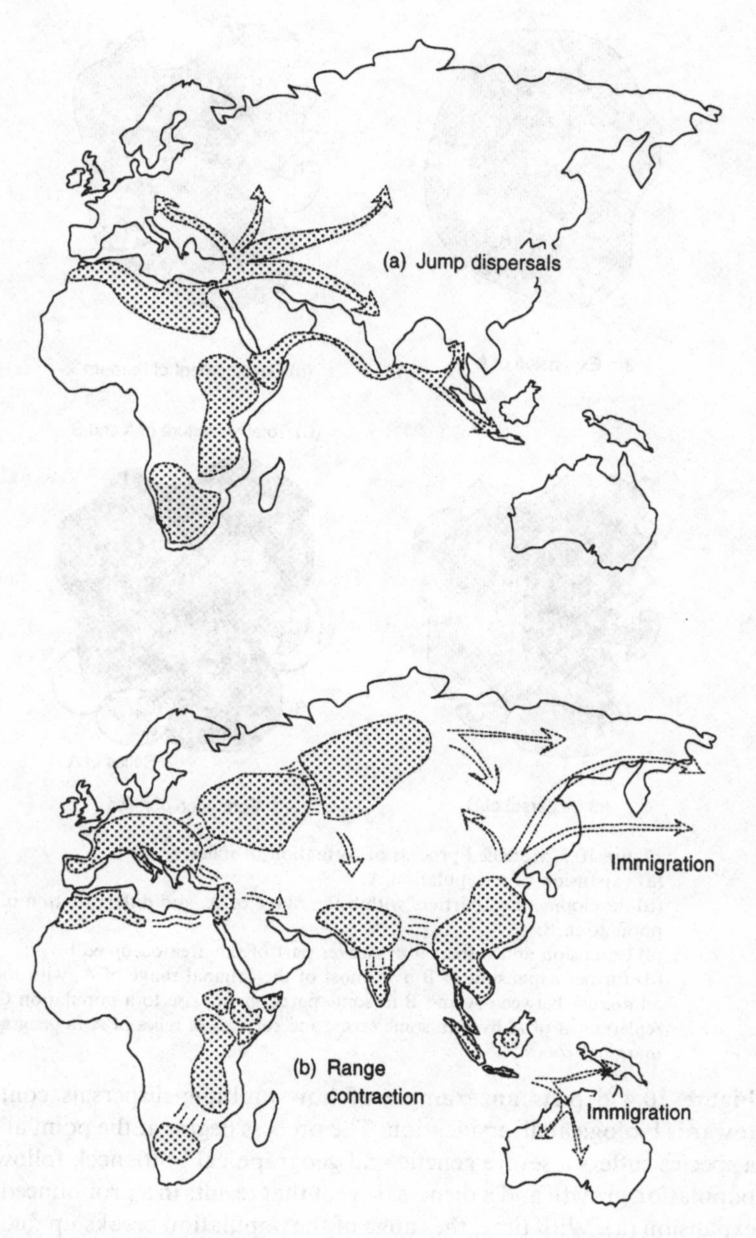

Figure 10.2 Modelled world expansion and secondary dispersals of early regional populations. (a) establishment of secondary sources of dispersal in North Africa and Southeast Asia, (b) dispersals from these secondary sources into Eurasia, Australia and the Americas.

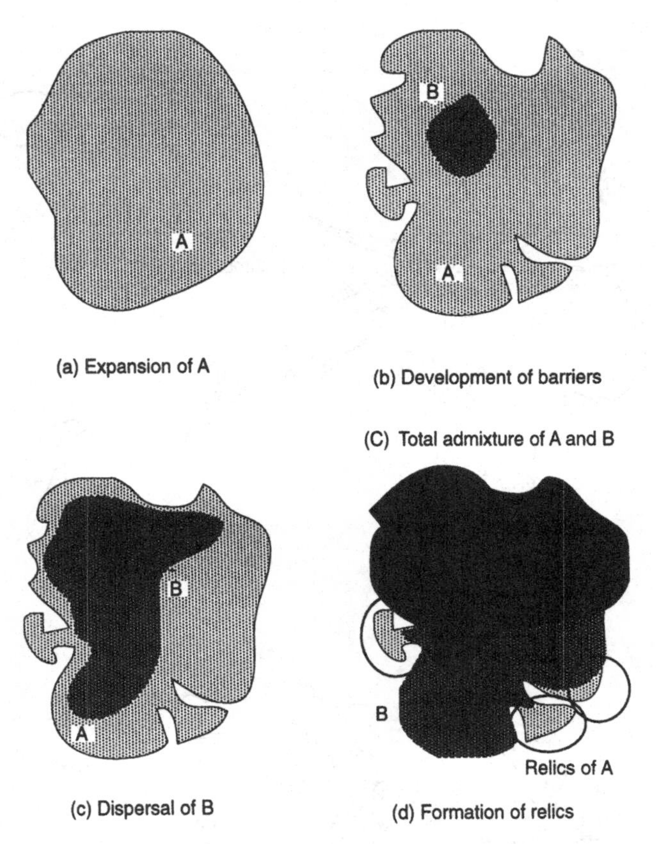

Figure 10.3 Modelled process of formation of relics:
(a) expansion of a population A;
(b) development of barriers within the range of A, and differentiation of a population B;
(c) expansion and dispersal of B over part of the area occupied by A;
(d) further expansion of B over most of the original range of A, with total admixture between A and B in some parts, giving rise to a population C, replacement of A by B in some areas, and survival of relics of A in geographically marginal zones.

Figure 10.4 depicts an example of how multiple dispersals contribute towards biological diversification. The process begins at the point at which a species suffers a severe genetic and geographical bottleneck followed by population growth and a dispersal event that results in a pronounced range expansion (a). With time, the range of the population breaks up due to the development of geographical barriers or to the immigration of peripheral groups, resulting in a number of smaller sub-populations (b). If contact between these groups is not renewed, the regional differentiation process

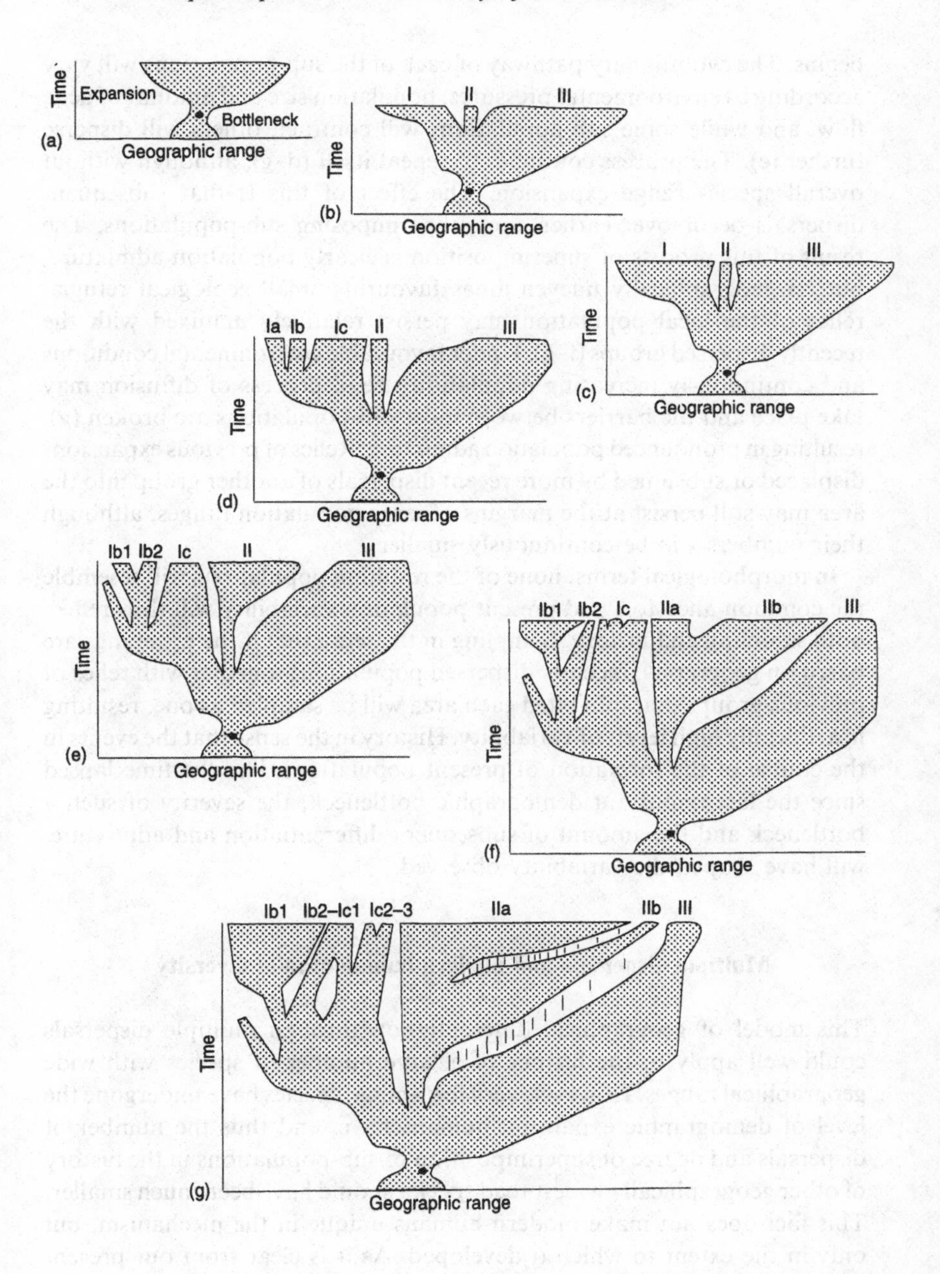

Figure 10.4 Modelled process of geographical differentiation of a single population through time, showing possible evolutionary events after a population bottleneck, involving population expansions, dispersals, extinctions and differentiation.

begins. The evolutionary pathway of each of the sub-populations will vary according to environmental pressures, population size and amount of gene flow, and while some sub-populations will contract, others will disperse further (c). The process continues to repeat itself (d–g), although without overall species range expansion. The effect of this is that subsequent dispersals occur over earlier ones, superimposing sub-populations. The result of this process of superimposition is clearly population admixture, but on geographically uneven areas favouring small ecological refugia, relics of the local population may persist relatively unmixed with the recently dispersed groups (f–g). Under favourable environmental conditions and continuously increasing population size, a process of diffusion may take place and the barriers between most sub-populations are broken (g), resulting in pronounced population admixture. Relics of previous expansions displaced or subsumed by more recent dispersals of another group into the area may still persist at the margins of large population ranges, although their numbers will be continuously smaller.

In morphological terms, none of the resulting populations will resemble the common ancestor, and present population variability will thus reflect both sampling and history. Sampling in the sense that if the groupings are based on geography, recently dispersed populations together with relics of previous groups that inhabited each area will be sampled as one, resulting in artificially high levels of variability. History in the sense that the events in the course of the formation of present populations, like the time lapsed since the last important demographic bottleneck, the severity of such a bottleneck and the amount of subsequent differentiation and admixture, will have shaped the variability observed.

Multiple dispersals and modern human cranial diversity

This model of geographical differentiation through multiple dispersals could well apply to the history of a large number of species with wide geographical ranges. However, very few animal species have undergone the level of demographic expansion humans have, and thus the number of dispersals and degree of superimposition of sub-populations in the history of other geographically widespread species should have been much smaller. This fact does not make modern humans unique in the mechanism, but only in the extent to which it developed. As it is clear from our present occupation of the world, modern humans are unique in their ability to expand and occupy virtually all environments. The selective pressures from these different environments throughout the world, together with different degrees of gene flow between populations, are at the root of the origins of diversity. By applying the concept of multiple temporal dispersals with incomplete spatial superimposition, sampling biases as described above become an important interpretation of the levels of morphological variability within geographical units at any one time, and the biological character of present regional populations may be understood through the historical reconstruction of evolutionary events.

11 *Reconstructing population histories*

In order to apply the model of multiple dispersals to the evolution of modern human diversity it is necessary to reconstruct the main events that took place in the period concerned. Such a reconstruction cannot be complete, for the fossil record only provides patterns, whether morphological, spatial or temporal. Establishing evolutionary events from these patterns depends, of course, on the quality of the fossil record in each region, the quality of the chronological control of existing human remains and the quality of their anatomical description. This is clearly very inconsistent from case to case and region to region. However, the only means of testing whether multiple dispersals acted as a mechanism for the evolutionary process of modern diversification is by applying the model to the global picture. The appearance of modern humans may have been a localised event, but the development of diversity is intimately linked to global expansion. Therefore, any attempt at understanding human diversification has to do so on a cross-regional scale. The following sections are such an attempt, integrating some of the important points of the character and affinities of fossil modern crania in the Upper Pleistocene with the theoretical framework of multiple geographical dispersals, so that a model for the evolution of modern human cranial diversity can be proposed.

Events in modern human evolution

Establishing all the events that took place in modern human history is beyond the scope of this book. The modern human record is very rich and intricately varied. Knowing and understanding these intricacies would be necessary to reconstruct the history of each population as far as the record allows, but again, this is beyond the scope of this book. Others, far more able and knowledgeable than myself have provided us with such reconstructions, most notably W.W. Howells with his early book *Mankind in the Making* (1959) and his recent volume *Getting Here* (1993). However, it remains true that few scholars have attempted to integrate the information on a global scale, and in not doing so, some important patterns in the data, only explained through broader regional comparisons, are lost. In the following sections, the integration of the modern human record with the model of multiple dispersals will be approached through the discussion of specific issues, so that both generalisations and certain specific problems can be addressed.

The African source

The first event in the history of modern human evolution is the appearance of the group itself. This may have been a true speciation event or the evolution of a distinct subspecies within the already existing species *Homo sapiens*. This issue is polemical and unresolved, and does not affect the processes described here. As mentioned before, several researchers have highlighted the fact that whether at the population level of species, subspecies or con-subspecies, the same kind of morphological differences, often involving the same features, characterise the diversification process (Groves, 1989a; Lewontin, 1974). Reproductive isolation is the only criterion for recognising the level of diversification between living species and subspecies, while the amount of morphological overlap differentiates subspecies from con-subspecies populations. Therefore, whether the evolutionary process that originated modern humans resulted in complete reproductive isolation from other hominid populations will not be addressed. As discussed in Chapter 2, most studies of evolutionary genetics, involving both mtDNA and nuclear DNA, find evidence of a single and relatively recent origin of all modern humans (Bowcock *et al.*, 1994; Cann *et al.*, 1987; Cavalli-Sforza *et al.*, 1988; Harpending *et al.*, 1993; Nei & Roychoudhury, 1982; Tishkoff *et al.*, unpub. data; Vigilant *et al.*, 1991; Wainscoat *et al.*, 1986). Furthermore, recent research by Harpending, Rogers and their co-workers (Harpending *et al.*, 1993; Rogers, unpub. data a,b; Rogers & Jorde, 1995; Sherry *et al.*, 1994) found strong evidence that a severe bottleneck occurred at the origins of modern humans, followed by a very early differentiation of groups from which modern populations later expanded. These palaeo-demographic reconstructions are consistent with a founding model of diversification. However, fossil and ecological data suggests that such an event did not occur at the periphery of the 'archaic' *H. sapiens* range as would be expected by the classical principles of a bottleneck resulting from population isolation (Eldredge & Gould, 1972; Stanley, 1979), but rather at the centre of that range – within eastern Africa.

The early modern fossil record is not sufficiently complete to make this assertion on morphological grounds. Some early modern fossils are found in East Africa (LH18, Omo 1 and 2), while others come from southern Africa (Florisbad, KRM, the later Border Cave and Tuinplaas). Kingdon (1993) argues on biogeographical grounds that East Africa was the focal area. Based on patterns of fragmentation and secondary expansions, Kingdon (1993) suggests that there was differentiation at the centre of the range. In terms of number of animal species, diversity and abundance, East Africa is the richest region in the world, a region ecologically unstable and thus promoting differentiation through a spatial and temporal mosaic of ecological barriers. The character of the fauna dispersing towards North Africa in the early Upper Pleistocene shows that eastern Africa was the geographical source of such dispersals (Tchernov, 1992a,b), and therefore consistent with Kingdon's view that the regions or ecosystems in which successful adaptive types first appear remain significant even after the dispersal of forms (Kingdon, 1993). In this case, the founder–flush model proposed by Carson (1975), by which the bottleneck is due to a demographic fluctuation rather than geographical isolation, would be most consistent. This may be an important explanation of how very high levels of polymorphism could have gone through the bottleneck. In any case, either founding model suggests a subsequent period of high variability due to selection release mechanisms that could account for the pronounced morphological variability in the early modern populations. If the genetic evidence for early separation of groups from this ancestral source is correct (Harpending *et al.*, 1993), the process of drift would have fixed not one, but several new morphological variants of the newly evolved population.

Early modern technologies and behaviour

Early modern human remains in Africa are found within Middle Stone Age (MSA) archaeological contexts. The African MSA covers the period from 200–40 ka (Clark *et al.*, 1984; Rightmire, 1978; Singer & Wymer, 1982; Vogel & Beaumont, 1972), and therefore not only the period of appearance of the earliest moderns, but also a large proportion of their African descendants. Although the MSA is technologically comparable to the European Middle Palaeolithic (Allsworth-Jones, 1986, 1993), certain assemblages show indications of the appearance of more complex behaviours. This complexity is reflected in incipient blade production at Howieson's Poort and pre-Aurignacian sites (Clark, 1982; Gowlett, 1987; McBurney, 1967), and the increased use of foreign raw materials, manufacture of some ornaments, specialised resource exploitation and other traits (Beaumont *et al.*, 1978; Brookes & Yellen, 1989). The meaning of these progressive traits in the MSA material culture is greatly disputed (Binford, 1984; Deacon,

1989; Deacon & Shuurman, 1992; Klein, 1992), and in most cases these 'progressive' traditions were temporally limited, and followed by a flake-based Middle Palaeolithic industry.

Many researchers consider the transition towards the Upper Palaeolithic to be the main technological-behavioural revolution associated with modern humans. However, Gamble (1993) contests this view, arguing that the behaviourally important technological change was the appearance of the Middle Palaeolithic-MSA technological level at around 200 ka. The changes associated with the Middle Palaeolithic are reflected in the tool types, the use and re-use of tools, the raw materials employed, the presence of hearths, burials and incipient art. The earliest evidence of art is from the site of Karolta, South Australia, with rock engravings dated to 32 ka, Apollo Cave, Namibia, with coloured images on two stone slabs dated to *c.* 26 ka, and in the Stadel cave, Germany, where a lion-headed figurine dates to *c.* 31 ka (Gamble, 1993). The fact that art was equally expressed within and without Upper Palaeolithic contexts (as is the case of Australia), supports Gamble's view.

Dispersals within Africa

As it was mentioned above, recent genetic studies suggest that after a severe bottleneck modern humans expanded rapidly to establish groups which remained unexpanded and relatively isolated for a long period of time (Harpending *et al.*, 1993; Rogers, unpub. data a,b; Rogers & Jorde, 1995). This rapid population expansion is the expected consequence following a founding effect (Carson, 1975; Eldredge & Gould, 1972), and given the intermittent ecological barriers in Africa, a later range reduction could have easily led to the establishment of a number of relatively isolated populations within the continent as discussed in Chapter 10.

The variation in MSA traditions in Africa may reflect such an early diversification. Masao (1992) has stressed the environmental specialisations of two partly contemporaneous traditions in the late Middle Pleistocene of Tanzania, the more archaic Stillbay in open sites, and the Sangoan and Charaman assemblages, associated with riverbeds and more forested coastal areas. This southeastern African archaeological diversity increases throughout the Upper Pleistocene. In Zimbabwe and Zambia, the Sangoan and Charaman give way to the Bambatan. It is in the Bambatan-related industries of southern Africa that the first widespread evidence of backed-microlith manufacture is observed (Phillipson, 1985). The Charaman industries of the Zaire basin later developed into the Lupemban tradition

(Phillipson, 1985), and in southern Africa, the late Upper Pleistocene increase in population numbers may be reflected in a number of differentiated industries, like Tshitolian, Nachikufan and Kintampo. Therefore, African archaeology discloses a pattern of continuous occupation from the Middle Pleistocene, with increasing cultural diversification and specialisation. It also shows important fluctuations in the intensity of occupation of certain areas (Klein,1992).

The fossil record of modern humans in Africa is unfortunately too incomplete to provide a morphological dimension to early African diversification. After the early modern specimens, there are almost no human remains until very late in the record. The only material that could testify towards this early diversification are the Border Cave (BC) remains, recently dated to 80–70 ka (Grün *et al.*, 1990). The well-preserved BC1 specimen has been intensely studied (Beaumont *et al.*, 1978; Rightmire, 1981, 1984; Tobias, 1972), and it has been considered to show Negro-Khoisan affinities (Beaumont *et al.*, 1978), Khoisan affinities (Rightmire, 1981, 1984) or no affinities towards any recent population (Ambergen & Schaafsma, 1984; Campbell, 1984; De Villiers, 1973; De Villiers & Fatti, 1982; Morris, 1992; van Vark, 1986). Therefore, we can neither establish the degree of morphological diversification between regional populations of modern humans in Africa in the early Upper Pleistocene, nor whether recent African groups are directly descended from any of these populations.

Genetic studies, both mtDNA and Y-chromosome, indicate that the San and Pygmies are genetically very distinct from other populations, suggesting a long-standing period of differentiation for these populations (Lucotte, 1992; Vigilant *et al.*, 1989). However, the earliest fossil evidence of people with clear San/Khoi affinities is late in the record. The late Pleistocene remains from Fish Hoek, Mjates River, Boskop and Cape Flats, generally grouped under the name 'Boskopoid', have been found to show affinities among themselves and possibly to the San and Khoi (Rightmire, 1975a). If the identification of these affinities is correct, these data show that, although their geographical range may have varied in time, these hunter–gatherer groups have been a regional southern African population since the Pleistocene, although since when in the Upper Pleistocene remains uncertain. Of these remains, Boskop and Fish Hoek are associated with MSA (Protsch, 1975), and possibly earlier in time (35 ka – Fish Hoek 4, Protsch, 1975). This date would be consistent with the time of increased archaeological density in southern Africa (Klein, 1992), but much later dates of 12–13 ka have also been suggested (Howells, 1989). In other parts of Africa there is much more variation, disclosing a mosaic of forms, some unrelated to recent groups (Lukenya Hill – Gramly & Rightmire, 1973), others with possible Khoi-San affinities (Neolithic crania associated with the Wilton tradition of Kenya), others with clear Negro traits (Ishango, Congo – Ferembach, 1986c; Howells, 1959; Rightmire, 1975b; Chad, Tamaya Mellet in Niger, and El Guettara in Mali – Chamla, 1968; Asselar, Ibalaghen, Tin Lalou sites – Chamla, 1968), and yet still others suggesting trans-Saharan movements (Wadi Halfa, Jebel Sahaba – Anderson, 1968; Greene & Armelagos, 1972). Furthermore, levels of robusticity and dimorphism also seem to have been extraordinarily variable (Gwisho, Zambia – Phillipson, 1985). Most of these remains are late Pleistocene to early Holocene, and associated with Late Stone Age technologies.

'Caucasoids' in East Africa and African variability

In Kenya, the remains from Gamble's Cave (10–8 ka) and Bromhead's site (12 ka?) have been interpreted as showing 'Caucasoid' features (Tobias, 1972) and possible archaeological affinities with the Mediterranean Caspian industries (Ferembach, 1979). As we have seen in previous chapters, recent sub-Saharan Africans are cranially more gracile than Europeans, and therefore fossil African specimens of greater size and robusticity have been traditionally considered non-African in character. However, Rightmire (1975b, 1981) found that these East African remains, as well as those from the related sites of Willy Kopje, Nakuru and Makalia, cluster with one or other sub-Saharan population in multivariate statistics, and not with either Egyptians or San/Khoi. Similar results were obtained by Bräuer (1978), and Rightmire (1975b) has suggested that these fossils may represent Nilotic peoples, non-Bantu morphologically and linguistically. These findings are very important, for they suggest that not only late Pleistocene to early Holocene remains like Gamble's Cave and Elmenteita should not be interpreted as Caucasoid immigrants, but that the great levels of cranial variation observed today in sub-Saharan Africa were probably even greater in the late Pleistocene.

With the exception of the San-Khoi lineage in southern Africa, other late Pleistocene sub-Saharan African fossils are so variable that no distinct regional or temporal morphological patterns can be identified at that time, with the possible exception of a reduction in cranial breadth. This is consistent with high levels of fragmentation of populations, allowing the development and maintenance of a number of specialised forms in intermittent contact with each other. This is in agreement with the results of Chapter 8 on the degree of variation within recent African crania. Recent sub-Saharan African crania show a mixed morphology that can be interpreted as the combination of derived characters of the vault in terms of small size and large lateral expansion resulting in rounded gracile crania, and primitive features in the face, seen in the wide interorbital areas, retracted upper faces, long palates and prognathic alveolar regions. However, they also exhibit a large number of facial variables with maximum within-group variation, suggesting that, at least in some groups, these facial traits are undergoing differentiation. Although African crania show the largest number of variables with mean values closest to the Australian mean, these relate mostly to the retention of some facial and cranial plesiomorphies in both groups, rather than a shared pattern of differentiation. The homogeneity in small dental and temporal muscle size is consistent with the shared pattern of gracility in an otherwise very variable group. A case could be made that morphological variability in sub-Saharan African, during the late Pleistocene to early Holocene, Late Stone Age populations was at its maximum, and included levels of robusticity largely lost in present populations.

Multiple dispersals in Africa?

The African continent has been continuously inhabited by modern people for over 100,000 years. The genetic evidence suggests an early diversification of groups, and establishment of semi-isolated population foci. However, did these early African populations later disperse, overlapping their ranges on each other with great genetic admixture, or did they maintain relatively discrete geographical ranges through time? Morphological data indicates much diversification, but in recent periods. It is again, the more complete archaeological record which points to important fluctuations in both ranges and population numbers (Klein, 1992; Phillipson, 1985). These fluctuations probably reflect the expansions and contractions of populations, which may have given rise to a number of relics reflected by the mosaic morphological pattern in the late Pleistocene of eastern Africa. However, superimposed on these Pleistocene events is the recent expansion of agriculturalists. It is known that Bantu farmers dispersed throughout large parts of sub-Saharan Africa in the past few thousand years (Phillipson, 1985), carrying with them the Niger-Kordofanian languages (Greenberg, 1963) and possibly the Chifumbaze archaeological complex (Phillipson, 1985). This dispersal was accompanied by gene flow, which would have had a homogenising effect on sub-Saharan African diversity (a possible reason for the greater Pleistocene–Holocene levels of variability). This dispersal from a northwestern focus was superimposed on earlier non-agricultural populations, of which relics survive in East Africa (the Hadza, a hunter–gatherer group speaking a Khoisan language), West Africa (the varied Pygmy populations, which are believed to have 'borrowed' the farmers' languages), and southern Africa (the San and Khoi, also speaking Khoisan languages). In northeastern Africa, the Nilotic groups form another major population cluster.

Dispersals out of Africa

One of the most significant aspects about the findings of Harpending and co-workers is the suggestion that the dispersals, from which the main world populations originated, occurred after the ancestral source had diversified into a number of semi-isolated groups, so that different world populations derive from already differentiated sources (Harpending *et al.*, 1993). In other words, the differentiation process between main world population lineages was initiated during a period of relatively small population sizes and before the main geographical and demographical expansion of these groups. It is possible that the early differentiated groups postulated by Harpending and co-workers represent sub-divisions within Africa, but the relatively long-standing separation between these groups indicated in the genetic data would suggest that the first jump dispersals out of Africa were part of this early process of differentiation from the ancestral source. As discussed below, the extent of archaeological and fossil remains outside Africa during the first half of the Upper Pleistocene show that although early movements out of Africa established human groups outside sub-Saharan Africa, they were not followed by marked population expansion. If this was the case, the contraction of geographical ranges due to environmental conditions, and thus separation from the ancestral sub-Saharan source, could account for the levels of relative isolation of these population lineages as postulated by Harpending *et al.* (1993).

A northern dispersal route

The biogeographical position of northern Africa gives it a pivotal role in the dispersal of modern humans out of Africa. Most of the time this region is part of a Mediterranean biogeographic zone, but during specific periods within glacial cycles it becomes part of the range of sub-Saharan African animals. Therefore, this region has been intermittently linked to sub-Saharan Africa at specific moments during which rainfall is such as to allow movements across the Sahara. These fluctuations in the ecological conditions of the Sahara are best known for the last 20 ka (last glacial maximum, deglaciation and settling of interglacial conditions), and can be used as a model for earlier periods.

At present, about 10% of the land area between 30°N and 30°S is covered by sand deserts. At 15 ka it may have been as much as 50% (Sarntheim, 1978). In North Africa, a belt of fossil dunes extends to 10–12°N in Senegal, Mali, Niger, Nigeria, Chad and Sudan, over an east–west distance of 5000 km (Williams, 1975). These dunes blocked the Senegal river in Mali (Mitchell, 1973), covered palaeolake basins, altered the course of the Niger river, and savanna expansion in Nigeria produced a gap between the rainforest of Upper Guinea and Cameroons (Goudie, 1977). In East Africa, west of the White Nile, fossil dunes extend to 10°N, covering the landscape up to the Jebel Marra, and merge northwards with mobile dunes at 16°N. At some stage, dunes crossed the Nile, which probably dried up locally at the time of their formation (Goudie, 1977). Evidence of greater humidity than at present also exists, related to the high precipitation levels that accompany the deglaciation process. According to Hamilton (1978), at 8 ka lowland forest vegetation covered an area 15 times larger than at 20 ka reducing afterwards to present conditions (Figure 11.1). The expansion of lowland forests coincides with a northward shift of the semi-arid grass-dominated savanna zone (Hooghiemstra & Agwu, 1988), with the expansion of herbivores into now desertic regions. Of major importance during periods of expanded savanna and tropical forests was the distribution of riverine and lacustrine environments, providing water connections across the entire sub-Saharan region, from the Nile to Senegal. There is much biogeographical evidence of these past river connections in northern Africa: the Eurasian green frog (*Rana ridibunda*) has been found in the streams of the Ahaggar Mts., a Nile crocodile (*Crocodilus niloticus*) was found in a pool in Tassili-Ajjer Mts. (Beadle, 1974; Seurat, 1934), tropical fish have been found in permanent, but isolated, water holes from Biskra to Tibesti (Goudie, 1977), and there are some species of fish which are common to all major basins of the Sudan belt, Senegal, Gambia, Volta, Niger, Chad and Nile, even though there are now more than 1600 km of desert between Lake Chad and the Nile (Bradley, 1985). Furthermore, Holocene pollen records in the eastern Sahara show that during the period 9.5 to 4.5 ka savanna and desert grasslands occupied regions that are plantless hyper-arid deserts today (Ritchie & Haines, 1987).

By analogy, these palaeoclimatic reconstructions of the last 20,000 years allow to suggest that at the end of stage 6 (160–125 ka) North and sub-Saharan Africa must have been completely isolated from each other; that during the period of deglaciation, *c.* 125 ka, savannas expanded into the Sahara, allowing faunal dispersals; that these phases of savanna expansion and contraction were also affected by smaller events (like the Younger Dryas of the late Pleistocene), so that during the sub-stages 5e to

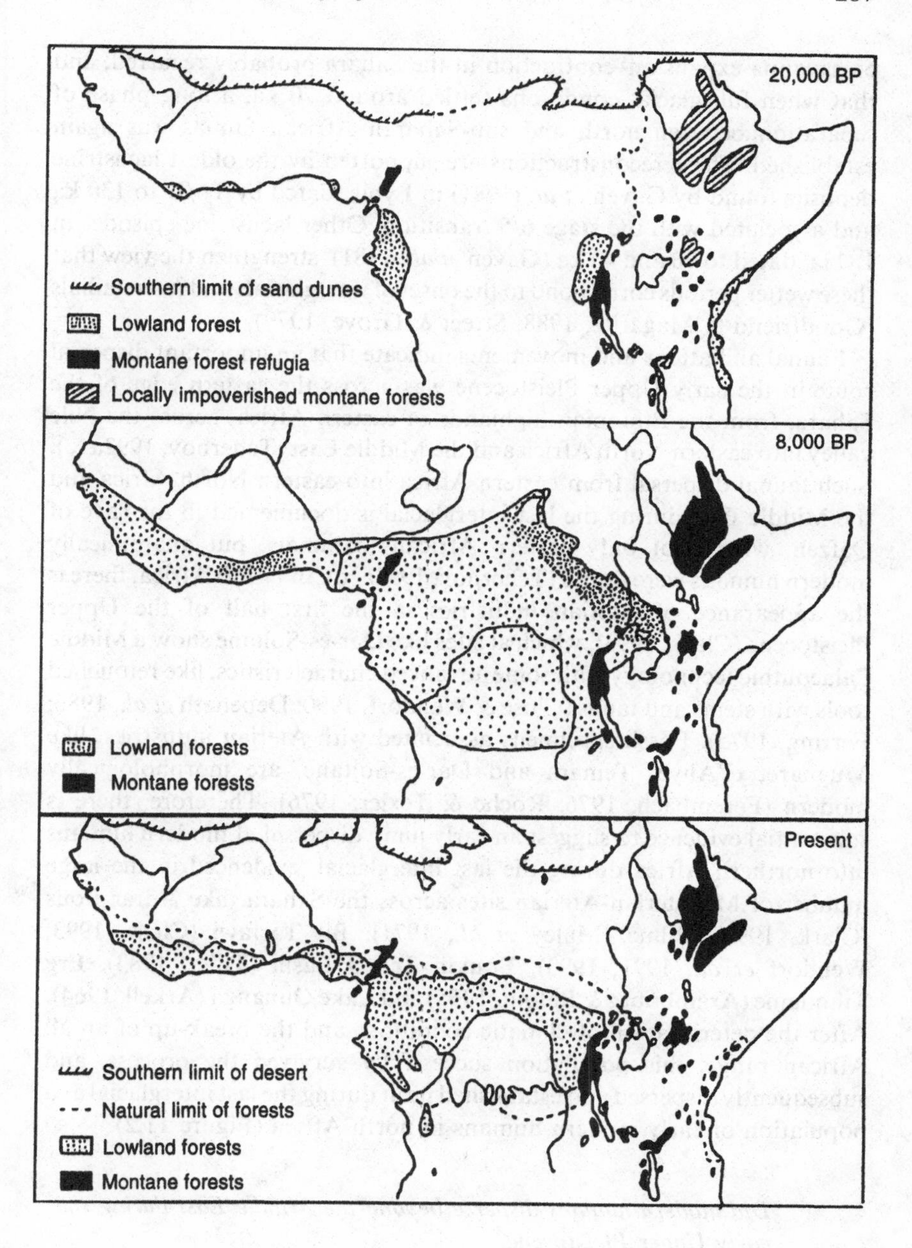

Figure 11.1 Expansion and contraction of tropical forests in sub-Saharan Africa between 20,000 yrs ago and the present. Reconstruction based on biogeographical, palynological and geomorphological data (redrawn from Hamilton, 1978).

5a savanna expansion–contraction in the Sahara probably recurred; and that when full glacial conditions settled around 70 ka, a long phase of separation between north and sub-Saharan African faunas was again established. These reconstructions are supported by the oldest lacustrine deposits found by Gaven *et al.* (1981) in Lybia, dated by Th/U to 130 ka and associated with the stage 6/5 transition. Other lacustrine episodes in Lybia, dated to 90 and 40 ka (Gaven *et al.*, 1981), strengthen the view that these wetter periods correspond to the onset of interglacials and interstadials (Goodfriend & Magaritz, 1988; Street & Grove, 1979).

Faunal affiliations and movements indicate that an important dispersal route in the early Upper Pleistocene was across the eastern edge of the Sahara, from the Ethiopian highlands of eastern Africa, across the Nile valley into eastern North Africa and the Middle East (Tchernov, 1992a–c). Such faunal dispersal from eastern Africa into eastern North Africa and the Middle East during the last interglacial is documented in the cave of Qafzeh, where not only eastern African mammals, but anatomically modern humans were found (Tchernov, 1992a–c). In North Africa, there is the appearance of Aterian industries in the first half of the Upper Pleistocene (Clark, 1993), which at sites like Dar-es-Soltane show a Middle Palaeolithic technology with some advanced characteristics, like retouched tools with stems and tangs (Close & Wendorf, 1990; Debenath *et al.*, 1986; Ferring, 1975). Fossil specimens associated with Aterian industries, like Mugharet el'Aliya, Temara and Dar-es-Soltane, are morphologically modern (Ferembach, 1976; Roche & Texier, 1976). Therefore, there is substantial evidence to suggest an early jump dispersal of modern humans into northern Africa during the last interglacial, evidenced in the large number of Mousterian-Aterian sites across the Sahara, like Adrar Bous (Clark, 1993), Bilma (Maley *et al.*, 1971), Bir Tarfawi (Close, 1993; Wendorf *et al.*, 1991, 1993), Bouku, Tchad Basin (Tillet, 1983), Erg Tihodaine (Arambourg & Balout, 1955) and Lake Ounanga (Arkell, 1964). After the deterioration of climatic conditions and the break-up of an all African range, this population successfully survived the process, and subsequently dispersed and established itself during the last interglacial as a population of early modern humans in north Africa (Figure 11.2).

Did modern humans disperse beyond the Middle East during the early Upper Pleistocene?

Even though northern Africa and the Middle East were at times an extension of sub-Saharan Africa, these periods were very short. Furthermore, Tchernov has called attention to the fact that the Middle East. an area

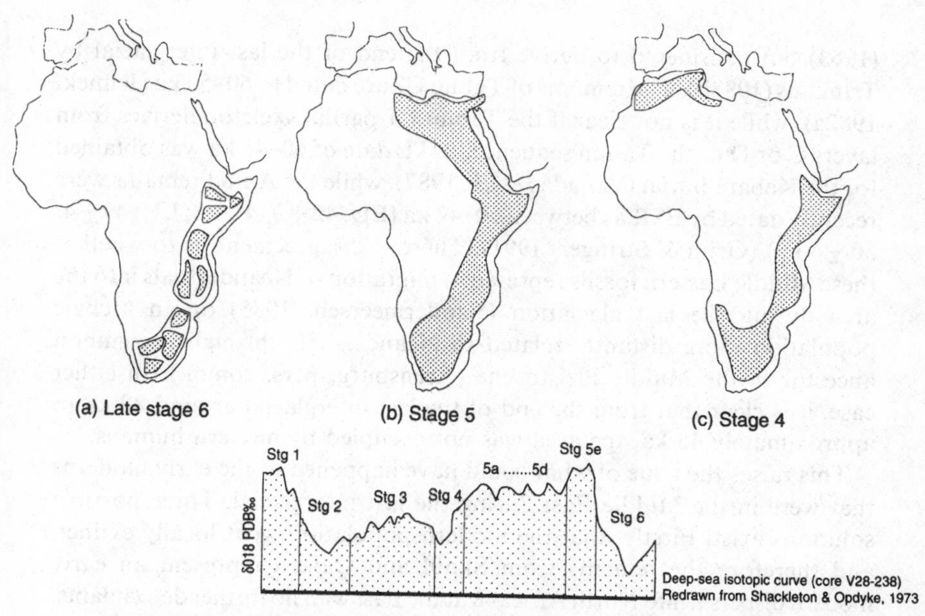

Figure 11.2 Modelled variations in the ancestral population geographical ranges in Africa, north and south of the Sahara during isotopic stages 6 (late glacial), 5 (interglacial) and 4 (onset of last glacial).

which is usually called a 'corridor', actually behaves in biogeographical terms like a *cul-de-sac* (Tchernov, 1992a). During pluvial periods due to fast deglaciation at the onset of interglacials, the Middle East was, faunistically, the northern fringe of Africa. Soon after the end of interglacial conditions, the Middle East was the southern fringe of Eurasia. This latter event is recorded in the movement of palaearctic faunal elements into the area at the beginning of stage 4 (Tchernov, 1992a–c).

These faunal reconstructions are fully consistent with the hominid fossil record. As was discussed in Chapter 2, the Middle East has a very complex and seemingly continuous history of occupation during the early Upper Pleistocene. The earliest modern humans in the area are the remains from Skhūl and Qafzeh, dated to between 100 and 80 ka (see Chapter 2). These fossils exhibit a large and robust cranial morphology, with a relatively high degree of variability (see Chapter 9). This early modern group is followed by an archaic population of clear Neanderthal affinities, composed by the fossils of Kebara, Tabun, Amud and Shanidar in Israel and Syria, and Kiik-Koba, Teshik-Tash and Zaskalnaya VI in Central Asia. These specimens are variously dated to the last glaciation, but the quality of the dating varies markedly. The earliest material could be a series from Shanidar (2, 4, 6, 7, 8 and 9), dated to approximately 60 ka by Solecki

(1963) but considered to derive from the end of the last interglacial by Trinkaus (1983b). The remains of Tabun C2 are dated to 60–50 ka (Jelinek, 1982a), while it is not clear if the Tabun C1 partial skeleton derives from layers C or D of the Tabun sequence. A TL date of 60–48 ka was obtained for the Kebara burial (Valladas *et al.*, 1987), while the Amud remains were recently dated by ESR as between 50–42 ka (EU: $42 \pm 3/41 \pm 3$; LU: $49 \pm 4/50 \pm 4$ ka) (Grün & Stringer, 1991). There is disagreement as to whether these Middle Eastern fossils represent a migration of Neanderthals into the area during the last glaciation (Vandermeersch, 1985) or an archaic population more distantly related to Neanderthals through a common ancestor in the Middle Pleistocene (Arensburg, pers. comm.). In either case, it is clear that from the end of the last interglacial around 80 ka to approximately 45 ka, the area was not occupied by modern humans.

This raises the issue of what could have happened to the early moderns that were in the Middle East during the last interglacial. Three possible solutions exist. Firstly, that the modern population went locally extinct, and therefore the remains from Skhūl and Qafzeh represent an early modern dispersal into North Africa–Middle East with no further descendants. Secondly, that under pressure (either environmental or competitive), the population's geographic range decreased, retreating from the Middle East back into North Africa. And thirdly, that early moderns dispersed further into southern Asia, and because of a southern dispersal of European Neanderthals into the Middle East, were cut off from other modern populations in Africa. There is no evidence to support further expansions of modern people into Asia at this time, and the archaeological and morphological evidence of occupation of northern Africa by modern people throughout the last glaciation would suggest that the second hypothesis raised above is the probable answer to the problem. In this case, Upper Pleistocene North African populations, and all the populations later associated with the Upper Palaeolithic dispersals, could be descendants from the Skhūl and Qafzeh lineage, although not of the Skhūl and Qafzeh populations themselves.

A southeastern dispersal route

As we have seen, the northern dispersal route was open during very specific periods of time, after which the trans-Saharan corridor was closed by the onset of arid conditions. After an early dispersal during the last interglacial, no further movements of people into Eurasia through the Middle East took place between 80 and 45 ka, a period when the Middle East was inhabited by an archaic population. Three lines of evidence suggest that a different route out of Africa must have existed. Firstly, the genetic data indicate that Asian and European populations expanded from already differentiated groups (Harpending *et al.*, 1993), which we have hypothesised were the first jump dispersals out of Africa. Secondly, the dates of colonisation of Australia fall during the time-span in which the northern route was closed. And lastly, the southeastern Asian–Australian populations have been considered to form a morphological complex relatively distinct from Eurasian groups (Brace *et al.*, 1989, 1990; K. Hanihara, 1985; Howells, 1984, 1986, 1989; Pietrusewsky, 1984, 1990b; Pietrusewsky *et al.*, 1992).

It is possible that a second route of dispersal out of Africa existed across the strait of Bab el Mandeb, from eastern Africa to southern Arabia. This southern route is also a faunal corridor (Kingdon, 1993), mainly constrained by sea-level fluctuations and climate. The strait of Bab el Mandeb apparently never dried during eustatic sea-level changes, but its width fluctuated markedly. Palaeoceanographic data indicate that both the Red Sea and the Persian Gulf became very shallow, with internal circulation of water, during the last glacial maximum (Reiss *et al.*, 1980). Interestingly, the moments that climatic constraints were acting on faunal movements from the Horn of Africa to Arabia were different from those acting on trans-Saharan dispersals. Palaeoclimatic reconstructions in Somalia show lacustrine episodes at glacial maximum (Karin, dated to 17.7 ± 300 BP, Gabriel *et al.*, 1993), while the rest of central and eastern Africa was characterised by cool and very dry conditions (Bonnefille & Hamilton, 1986; Flohn & Nicholson, 1980; Gasse *et al.*, 1980 Hurni, 1981; Rognon, 1976; van Zinderen Bakker, 1982; Williams, 1985), highlighting the different localised climatic regime. The lowering of sea-level would have occurred with the onset of glaciation, at the time when arid conditions settled across the Sahara. But as it was the case with the northern route, this southern corridor would not have remained open for a long time, for although there are Middle Stone Age (MSA) assemblages in India with affinities towards the East African MSA (Allchin *et al.*, 1978), the Ethiopian and eastern African Late Stone Age traditions did not reach southern Arabia. It is possible that the progressive desiccation of Arabia during the last glaciation prevented further occupation of this area.

Australian evidence of an early southern dispersal

Although it is obvious that any dispersal into Australia had to go through Southeast Asia, the archaeological record of tropical southern Asia in general is not well documented. There is also the possibility that if these early populations exploited mainly the then extended coastal margins and these margins are now submerged, that their vestiges will never be found. The sea-level changes in Southeast Asia were more pronounced than anywhere else in the world as a consequence of two main factors relating to the region's continental shelves: first, the shallow waters that characterise the sea connections between the various islands of Indonesia resulted, under low sea-level stands, in the emergence of an extensive landmass – the Sunda subcontinent (Figure 11.3). This in turn, influenced the area's vegetation, maintaining a tropical character even during glacial conditions (Pope, 1985), differing from most other intertropical areas of the world that showed maximum aridity and forest contraction during glaciations. The topography and biodistribution of Southeast Asia is and was influenced by north–south running rivers and intermontane valleys (Pope, 1988). Even during times of exposure of the Sunda shelf, faunal barriers continued to exist, since open dweller forms like camelids, equids and giraffids, known from other regions of Eurasia at the time, never reached Java (Pope, 1984, 1985). The main obstacle to these forms probably was the maintenance of tropical habitats, composed of swamp and forest environments (Bellwood, 1985; Pope, 1983; Tokunaga *et al.*, 1985), although with probably increased seasonality (Whitmore, 1975; Whyte, 1984). The second reason why the sea-level changes in Southeast Asia were so pronounced is because of the extension of the continental shelves, resulting in a sea-level drop in the order of -150m to -175m (Chappell & Shackleton, 1986; Veeh & Veevers, 1970). Furthermore, the post-glacial relative rise in sea-level in the southwestern Pacific may have been 25% greater than in the North Atlantic due to elastic deformation caused by isostatic effects of off-loading of ice (Chappell, 1976). The Torres Strait land bridge, connecting northeastern Australia and New Guinea as the Sahul landmass, existed from 60 to 8 ka (Chappell, 1976). During the periods of maximum lowering of sea-level, the distance between Timor and the Australian coast was ~ 100 km, while the other main possible route of crossing, from Sulawesi through Sula Is. and Ceram to New Guinea involved not more that 50 km of sea crossing at a time (Kirk, 1981).

Australian prehistory has been intensely researched (see Smith *et al.*, 1993, for an up-dated review), and it indicates that modern people were inhabiting the eastern fringes of the Old World at the time when Neanderthals were inhabiting Europe. Current archaeological evidence in Australia indicates that the continent was first occupied earlier than other regions outside Africa except the Middle East. The Australian early dates are from the Kakadu region of western Arnhem land, from the sites of Nauwalabila I, Deaf Adder Gorge, dated by OSL to 55 ± 11 ka, and Malakunanja, dated to 61 ± 10ka (Roberts *et al.*, 1990; Roberts &

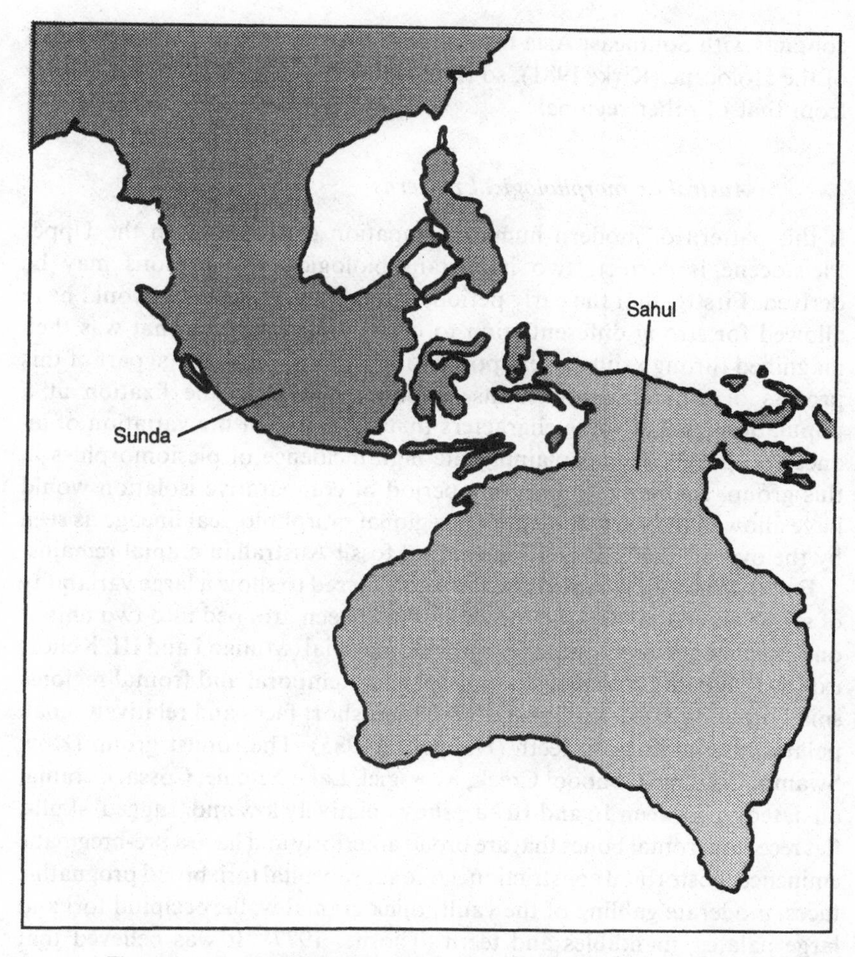

Figure 11.3 Exposed Sunda (mainland and island southeast Asia, extending along the eastern Asian coast to Japan) and Sahul (Australia, New Guinea and Tasmania) land-masses at low sea-level stands during periods of last glacial maxima.

Spooner, 1992). Other early dates from archaeological sites in Australia are from the Cranebrook Terrace, Blue Mountains (47–43 ka, Smith & Sharp, 1993), Upper Swan River, Southwestern Australia (^{14}C 39–37 ka, Pearce & Barbetti, 1981), Mammoth Cave, west coast (>37 ka, Smith & Sharp, 1993), and Devil's Lair (^{14}C 37–30 ka, Jones, 1992). Marked population expansion as reflected by a number of archaeological sites, seems to have occurred around 40 ka. This expansion was relatively fast, and by 30 ka all major ecological zones had been occupied (Smith & Sharp, 1993). Important contacts with Southeast Asia may not have occurred until the second half of the Holocene (Kirk, 1981), so the continent had a very distinct history from that of other regions.

Australian morphological patterns

If this pattern of modern human occupation of Australia in the Upper Pleistocene is correct, two important biological implications may be derived. Firstly, that the early period of small population size would have allowed for strong differentiation to occur, differentiation that was then magnified through subsequent population expansions. A large part of this process of differentiation seems to have resulted in the fixation at a population level of some characters that were part of the variation of an ancestral population, explaining the high incidence of plesiomorphies in this group. Secondly, that a long period of comparative isolation would have allowed the development of a regional morphological lineage as seen by the overall similarities of recent and fossil Australian cranial remains.

Fossil Australian crania have been considered to show a large variability of shape and form, and the specimens have been grouped into two units – one gracile, the other robust. The gracile material (Mungo I and III, Keilor) exhibits high and rounded skulls, expanded temporal and frontal regions, small brow ridge development, thin bones, short faces and relatively small palates, mandibles and teeth (Habgood, 1985). The robust group (Kow Swamp, Cohuna, Coobool Creek, Mossgiel, Lake Nitchie, Cossack crania, all dated to between 16 and 10 ka), show relatively low and 'rugged' skulls, flat receding frontal bones that are broad anteriorly and have a pre-bregmatic eminence, postorbital constriction, large supraorbital tori, broad prognathic faces, moderate gabling of the vault, thick cranial walls, occipital tori and large palates, mandibles and teeth (Thorne, 1977). It was believed that these two groups represented early and late morphologies (the gracile earlier than the robust), but the recent finding of an early robust specimen (WLH50) and the likely late Pleistocene date of the Keilor gracile specimen (Brown, 1990) discredit this view. Furthermore, Brothwell (1975), Brown (1981) and Kennedy (1984) have argued that some of the robust Australian specimens are artificially deformed (like Kow Swamp 5, Cohuna and some of the Coobool Creek material), creating the high degree of variability and some of the 'primitive' features like flat receding frontals, pre-bregmatic eminences and low position of biparietal breadth. According to the results of Anton (1989) on the anatomical effects of artificial deformation, the practice may have also affected some of the facial features.

Although there are spatial and temporal variations, Australian crania are very distinctive as a regional group. Recent Australians appear as relatively homogeneous in comparison to other regional samples, except for the high levels of within-group variance observed in tooth size and temporal muscle dimensions, suggesting that this population, or groups within it, are undergoing reduction of the masticatory apparatus. The number of variables with minimum variance that fall outside the range of variation of the other regional groups results in the distinctiveness of this population. As was discussed in Chapter 9, recent Australians have relatively small cranial size, associated with a very large dentition and very pronounced superstructures. In this way, Australians are differentiated from all other modern humans by expressing a unique relationship between cranial shape and the size-robusticity complex of co-variance.

Fluctuations in the occupation of Australia

It may be hypothesised that the colonisation of Australia was the result of a founding effect, itself a consequence of a jump dispersal from eastern Africa into Southeast Asia. The subsequent period of population expansion and release of pressure on the constraints acting on variation would account for the high levels of polymorphism among early Australian populations. Environmental conditions deteriorated between 22 and 12 ka (Smith, 1989), resulting in the break-up of the continental range, to create a number of ecological refugia (Veth, 1993), which would have subjected the refuged groups to the genetic processes of drift and founding acting on small populations (Figure 11.4). Increased pressure on resource exploitation at this time resulted in either wider ranges within the arid refugia, or reduced ranges in locally rich pockets (Hiscock, 1981; O'Connor *et al.*, 1993). Lakes in New South Wales (Willandra Lakes, Lake Mungo) rose to high levels around 45 ka, and remained high until about 26 ka, when the sequence at Lake Mungo records the first signs of progressive aridity (Bowler *et al.*, 1976). By 14 ka, the system was dry, with dune formation starting at 16 ka (Bowler, 1976). Wetter conditions seem to have persisted slightly longer in the temperate margin of Australia (Rognon & Williams, 1977). This period of increased aridity and localised ranges would have been crucial for setting new evolutionary trends, among which was possibly the reduction of cranial size with intensification of robusticity discussed in Chapter 9. Expansion and re-colonisation of areas took place between 15 and 7 ka (Horton, 1986; Lampert & Hughes, 1988; Veth, 1993), a time that also coincides with the appearance of the very robust Australian fossils in the record (Lake Tandou: 15 ± 160 BP – Freedman & Lofgren, 1983; Kow Swamp: 14 ± 9.5 BP, Cohuna: 14 ± 9.5 BP – Thorne & Wolpoff, 1981). This

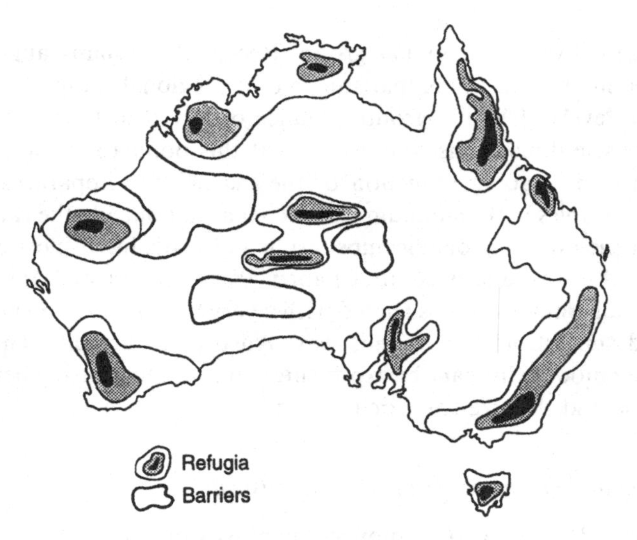

Figure 11.4 Formation of refugia and barriers in Australia during the last glacial maximum. (Redrawn from Veth, 1993.)

reconstruction of the history of occupation of Australia is very different from the traditionally fast or slow models of Birdsell (1977), Lourandos (1987) and Beaton (1983), resulting in a more complex process (O'Connor *et al.*, 1993). In evolutionary terms, these fluctuations in demography (Figure 11.5) are of major importance for interpreting the morphological patterns of diversity.

Close Australian affinities

The variability observed in Australian crania has led to various different phylogenetic hypotheses. Birdsell (1967, 1977, 1993) postulated three discrete waves of migrations into Australia, the first of oceanic Negritos (of undetermined source), the second of Murrayans (with an Ainu link), and the third of Carpentarians (from India). Freedman & Lofgren (1979a,b) and Thorne (1977, 1980a,b) argued for the existence of two sources for the Australians, one gracile from East Asia and one robust from Indonesia (archaic), followed by admixture to produce Holocene Aboriginal patterns. In contrast, Abbie (1966, 1968, 1975) suggested that Australians derive from a single undetermined source and that all observed variation is the result of local adaptation and drift. He considered the Tasmanians to represent a separate Melanesoid group. A similar conclusion was reached by Macintosh & Larnach (1976) and Habgood (1985, 1987), arguing that the observed prehistoric variation, originating from Javanese archaic

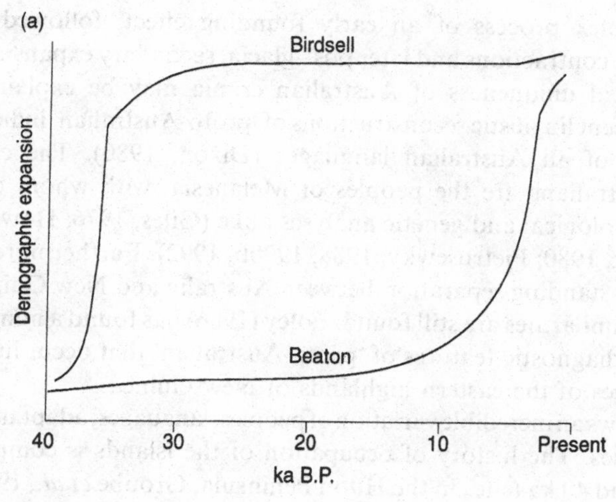

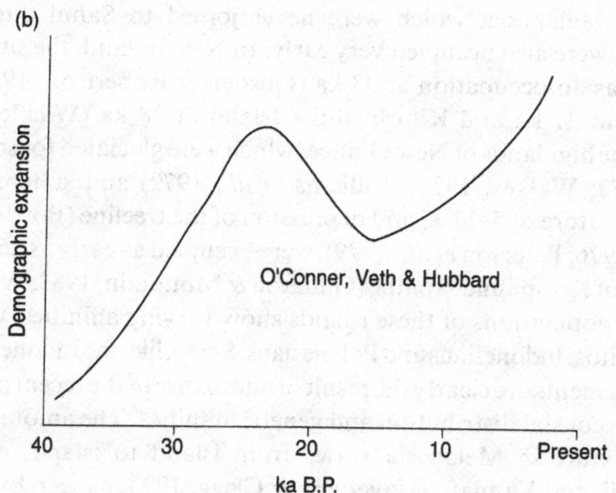

Figure 11.5 Different models of population expansion in Australia, showing (a) fast and slow models of population growth, and (b) important demographic fluctuations proposed by O'Connor *et al.* (1993). (Redrawn from O'Connor *et al.*, 1993.)

sources, can be subsumed within the modern range of features. Pietrusewsky (1990b) strongly argues that Australian crania (fossil and recent) are comparatively homogeneous, with a deep ancestry, and that the temporo-spatial patterns of variability may be explained through variations in the intensity of gene flow.

I have argued that Australians have a common origin with other modern humans in late Middle to early Upper Pleistocene Africans, and that through a complex process of an early founding effect, followed by expansion, range contractions and later post-glacial secondary expansions, the morphological uniqueness of Australian crania may be explained. Furthermore, recent linguistic reconstructions of 'proto-Australian' indicate a genetic unity of all Australian languages (Dixon, 1980). The clear relatives of Australians are the peoples of Melanesia, with whom they cluster in morphological and genetic analyses alike (Giles, 1976; Howells, 1973, 1989; Kirk, 1980; Pietrusewky, 1988, 1990b, 1992). Furthermore, in spite of the long-standing separation between Australia and New Guinea, some linguistic similarities are still found. Foley (1986) has found a number of conservative diagnostic features of 'proto-Australian' that occur in the Papuan languages of the eastern highlands of New Guinea.

Melanesia shows an incredible variation of peoples, languages, adaptations and morphologies. The history of occupation of the islands is complex, starting at around 40 ka (sites in the Huon Peninsula, Groube *et al.*, 1986). Other islands of Melanesia, which were never joined to Sahul during sea-level changes, were also occupied very early. In New Ireland, the site of Matenkupkum has an occupation at 33 ka (Gosden & Robertson, 1991), Buang Merabak at 31 ka and Kilu in Buka Island at 28 ka (Wickler & Spriggs, 1988). The highlands of New Guinea, which were glaciated for some time (Flenly, 1972; Walker, 1973; Williams *et al.*, 1972) and suffered a decrease of temperature of 5–11°C and depression of the treeline (Bowler *et al.*, 1976; Hope, 1976; Peterson *et al.*, 1979), were occupied as early as 26 ka, shown by the site of Kosipe and Nombe (Gillieson & Mountain, 1983; White *et al.*, 1970). The populations of these islands show varying affinities, with Australians, Negritos, Indonesians and Polynesians. Some, like the Indonesian and Polynesian elements are clearly the result of admixture in the recent past, as indicated by its coastal distribution and genetic affinities. The amount of Polynesian admixture in Melanesia varies from island to island, most pronounced in Fiji and Vanuatu (Hagelberg & Clegg, 1993) as seen by the distribution of the 9bp deletion, an Asian, Polynesian and Amerindian common trait. The 9bp deletion is present at low frequencies in coastal New Guinean groups, but completely absent in New Guinean highlanders, Australians and the inhabitants of Bougainville island (Friedlander & Merriwether, 1995). In certain islands, there are pygmy Negritos that seem to have had little admixture with other groups (like in New Guinea), while in New Britain and New Caledonian the morphological affinities towards Australians are strongest (Howells, 1959). There is genetic evidence of relatively small amounts of gene flow between linguistic groups (Boyce *et al.*, 1978) and the great within and between group mtDNA diversity in New Guinea suggest deep ancestral roots (Stoneking *et al.*, 1990).

Tasmania was part of the Sahul landmass during the period between 37 and 29 ka (Bloom, 1988), when archaeological evidence shows it was colonised (Cosgrove, 1989; Jennings, 1971; Jones, 1968). The earliest evidence of occupation comes from the southwestern area, intensely studied by the southern Forest Archaeological Project (SFAP). A number of sites have been discovered that cover the time span between 35 and 12 ka (Nunamira: 30–12 ka; Acheron: 30–13 ka; Bone cave: 29–14 ka; Kutikina cave: 20–12 ka; Mackintosh cave: 17–15 ka; Warreen: 35–16 ka), after which time the area seems to have been completely abandoned (Allen, 1989; Cosgrove, 1989; Kiernan *et al.*, 1983; Jones, 1992; McNiven *et al.*, 1993). The Pleistocene occupation of southwestern Tasmania was closely associated with the specialised hunting of Benett's wallaby (McNiven *et al.*, 1993), while the Holocene sites disclose intense exploitation of coastal resources (Bowdler, 1988; Dunnett, 1993). The morphological differences between Tasmanians and Australians have puzzled anthropologists for decades, and are generally attributed to genetic drift and isolation (Macintosh & Barker, 1965; Pardoe, 1991). Some workers would consider the Tasmanians as derived from southeastern Australia (Adam, 1940; Hrdlička, 1928; Larnach & Macintosh, 1966; Pardoe, 1984; Pietrusewsky, 1979, 1981, 1984), while others favour a more generalised Australian or even Melanesian ancestry (Giles, 1976; Howells, 1970, 1973, 1976).

Distant Australo-Melanesian affinities

Given the early dates of occupation of Australia (Roberts & Spooner, 1992; Roberts *et al.*, 1990), it may be assumed that a population of robust modern humans, with an unspecialised flake-based technology, from which Australians derive, inhabited the Sunda landmass between 70 and 60 ka. There are however, no traces of this early Sunda population, and its existence is only inferred on biogeographical grounds (the first Australians had to go through Southeast Asia to occupy Sahul), the shared dental similarities between Southeast Asians and Australians (T. Hanihara, 1992a; Turner, 1992), and by the morphological relations between certain Southeast Asian (Negritos) and Indian (Veddas) groups and Australo-Melanesian populations (Bellwood, 1985; Birdsell, 1949, 1977, 1993; Brues, 1977; Coon, 1962; Garn, 1971; Glinka, 1981; Howells, 1976; Jacob, 1967; Kennedy, 1977).

Greenberg (1971) has advanced the 'Indo-Pacific' hypothesis, which is the strongest available evidence for a relationship between all non-Austronesian speakers outside Australia. Greenberg recognises a number of non-Austronesian linguistic families that share a distant common ancestor: Andamanese (AN), the Timor-Alor group (TA), Halmahera (HA), Western New Guinea (WNG), Northern New Guinea (NNG), Southwestern New Guinea (SWNG), Southern New Guinea (SNG), Central New Guinea (CNG), Northeastern New Guinea (NENG), Eastern New Guinea (ENG), unclassified New Guinean languages (UNG), New Britain (NB), Bougainville (BO), Central Melanesian (CM) and Tasmania (TS). Among these, three tentative 'super-groups' further uniting some of these language families are also identified (1: HA, TA, WNG; 2: NNG, CNG, SWNG, SNG; 3: NB, BO, CM and Rossel Is.). Such a relationship, and its geographic distribution as outliers of the more recently dispersed Austronesian languages, are fully consistent with the model presented here of an original Southeast Asian-Melanesian and Australian population, before the expansion of 'southern' and 'typical' Mongoloids.

There are few fossil remains in Southeast Asia, which in general terms, support the hypothesis of shared ancestry between aboriginal Southeast Asian and Australian populations. The fossil of Niah, Sarawak, was found in association with Middle Soan-Indian MSA (Harrison, 1959), and has been dated to 41.7 ± 1000 BP (DeVries & Waterbolk, 1958; Majid, 1982; Vogel & Waterbolk, 1972; Zuraina, 1982), but is considered of uncertain age by Kennedy (1977). An early study by Brothwell (1960) suggested Tasmanian or general 'Australoid' affinities. The Tabon remains from Palawan, Philippines, have been dated to 26–24 ka (Bellwood, 1987; Fox, 1970; Jones, 1989), and also have been considered to show some Australian affinities (Macintosh, 1978). The Wajak remains from Java have no certain date (Storm & Nelson, 1992). These crania have been thought to be similar to both Australians (Howells, 1959) and Mongoloids (large and flat face), but Storm & Nelson (1992) argue strongly that their affinities should remain open. The relationship between Wajak and later fossils in Java, like Hoekgrot and Goea Djimbe, is not clear (Storm, 1990; Storm & Nelson, 1992). A number of archaeological sites in the area have early dates – ~37 ka at Long Rongrien, in Peninsular Thailand (Anderson, 1987), 31 ka at Lean Burung 2 cave, Sulawesi (Glover, 1981), 31 ka at Kota Tampan in Perak, peninsula Malaysia (Zuraina & Tija, 1988), 28 ka at Ting Kayum in Sabah, island Malaysia (Bellwood & Koon, 1988), 28 ka at Pilanduk cave (Fox, 1978), *c.* 25 ka at Da-phuk, Vietnam, 25 ka at Sai-vok, Thailand, 14 ka at Uai Bobo 2 cave, Timor (Glover, 1987; Glover & Presland, 1985) and 18–23 ka at the sites of Mài Dà Nguòm, Mài Dà Dieu, Ong Quyen and Xom Trai, Vietnam (Ha Va Tan, 1980; Hoàng Xuân Chinh, 1991) – which confirm the widespread human occupation of the area as from 40 ka.

Genetic and morphological analyses place Australo-Melanesians as a distinct cluster. either as the first bifurcation from an African cluster or as an early division of a non-African group (Hedges *et al.*, 1991; Howells, 1989; Nei & Roychoudhury, 1993; Vigilant *et al.*, 1991; Wright, 1992). These results, together with the morphological distinctiveness discussed before, indicate that because of their early geographical dispersal and subsequent relative isolation, Australo-Melanesians are distantly related to all other populations in Southeast Asia.

Are any recent Southeast Asian populations related to a dispersal out of East Africa?

Recent Southeast Asian crania appear as a derived group, both in terms of number of significant differences and the character of this differentiation. They exhibit rounded globular crania and a special reduction of the occipital region. These specialisations most certainly relate to the overall gracilisation observed in this population, especially post-cranial, with its consequent effect on the reduction of the nuchal muscles. In terms of dental size, they show a wide range of variation for M1 and M2 values, and a marked reduction in M3 dimensions. In spite of these few generalisations, Southeast Asian crania show very high levels of variability on a regional scale. Most researchers identify at least two, and possibly three, distinct human groups in Southeast Asia. The two clear groupings are the Negritos and the 'southern Mongoloids' or Indo-Malay, while some researchers argue for a third aboriginal population with neither Negrito nor Mongoloid affinities. In Malaysia, three groups (Negrito, Senoi Semai and Proto-Malay) have been recognised (Saha *et al.*, 1995). At the moment this issue cannot be settled.

The Negritos

The Negrito populations today are small and geographically restricted. They are found in the Andaman Islands, the Semang from the mountains of the Malay peninsula, the Aetas from the island of Luzon in the Philippines and 12 tribes in the high Atherton and Herberton Tablelands with Indo-Malayan rainforests of northeastern Queensland, Australia (Figure 11.6) (Birdsell, 1972; Howells, 1959; Kingdon, 1993). Other groups have been claimed to show Negrito affinities, like certain Melanesian groups, certain Indonesian horticulturalist groups and Tasmanians (Birdsell, 1972) supported by the linguistic reconstruction of Indo-Pacific of Greenberg (1971). Further Negrito associations have also been made with groups in India and the Hadramaut of southern Arabia (Howells, 1959). However, these relationships remain unconfirmed. The languages spoken in the

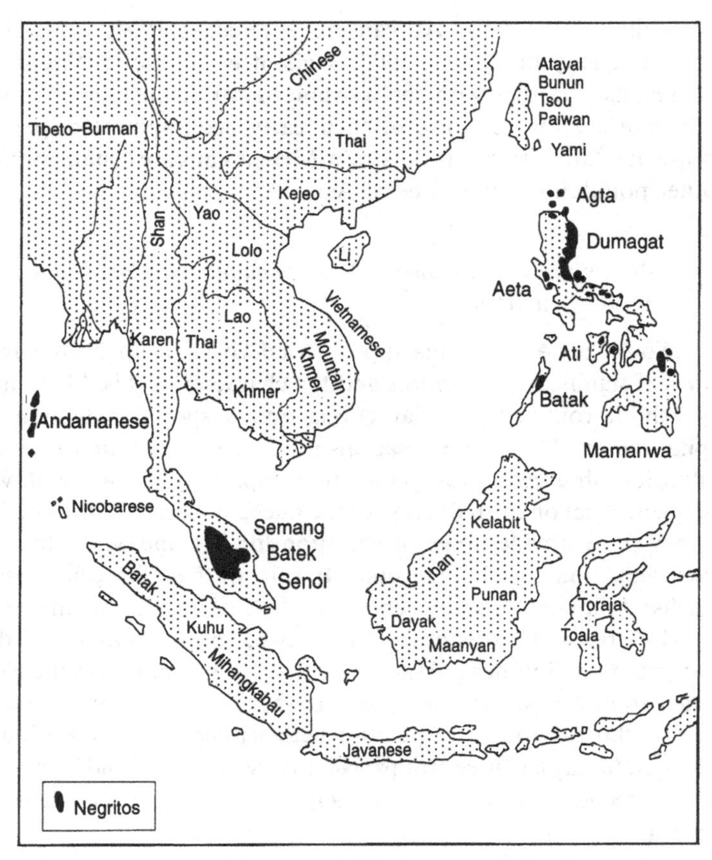

Figure 11.6 Distribution of Negrito populations in Southeast Asia. (Redrawn from Kingdon, 1993.)

Andaman Islands are not related to any language family of mainland southern Asia (Greenberg, 1953). Within the Andamans, two separate, totally unrelated families of languages exist (Greenberg, 1971), indicating high levels of isolation. In the Philippines and Malaysia, the Negrito tribes are considered to have 'borrowed' Austronesian languages from their agriculturalist neighbours (Griffin, 1985), as did the Pygmy groups in Africa. In Malaysia they speak a dialect belonging to the Mon Annam linguistic group, although it contains a number of words that are not Mon Annam or Malay or Andamanese (Evans, 1937).

Two hypotheses have been proposed for the origins of the Negritos. The first of these considers that their small body size, gracility and distinctive features are the result of local adaptation to the rainforests. Average heights among Negrito groups vary between 140 and 150 cm (Lanoh Negritos of Lenggong: $\bar{x} = 148.6$ cm; Ulu Lebin, Kantan: $\bar{x} = 149$ cm (men), $\bar{x} = 140.8$ cm (women); Semang-Pagan, Malaysia: $\bar{x} = 150$ cm (men), $\bar{x} = 140$ cm (women) – Evans, 1937; Schebesta, 1927, 1928). However, although this view of extreme gracilisation as a rainforest adaptation explains the small body size, it does not account for the other distinctive features in this population, like hair form, hair pigmentation, skin colour and steatopygia, which are clearly not found in small forest dwellers in the Americas who may also have reduced in size as a specific rainforest adaptation (Birdsell, 1972). The second hypothesis suggests that Negritos share a common ancestor with the San, Khoi and some Congo pygmy tribes of Africa. This hypothesis is based on shared expression of the features mentioned above, especially steatopygia, which is only observed in Negritos and San-Khoi populations (Howells, 1959). There is no fossil evidence to support or refute either of these views. However, genetic studies show that African Pygmies and Southeast Asian Negritos are more closely related to their near neighbours than to each other, suggesting that pygmy populations appeared independently in different regions of the world (Boyd, 1963; Cavalli-Sforza, 1986). Nevertheless, a relationship between Australian and Melanesians and Negritos has long been argued (Bellwood, 1985; Birdsell, 1949, 1977, 1993; Brues, 1977; Coon, 1962; Garn, 1971; Glinka, 1981; Greenberg, 1971; Howells, 1959, 1976; Jacob, 1967; Kennedy, 1977). Melanesians and Australians appear consistently as the first branching event from an African source (Hedges *et al.*, 1991; Stoneking *et al.*, 1992; Vigilant *et al.*, 1991), especially San-Khoi and Pygmy groups, and the remote association of African and Australo-Melanesians could be extended to the Negritos of Southeast Asia. The fact that there is no evidence of extinct widespread rainforests in southern Africa, across East Africa or in southern Arabia, suggests that if small body size is ancestral to the southern dispersed population, it did not result from rainforest adaptations, for to expand into these areas, populations with such a specific forest adaptation would have had to overcome a period of seriously decreased fitness. It is possible that selection for small body size was not associated to forest adaptations in the first place, but to the stringent arid conditions during certain periods in the Upper Pleistocene of Africa. In this case, small body size as a successful feature for forest exploitation would have been an exaptation. Extreme small body size may have thus favoured the colonisation of the rainforests by certain populations of Africa and Southeast Asia, and under further positive selection, been intensified in the forest dwellers. Of course, even if this scenario were correct, it does not need to be universal, since certain populations in South America seem to have evolved small body size as an adaptation to rainforests, although not as extreme as either the African San-Khoi-Pygmy or Southeast Asian Negritos. It is also possible that, rather than reflecting a distinct relationship between San-Khoi and Negritos, the few shared features between the groups are plesiomorphic for early modern humans, and that the distribution of fat in the body is functionally constrained by body size, thus converging into steatopygia in pygmy populations.

Are any populations of India related to a dispersal out of East Africa?

The history of the prehistoric modern population of the Indian subcontinent is poorly known. There are Middle Stone Age deposits in India (central Rajasthan) during a moist period (130–80 ka?), and the material culture has been considered as showing East African affinities (Allchin *et al.*, 1978). The Indus delta appears to have been occupied around 50 ka, and in the very south, the site of Batadomba Lena, Sri Lanka, has yielded modern human fossils associated with a sophisticated tool industry dated to 28 ka (Kennedy & Deraniyagala, 1989). The foraging tribes still inhabiting southern India, like the Veddas, and speakers of Dravidian languages, although very mixed with other Indian groups, show morphological affinities towards Southeast Asian Negritos and Australian aborigines (Howells, 1959, 1993), suggesting that aboriginal populations of Southeast Asia once reached and occupied the southern Indian subcontinent. However, recent genetic studies among the Senoi Semang of Malaysia do not show any relationship between this population and Veddah or other Indian groups (Saha *et al.*, 1995). A large portion of the population today is of Caucasian affinities, speakers of Indo-European languages, although there is an important Mongoloid component in the area immediately south of the Himalayas (Bhasin *et al.*, 1994). Howells (1959) has argued that the first Caucasian groups to enter India from the west (Middle East ?) were Upper Palaeolithic populations, and not the Neolithic farmers who, more than 20,000 years later, would have already found a locally inter-mixed population. This is consistent with the archaeological evidence from Batadomba Lena, which shows Upper Palaeolithic affinities (Kennedy & Deraniyagala, 1989).

A measure of the complexity of populations in India may be obtained by the linguistic diversity in the country. Four language families are currently spoken in the Indian subcontinent (Austro-Asiatic; Tibeto-Chinese; Dravidian; Indo-European), subdivided into 12 major languages, several less important ones, and a number of tribal languages, some belonging to unclassified Dravidian tongues, others to unspecified family affinities (Bhasin *et al.*, 1994). It is not clear from genetic and morphological studies what are the underlying sources for the observed patterning of variation among Indian populations. Some studies point towards a main genetic division between caste Hindus and non-caste tribes, with a possible further division within these groups along linguistic and geographical boundaries (Balakrshnan, 1978; Malhotra, 1978; Walter, 1983). However, dental studies suggest that Hinduization and genetic admixture of some tribal groups make these distinctions less clear (Lukacs & Hemphill, 1993).

Secondary sources of dispersal

As argued in Chapter 10, the establishment of successful independent populations outside the range of the original source would occur by jump dispersals followed by range contraction. We have argued that the occupation of North Africa by modern humans during stage 5 and southeastern Asia during stage 4 were the results of such dispersal events which led to the establishment of early modern populations in those areas. It is dispersals from these secondary sources, northern Africa and Southeast Asia, that will be discussed here. The character of the colonisation of Australia would indicate it was an immigration from Southeast Asia, but its early date would suggest that it was part of the early dispersal of modern humans throughout southern Asia.

Secondary dispersals resulting from the early northern route

As already mentioned, the Middle East, Central Asia and Europe were occupied by Neanderthal and Neanderthal-related groups until 45–35 ka, and modern populations remained restricted to North Africa during the first half of the Upper Pleistocene. However, around 45 ka, Upper Palaeolithic assemblages are found at Boker Tachtit in the Middle East and Ksar Akil in Israel, the latter associated with a modern human skeleton (Bergman & Stringer, 1989). The origin of this palaeolithic tradition seems to be local to the area (Marks, 1990), and although there may have been a period of contact with sub-Saharan Africa (Fontes & Gasse, 1989; Petit-Marie, 1989; Wendorf *et al.*, 1976), the dispersal of modern people into the Middle East at this time was not a faunal event (Tchernov, 1992a–c). Therefore, the second phase of modern occupation of the Middle East results from an expansion of the range of modern humans in northeastern Africa, starting around 50–45 ka (Klein, 1992; Mellars, 1992;

Strauss, 1994). However, in contrast to the similar event during the last interglacial, modern humans dispersed further into Eurasia at this time. In evolutionary terms, all the subsequent populations of North Africa, the Middle East, India, Central Asia and Europe share a common ancestor at the base of the Upper Palaeolithic.

Northern Africa

In North Africa, the Aterian industries give way to a number of later specialised Upper Palaeolithic traditions, like the Iberomaurusian and the Caspian of the Mediterranean coast. Iberomaurusian sites appear 22–20 ka (early levels at Taforalt, Maroc; Tamar Hat, Algiers – Camps *et al.*, 1973; Roche, 1976), while the site of Columnata at 7 ka shows the recent range of this tradition. Palaeoclimatic reconstructions show that the Maghreb had a different climatic regime from that of other regions. Studies by Rognon (1987) have shown that the temperate border of the Sahara did not experience maximum aridity until the end of the last glaciation, i.e. the Iberomaurusian expands during the glacial maximum but not under conditions of extreme aridity, rather in a refuge of humid climate. During maximum glaciation, strong trade winds in the western Sahara (Sarntheim *et al.*, 1982), also evidenced by sediments in deep-sea cores (Hooghiemstra & Agwu, 1988; Parkin & Shackleton, 1973), together with the southwards displacement of the Mediterranean winter belt (Diester-Haas, 1976; Street & Grove, 1976), promoted more frequent incursions of cold northerly winds and maintained moister conditions. The period 40–20 ka in the region is characterised by wetter environments (Conrad, 1969; Rognon & Williams, 1977), and pollen studies in northern Tunisia indicate that before 20 ka the region was covered by forests, while in southern Tunisia two wet episodes are represented by pollen and palaeosols at 27 ka and 21 ka. Near the Gulf of Gabes, at Wadi El Akarit, steppe conditions are indicated between 30–20 ka, where several Rhino skeletons were found. Between 20–15 ka rainfall was more irregular, and by 12–10 ka maximum aridity settled, when most of intertropical and tropical Africa had wetter conditions (Rognon, 1987).

Ferembach (1976, 1986c) found that the fossil of Dar-es-Soltane (Aterian, Morocco) shows affinities towards Ibero-Maurusian remains (Mechta, Afalou, Taforalt) and the related Wadi Halfa series. The similarities between the Afalou-Taforalt and the Wadi Halfa series from Sudan (Greene & Armelagos, 1972) show that the late Pleistocene-early Holocene was a period of important movements across the Sahara. Palaeoclimatic data indicates that the Sahara was crossable at this time. as does the distribution of Ounanian and later archaeological sites in the Sahara (Smith, 1993). However, the strongest relationships of the Iberomaurusian samples are with early European Upper Palaeolithic crania (Chamla, 1978; Ferembach, 1986c), while the more gracile skeletal remains associated with the Caspian industries could be related to the Natufian people of the Levant (Ferembach, 1986c). Indeed, for many variables in Figures 9.2 to 9.4, the Afalou and Taforalt samples show different values to the contemporary Natufians, but similar to early European modern fossils even though they are some 20 ka apart.

Contact between North African and southern European populations across the Mediterranean has been suggested by several authors (Alimen, 1975; Ferembach, 1986c). The deepest point between Spain and Morocco is too deep to have ever been exposed by glacio-eustatic regressions of the sea (British Admiralty chart, 1988). Between Sicily and Tunisia, even at −200 m sea-level, there was a narrow corridor of water separating the two bodies of land, but with islands in between. This corridor was approximately 19 km wide, but through island connections, sea passages of ~8 km and less were formed (distances calculated from Admiralty charts, 1988, 200 m isobath lines). Herman *et al.* (1969), studying the upper half (last 50,000 years) of core V10-67 in the eastern Mediterranean, found that the record is marked by repeated layers of black mud resulting from periods of stagnation of the bottom waters, suggesting that the eastern Mediterranean basin must have been nearly cut-off from the western basin and the Atlantic Ocean by the Strait of Sicily reduced to a narrow passage (Figure 11.7a–e). Considering that humans with a less sophisticated technology made larger crossings into Australia well before the last glacial maximum, the possibility of cross-Mediterranean dispersals, specially in view of the affinities of the Iberomaurusian and European populations, should not be discarded.

As discussed in Chapter 8, the main characteristic of the fossil samples from Afalou and Taforalt is large cranial size. They have also retained a large posterior dentition, with its related wide and retracted upper faces, long palates, projecting supraorbitals, minimum frontal breadth positioned away from the face and flat frontal bones. It is the retention of large teeth and facial superstructures that make this group appear morphologically similar to the Australians. However, contrary to the latter group, the fossils from Afalou and Taforalt show strong homogeneity in their large dental dimensions. Furthermore, the maintenance of large molars in combination with such a large cranium may be associated with the development of very large temporal muscles, also reflected in the wide bizygomatic breadths and the eversion of the lower border of the malar that give the lower face an East Asian appearance. Like recent sub-Saharan Africans, they have very

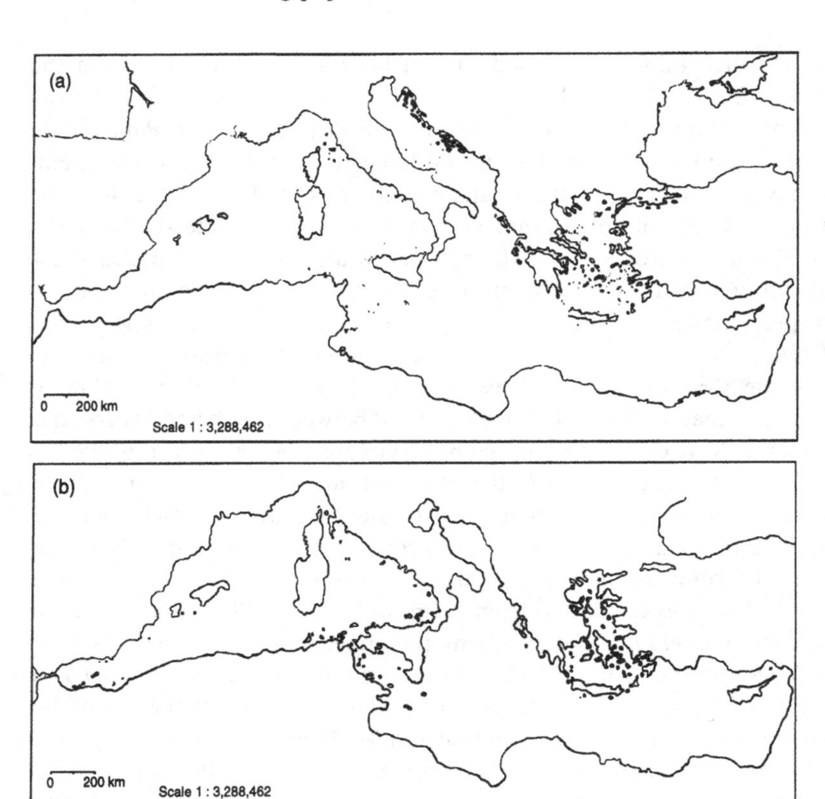

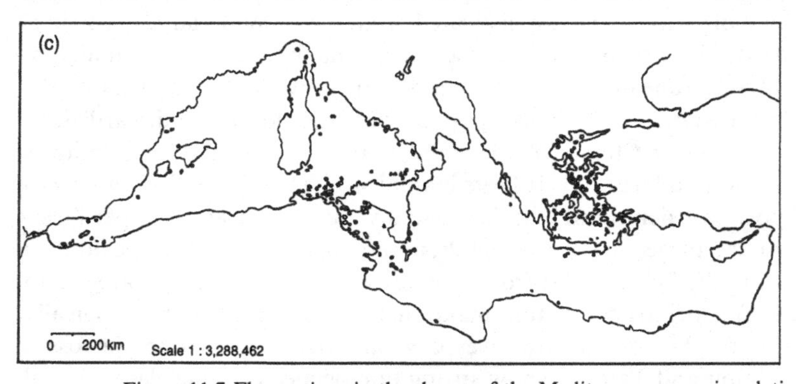

Figure 11.7 Fluctuations in the shores of the Mediterranean sea in relation to eustatic changes in sea-level:
(a) Present day shorelines (map drawn from the British Admiralty chart, based on the chart produced by the Italian Hydrographic Institute, 1972).
(b) Shorelines assuming a sea-level drop of 100 m, calculated from the depths shown in the British Admiralty Chart.
(c) Shorelines assuming a sea-level drop of 200 m, calculated from the depths shown in the British Admiralty Chart.

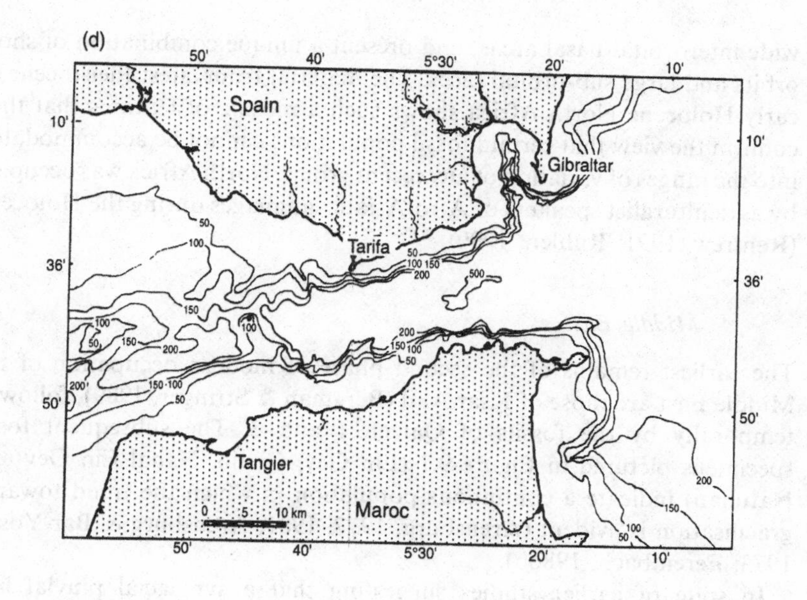

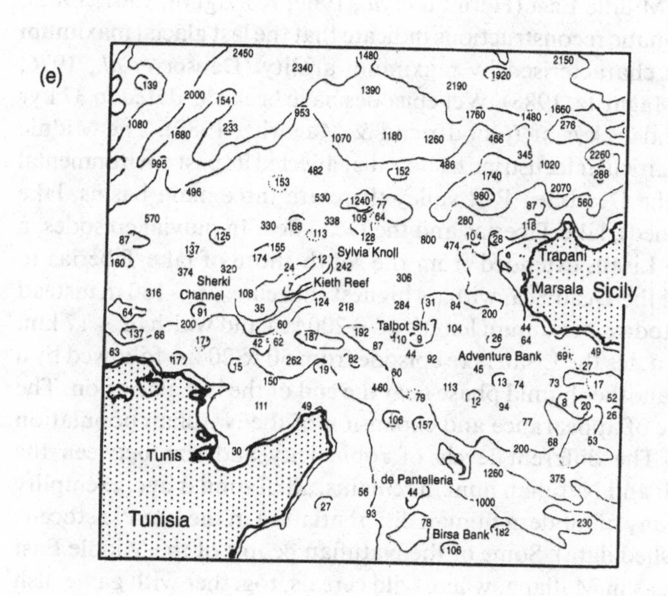

(d) Detailed map of the Strait of Gibraltar, showing the progressive depths between Tangier and Tarifa.
(e) Detailed map of the Strait of Sicily, showing the complicated and irregular, but largely shallow depths between Tunis and Marsala.

wide interorbital-nasal areas, and present a unique combination of short orbits and large sub-orbital faces. The material from latest Pleistocene to early Holocene North Africa shows such a mosaic of features that they confirm the view that certain fossil populations cannot be accommodated into the ranges of variation of any recent group. North Africa was occupied by agriculturalist speakers of Afro-Asiatic languages during the Holocene (Renfrew, 1991; Ruhlen, 1990).

Middle East

The earliest remains of the second phase of modern occupation of the Middle East are those of Ksar Akil (Bergman & Stringer, 1989), followed temporally by the fossils of Qafzeh 1 and 2. The subsequent fossil specimens pictured in Figure 9.1 (Ein Gev, Ohalo, Nahal Ein Gev and Natufian) indicate a continuous population in which the trend towards gracilisation is evident (Arensburg, 1977, 1981; Arensburg & Bar-Yosef, 1973; Ferembach, 1986b).

In spite of earlier studies suggesting that a synglacial pluvial had occurred in the Middle East (Herman *et al.*, 1969; Rossignol, 1961, 1962), recent palaeoclimatic reconstructions indicate that the last glacial maximum in the area was characterised by maximum aridity (Deuser *et al.*, 1976; Goodfriend & Magaritz, 1988). Wet episodes have been ^{14}C dated to 37 kyr BP, 28 kyr BP and 13 kyr BP (Goodfriend & Magaritz, 1988). The Middle East also has a particular lacustrine history that affected its past environmental conditions. In the Dead Sea Rift valley there are three main basins, lake Hula (now drained), lake Tiberias and the Dead Sea. In pluvial episodes, a great lake, lake Lisan, extended from the south shore of lake Tiberias to 335 km south of the Dead Sea, with its highest shoreline at −180 m instead of the −400 m today, maximum length of ~200 km and width of ~17 km. Farrand (1971) dates the Lisan lake episode from 50 to 20 ka, followed by a dry period and another humid phase near the end of the last glaciation. The latter is the time of appearance and expansion of the Natufian population of the Levant. The different levels of robusticity and size between the Afalou-Taforalt and Natufian human remains, all of similar age, exemplify well the anachrony of modern human differentiation in the late Pleistocene (Lahr, unpublished data). Some of the Natufian people of the Middle East were sedentary, as in Mallaha, where wild cereals, together with game, fish and shellfish were exploited (Perrot, 1968), while others seem to have maintained a hunter–gathering existence, as in Erq el Ahmar. As a group, the Natufians foreshadow several of the trends of the Neolithic agricultural goups (Bar-Yosef & Valla, 1991).

Europe

The dispersal of modern humans into Europe occurred relatively fast, as it is apparent by the spatial distribution of early Aurignacian sites (Strauss, 1994). In three areas of Europe, southwestern France, central Europe and Italy, sites with progressive Mousterian assemblages are found (Chatelperronean, Szelettian and Ulluzzian). These sites, which are broadly contemporaneous with the early Aurignacian, were occupied by Neanderthals, as evidenced by the St. Cesaire skeleton. The progressive aspect of these industries has been interpreted as resulting from the influence of contact with Upper Palaeolithic assemblages (Mellars, 1989, 1992), yet some Neanderthal groups seem to have survived for a long period without evident modern cultural influence, as in Zafarraya, Spain (Hublin, 1994).

The area of origin of the European Aurignacian is not clear. Although the earliest Upper Palaeolithic industries are found in the Middle East, and an east to west movement of the Aurignacian into Europe has been proposed (Mellars, 1989, 1992), the Middle Eastern Upper Palaeolithic is not Aurignacian (Marks, 1990). Besides the very pronounced technological shift between the Middle and Upper Palaeolithic in Europe, suggesting replacement of Neanderthal populations by immigrant modern humans (Harrold, 1989; Hublin & Tillier, 1992b; Klein, 1992; Mellar, 1989, 1992; Stringer *et al.*, 1984), other significant behavioural changes occur later (Chase, 1989; Clark, 1987; Rigaud, 1989).

The European Upper Palaeolithic is characterised by a number of cultural traditions reflecting the dynamics of population movements and environmental changes during the period. Around 28–26 ka, the Gravettian tradition disperses from present-day Russia into Europe (Svoboda, 1990), surviving alongside Aurignacian assemblages for a long period of time. Based on the percentage and sources of exotic raw material (several hundred kilometres) in the Gravettian assemblages, Svoboda concludes that the geographical ranges exploited by this group were much wider than those of the people associated with the Aurignacian (Svoboda, 1990). This greater vagility could explain both the intense rate of expansion and the cultural homogeneity across the vast area occupied by the Gravettian. In western Europe, both these industries disappear during the glacial maximum (Otte, 1990), although they continue to be manufactured in central Europe and the Italian peninsula. Maximum glacial conditions resulted in a major depopulation of western northern Europe (Schnider, 1990), but of very short duration, since the site of Hallines in northern France was occupied by 16 ka (Fagnart, 1984). This period of northwestern depopulation witnessed an increase in site density in Iberia and southwestern France, where the Solutrean and Magdalenian developed (Otte, 1990). The differences between the Badegoulian (or Early Magdalenian) and both the Solutrean and the final Gravettian suggest a possible late Aurignacian source for this tradition (Rigaud, 1976). There is an eastward dispersal of the Magdalenian in post-glacial times (Otte, 1990).

Interestingly, during this period the Italian peninsula seems to have become completely isolated from the rest of Europe by the alpine ice sheet (Figure 11.8). This is reflected archaeologically, since neither the Solutrean nor the Magdalenian of neighbouring regions made its way into Italy. Between 30–25 ka, the late Aurignacian disappears in Italy, and the area seems to have remained unoccupied until the appearance of Gravettian and subsequent Epi-Gravettian industries (Mussi, 1990; Mussi & Zampetti, 1987). In central Europe, large Gravettian sites, like Dolní Vestonice, Pavlov, Predmostí and Petřkovice, are characteristic of the period between 28 and 24 ka. There is sparse, and possibly only temporary task-specific occupation of the area during the last glacial maximum, as evidenced in the sites of Stránská Skála and Konevova (Svoboda, 1990). The local Gravettian (Pavlovian) continued to be manufactured throughout the period until the late appearance of Magdalenian assemblages from the west around 12 ka (Kozlowski, 1985). This area of Moravia seems to have been the eastern limit of the Magdalenian expansion, for Magdalenian tools are not seen across the Carpathian Mountains. The pronounced decrease in population size seen in Moravia during the pleniglacial did not occur in Bohemia, where microclimatic factors seemed to have allowed continuity of occupation of Gravettian peoples, with some technological specialisations (Kozlowski, 1990). These were eventually influenced by Epi-Gravettian traditions around 17 ka. Some of the Gravettian groups from Slovakia dispersed eastwards, related to the Kostenki-Avdeevo culture (Kozlowski, 1990).

The archaeological data of Europe at the last glacial maximum suggest important intra-continental differences for a short period of time. Northern Europe seems to have been largely depopulated, while two temporary isolated provinces were formed: southwestern Europe, with its intense and dense Solutrean and Magdalenian occupation, and Italy and eastern Europe, with surviving Gravettian sites. It is interesting to note that, during glacial maxima, the climate in the western and eastern Mediterranean was very different. The northwestern Mediterranean region experienced a marked fall in temperature because of the southward shift of the Gulf Stream that reached directly the Iberian Peninsula, while the eastern Mediterranean had a temperature decrease of only 1–2 °C. Therefore, while the temperature gradient today from the Balearic Islands to Crete is just 1–2 °C, at 18 ka it was about 10 °C (CLIMAP Project Members, 1976).

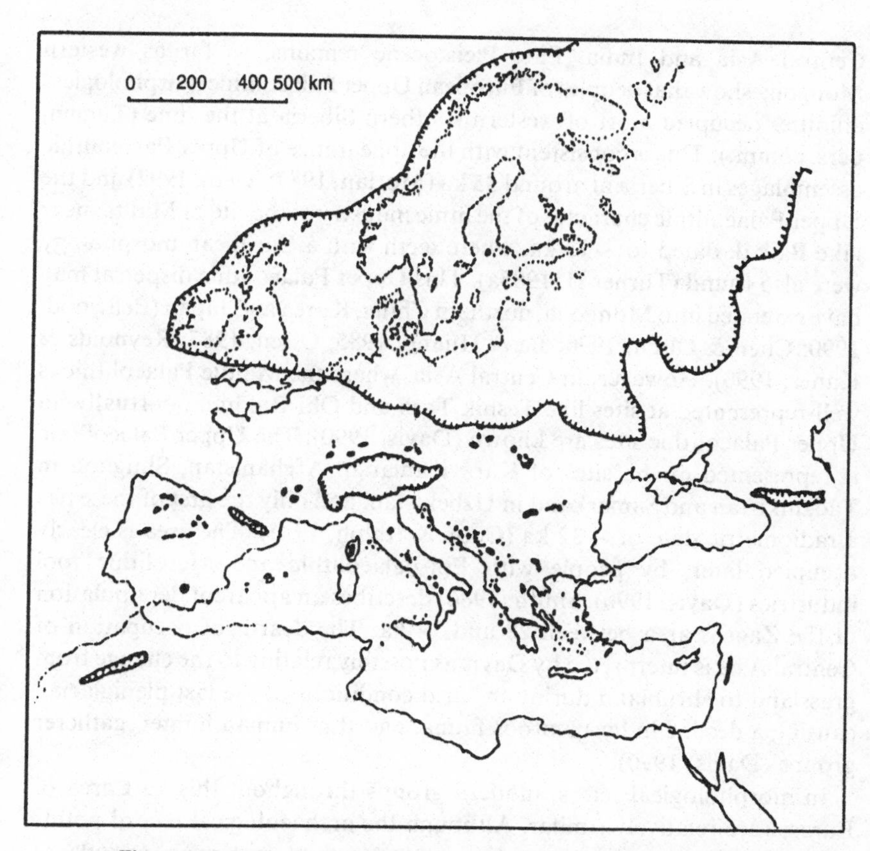

Figure 11.8 Ice limits in Europe and north Africa at the time of the last glacial maximum. Minor glaciated areas spotted in black. The shorelines are as present day limits, without showing the eustatic drop in sea-level during galcial maxima. (Redrawn from West, 1977.)

However, the eastern Mediterranean was significantly drier than the west, and during pleniglacial times there still were some *Pinus* forests in Spain, while Greece was essentially treeless (Florschutz *et al.*, 1971).

Siberia

Modern humans seem to have been the first hominids to occupy Siberia. Gamble (1993) argues that the vast Siberian plains could only be occupied by modern people because, although rich in resources, the food is found at densities that require storage, mobility over large distances and much forward planning to exploit. This occupation seems to have been part of the early Upper Palaeolithic expansion that dispersed into Europe, Siberia,

Central Asia and India. Late Pleistocene remains, as far as western Mongolia show that peoples of European Upper Palaeolithic morphological affinities occupied most of western southern Siberia at the time (Tumen, pers. comm.). This is consistent with the appearance of Upper Palaeolithic assemblages in Siberia at around 35 ka (Morlan, 1987; Klein, 1992) and the Upper Palaeolithic character of the lithic industry at the site of Mal'ta, near lake Baikal, dated to ~18 ka, where teeth with a European morphology were also found (Turner II, 1985a). This Upper Palaeolithic dispersal may have extended into Mongolia, northern China, Korea and Japan (Bellwood, 1990; Chen & Olsen, 1990; Jia & Huang, 1985; Olsen, 1987; Reynolds & Kaner, 1990). However, in Central Asia, where the Middle Palaeolithic is well-represented at sites like Teshik Tash and Obi-Rakhmat, virtually no Upper Palaeolithic sites are known (Davis, 1990). The Upper Palaeolithic is represented at the sites of Kara Kamar in Afghanistan, Shugnou in Tadzhikistan and Samarkand in Uzbekistan, and only the first of these has a radiometric date of +32 ka (Coon & Ralph, 1955). The area is clearly occupied later, by people with Epi-Palaeolithic and Mesolithic tool industries (Davis, 1990). Smith (1986) describes an apparent depopulation of the Zagros area between 28 and 14 ka. The dearth of occupation of Central Asia is interpreted by Davis as possibly relating to the change from grassland to shrubland during the arid conditions of the last pleniglacial, causing a decline in herbivorous fauna, and thus human hunter–gatherer groups (Davis, 1990).

In morphological terms, modern groups throughout this vast area of Eurasia are relatively similar. Although the archaeological record points towards quite extensive population movements, and even periods of regionalisation of traditions, the fast occupation of Eurasia by Upper Palaeolithic populations (approximately by 35 ka most regions seem to have been reached) deriving from a common source in North Africa–Middle East around 45–40 ka, indicates that all these populations were closely related. Early European Upper Palaeolithic remains can be described as large, robust, with strong development of the supraorbital ridges, not very pronounced occipital tori, with a broad and short face (Billy, 1975; Ferembach, 1986a; Vallois & Billy, 1965), a description that applies to Aurignacian and Gravettian fossils alike (Abri Pataud, Barma Grande, Baousso da Torre, Cro-Magnon, Grimaldi, Mladeč). A small number of crania were recognised as different from the above group, like Combe Capelle, Brno and Svitavka, characterised by hyper-dolichocephaly, high and keeled vaults, strong supraorbital ridges, prognathic faces and pronounced sexual dimorphism (Billy, 1975, 1986; Ferembach, 1986a; Vallois & Billy, 1965). However, the variable remains from Predmost seem to bridge the morphological gap between these two groups, as do those of Cioclovina, whose affinities were alternately described as Cro-Magnon (Rainer & Simionescu, 1942; Vlcek, 1970), Predmost (Necrasov & Cristescu, 1965) and Combe Capelle (Riquet, 1982). Henke (1991) in a statistical study of Upper Palaeolithic and Mesolithic fossils of Europe finds no distinct sub-division of the sample, the level of intra- and intergroup variability (with groups defined regionally and temporally) are of similar magnitude. He argues that intense gene flow maintained this relative homogeneity through time and space (Henke, 1991). Magdalenian human remains are described as particularly homogeneous (Ferembach, 1986a; Vallois, 1972), which is consistent with a period of localised geography and intense selection as would have been pleniglacial southwestern Europe.

There seems to be a decrease in size and robusticity, although maintaining the original levels of variability, from early to late Upper Palaeolithic to modern times (Billy, 1972; Ferembach, 1986a; Frayer, 1984; Chapter 9). This gracilisation is visible in the overall fossil assemblage, while individual mesolithic sites show more differentiated morphologies. Therefore, populations like Teviec, Gramat and Saint-Rabier show high levels of robusticity and size, interpreted as reflecting a lineage from Cro-Magnoid Magdalenian ancestors by Ferembach (1986a). On the other hand, remains like Staré Mesto in the Czech Republic, Muge in Portugal, or Vassilevska in the Crimea, are characteristically more gracile. It is possible that along the Mediterranean and associated river basins, gene flow from the Middle East accelerated a process of gracilisation (already observed to a large extent in the Natufian people of Israel), creating a morphological cline towards northern Europe. This is supported by the analyses of Henke (1991) which point towards clines rather than distinct groupings. Cavalli-Sforza (1991) and Cavalli-Sforza *et al.* (1994) have identified an east–west geographical cline in the expression of a number of genes, interpreted as resulting from the movement of Neolithic farmers from the Middle East. Relics of the Holocene population prior to this expansion would be represented by the genetically and linguistically distinct Basque people. Agriculture in Europe is first seen at Ile Riou, at 5.7 BP, while it took another approximately 2000 years to reach the north of the continent (Newell *et al.*, 1979).

Chapter 8 shows that, as a group, recent Europeans are relatively homogeneous in morphology, which is consistent with the fossil evidence discussed above. The few specialisations present in recent crania are all of a derived character, except the low position of maximum parietal breadth, as discussed extensively in Chapter 4. Since there is no evidence of a recent demographic bottleneck in the area, other processual explanations for this homogeneity are necessary. It is possible that the amount of inter-regional movements in Europe during the Upper Palaeolithic had a resulting homogenising effect over the population on a regional scale. However, as mentioned before, nuclear genetic data support a recent (Neolithic) influx of people into Europe (Cavalli-Sforza, 1991). This agricultural dispersal was neither the first nor the last, but the demographic dimensions of farming populations could have had a swamping effect over the sparser foraging populations. Therefore, while Palaeolithic expansions could have been realised through gradual displacement and admixture, with the possibility of surviving groups in relatively isolated pockets, the rapidly expanding Neolithic populations would have swamped local groups, acting as a homogenising mechanism. The reality of geographical isolates among the small human bands in glacial Europe is exemplified by the late survival of a Neanderthal group in Zafarraya, southern Spain, some 15 ka after modern humans were present in the region. However, the mechanism of genetic diffusion of Neolithic agriculturalists into Europe argued by Cavalli-Sforza and co-workers was not universal, as shown by the studies of genetic diversity in the Caucasus by Barbujani *et al.* (1994). These studies are consistent with the views advanced here, that specific topographic, linguistic and demographic parameters will determine the pattern and rate of genetic admixture of particular groups.

In terms of archaic–modern continuity, neither earlier nor later modern European crania show any of the facial and occipital specialisations observed in the Neanderthal group, such as mid-facial prognathism, inflated maxillas with lack of canine fossae, large nasal aperture, flattened cranial base, small mastoids, occipital bunning and supra-iniac depressions. The level of difference between Neanderthal and modern European populations (Upper Palaeolithic and recent) has been clearly shown by the multivariate analyses of Howells (1989), van Vark (1984) and van Vark *et al.* (1992), among others.

Secondary dispersals resulting from the early southern route

The study of the main population of Southeast Asia is intimately linked to concepts of the evolution of Mongoloids. This term is currently used to refer to a very wide range of populations of variable features (Howells, 1986), and two sub-groups are commonly identified, the 'southern Mongoloids' from southeastern Asia and the 'typical Mongoloids' from northeastern Asia. Because of the more intense expression of certain features, like the epicanthic fold, facial flatness and shovel-shaped incisors (among others) in the 'typical Mongoloid' populations, the traditional view has been one of north to south gene flow along eastern Asia resulting in the

'Mongolisation' of aboriginal Southeast Asian populations of Australo-Melanesian-Negrito affinities (Coon, 1962; Coon *et al.*, 1950; Howells, 1960; Macintosh & Larnach, 1976; Thoma, 1973; Wu, 1959; Wu & Wu, 1984). However, the 'typical Mongoloids' of Northeast Asia are very specialised morphologically (the reason why they are so easy to characterise), and in evolutionary terms represent a derived form and not the likely ancestral condition to generalised groups such as Southeast Asians.

The view of Southeast Asia as a more generalised group from which the East Asian patterns derived has been developed by Turner II's work on dental morphology (1979, 1983, 1985a, 1986a,b, 1987, 1989, 1990a,b, 1992). Turner found that it is possible to recognise two dental complexes in southern and eastern Asia. The first of these is characteristic of the people of Southeast Asia, Polynesia and southern China, which he called sundadonty. It represents a general morphology, characterised by low frequencies of incisor shovelling, double-shovelling, lower first molar cusp 6, lower second molar cusp 5 and 3-rooted first molars. The second pattern is characteristic of Northeast Asian populations, as far west as lake Baikal. It represents a more specialised morphology, characterised by an intensified expression of the traits mentioned above. Turner proposed that sinodonty arose from a sundadont pattern, and therefore reversed the traditional view of a north to south direction of influence. This suggestion is based on two facts. Firstly, the polarity of the features involved which shows that the East Asian sinodont condition is the derived form; secondly, the existence of sundadont relics within the range of sinodont populations, which shows that the original inhabitants of East Asia were part of a wider sundadont range over which sinodonts later expanded. A Southeast Asian origin of all eastern Asian populations is also supported by genetic data (Ballinger *et al.*, 1992).

The origin of 'southern Mongoloids'

The findings of Turner II are fully consistent with those of multivariate analyses of cranial form, which show that Southeast Asians are less specialised morphologically and more closely related to a number of marginal Mongoloid populations, such as the Pacific islanders (Brace *et al.*, 1992; Hanihara, 1993a; Pietrusewsky, 1992). Therefore, it is among the southern Mongoloids that the origins of Mongoloids in general lie. However, the distinctions between the 'southern Mongoloids' and the Negritos, Melanesians and Australians, also indicate that this population itself is derived in relation to the original occupation of Southeast Asia by modern humans. Unfortunately, the fossil evidence is insufficient for tracing this evolution. It may be hypothesised however, that somewhere within Sunda a group differentiated, acquiring certain dental features (sundadonty), upper facial flatness (although maintaining a relatively large dentition and alveolar prognathism) and skeletal gracility. The dental pattern must have been acquired before 17 ka for it is present in several fossil remains (Turner, 1987), but the skeletal gracilisation seems to have occurred later (Bellwood, 1985, 1987). This population, ancestral to the 'southern Mongoloids', called 'proto-Malay' by T. Hanihara (1990a,b, 1991a, 1992c, 1993a,b) and Omoto (1984), subsequently dispersed over the range of the original group, of which a small number of relics survive in the Negritos. However, an important factor in this differentiation of Southeast Asian populations, away from an ancestral form shared with Australo-Melanesians, could have been gene flow from the west. It is possible that the dispersal of people from the northern route (North Africa–Middle Eastern secondary source) reached beyond India into Southeast Asia, and influenced the morphological changes in the region. Archaeological evidence would argue against this dispersal, for there is no evidence of Upper Palaeolithic technologies in Southeast Asia (Pope, 1988). Cranial multivariate analyses also suggest a clear separation between Mongoloid and Eurasian-North African populations (Howells, 1989). However, genetic data indicate a clustering of Eurasian groups that would point towards such gene flow (Nei & Roychoudhary, 1993).

The origin of typical Mongoloids

The derived morphology of sinodont populations appears in northeastern Asia in the late Pleistocene, before the expansion of rice agriculturalists. The Upper Cave remains have long been recognised as not showing typical Mongoloid features (Brace *et al.*, 1984; Coon, 1962; Han & Pan, 1984; Kamminga & Wright, 1988; Weidenreich, 1939; Weiner, 1971; Wu, 1961; Wu & Zhang, 1985). The results of the recent study by Kamminga & Wright (1988) show that the Upper Cave material has its closest affinities with recent Australians, which is consistent with the view that they were part of the widely dispersed Southeast Asian–Eastern Asian population descended from a group ancestral to both Australian and Asian populations. It has been argued that the features of typical Mongoloids reflect specific cold adaptations (Coon, 1962). As it happened in other regions of the world, it is possible that with the onset of glacial conditions the widespread population of eastern Asia contracted its range in its northern latitudes, resulting in a number of temporarily isolated groups. Under strong environmental pressure, morphological change could have become rapidly fixed in a population of small size. As conditions ameliorated, this population dispersed north into eastern Siberia and the Americas, west into Mongolia (where typical Mongoloids appear at the beginning of the Holocene displacing a local population of Eurasian affinities, Tumen pers. comm.), and south into northern Southeast Asia. Holocene population movements across northern Eurasia have been extensive, and this must have been an important area of genetic admixture between people derived from Eurasia and those from eastern and southeastern Asian origins.

In China, there is morphological continuity from the early Neolithic (7000 BC) to the present (Han & Pan, 1984; Howells, 1986, 1989). In terms of present morphologies, although the East Asians show a relatively large number of features significantly different from other regional populations, these are largely correlated among themselves. They have two main specialisations, one in terms of absolutely large temporal muscles, that result in small distances between left and right temporal muscles across the vault, wide bizygomatic breadths and eversion of the lower border of the malars, and the other in terms of tall faces and noses, with small interorbital distances. The curvature of the posterior alveolar plane of the maxilla, absent in the fossils with vertically large faces, indicates that the tall face of East Asians is not a primitive retention, but rather a secondary specialisation in this group. With the exception of the lack of prognathism among Europeans, the East Asians are the only group to show facial specialisations.

The Ainu as a sundadont relic

The Ainu population of the northern Japanese islands of Hokkaido and neighbouring Russian island of Sakhalin have been long recognised as distinct from the rest of the Japanese population (Brace *et al.*, 1989, 1990; Howells, 1966, 1986; Pietrusewsky, 1992; Turner, 1979; Yamaguchi, 1982). It is known that around 2 ka, a sinodont population from mainland Asia entered the island of Japan and displaced the local population, of which only the Ainu survived. The Ainu seem to be descended from the Jomon population of prehistoric Japan (Brace & Nagai, 1982; Dodo, 1986; Dodo *et al.*, 1992; T. Hanihara, 1989a,b,c, 1990c, 1991b,c, 1992d; Mouri, 1986, 1988; Ossenberg, 1986; Turner, 1976, 1986a, 1989), which is also different from other recent Japanese (Brace & Nagai, 1982; Brace & Tracer, 1992; Brace *et al.*, 1989; Dodo *et al.*, 1992; Gates, 1956; Howells, 1983; Turner, 1990a). Furthermore, the Minatogawa remains, dated to 17 ka (Matsu'ura, 1982; Suzuki & Hanihara, 1982) have been found to show affinities towards the Jomonese–Ainu line (Brace & Tracer, 1992; Suzuki & Hanihara, 1982; Turner, 1983, 1987). The Minatogawa fossils, Jomonese and Ainu crania all exhibit sundadonty, and probably represent the northern extent of the early sundadont population over which sinodont peoples expanded. Japan was connected to the mainland during pleniglacial times. The sinodont population that entered Japan between 300 BC and AD 700 were the Yayoi and Kofun groups, who are the main ancestors of current Japanese (Dodo & Ishida, 1990; K. Hanihara, 1985, 1987, 1991; Kanaseki, 1956, 1966; Mizoguchi, 1988; Yamaguchi, 1982, 1985).

The agricultural dispersals into Southeast Asia

Bellwood (1985, 1986, 1987) has argued that there has been important gene flow from southern China into Southeast Asia in recent years (within the past 5000 years) associated with the spread of farmers speaking of Austronesian languages. He has suggested that this flow of people originated around the area of the Yangtze delta, where large scale rice cultivation is observed by 5000 BC, reaching Thailand about 1000 years later (Yen, 1977). This early farming population is believed to have first dispersed into Taiwan, taking a branch of the Austro-Thai languages and agriculture-related vocabulary, which developed into the proto-Austronesian languages. This population may be represented in the Ta-p'en-'eng culture of Taiwan around 4000 BC. These early farmers later dispersed into Southeast Asia and the Pacific, with the Malayo-Polynesian branch of Austronesian languages, which at this stage acquired a number of words related to the use of tropical plants. This agricultural dispersal would have been superimposed over the range of foraging groups, and the resulting admixture has created a clinal pattern of morphology. Such clinal variation has also been identified by Turner in the frequencies of sundadont and sinodont traits in the populations of Hong Kong, southern China and Taiwan (Turner, 1990a). The Negritos of Malaya and Philippines are considered to have 'borrowed' Austronesian languages from their agriculturalist neighbours, with whom they trade extensively (Griffin, 1985).

The spread of rice agriculturalists had no penetration in the island of New Guinea. It has been suggested that because the inhabitants of New Guinea were already horticulturalists and supported large population numbers in comparison to foraging societies (Golson, 1977, 1981), they were less susceptible to the influences of new subsistence strategies (Bellwood, 1987).

Polynesia

The origins and affinities of the Polynesian islanders have been much investigated. In recent years, Brace and co-workers (Brace & Hunt, 1990;

Brace *et al.*, 1989, 1990) and Katayama (1990) have suggested that the Jomon population of Japan is the ancestral source of Polynesians, based on cranial similarities between the two groups. However, other cranio-dental studies indicate that Polynesians and Micronesians represent a late dispersal from Southeast Asia (T. Hanihara, 1992b, 1993a; Harris *et al.*, 1975; Howells, 1970, 1976, 1989, 1990; Kirch *et al.*, 1989; Pietrusewsky, 1983, 1984, 1985, 1988, 1990a; Turner, 1987, 1989, 1990a,b; Turner & Scott, 1977; Turner & Swindler, 1978). T. Hanihara has suggested that the close similarities between Pacific and Jomonese crania may result from the fact that these two groups share a close common ancestor in Southeast Asia (T. Hanihara, 1993a).

It is believed that the colonisation of Polynesia started between 6 and 3.5 ka with a dispersal from Southeast Asia to Melanesia (Bellwood, 1975, 1978, 1985; Serjeantson, 1984), from where (New Ireland, New Britain) the Lapita cultural complex dispersed further into the Polynesian islands (Green, 1979; Kirch, 1986; Kirch & Green, 1987; Kirch *et al.*, 1989; Pietrusewsky, 1985). However, Melanesian traits have been identified in some Lapita skeletons, suggesting that the Lapita population could have been of Melanesian and not southern Mongoloid origin (Pietrusewsky, 1985, 1989a, 1989b; Turner, 1989), also supported by the lack of the mtDNA 9bp deletion in Lapita samples (Hagelberg & Clegg, 1993). Micronesia was apparently colonised from another source within Southeast Asia (Bellwood, 1978, 1985; Howells, 1990; Pietrusewsky, 1990a, Turner, 1990b). Recent mtDNA studies of a variety of Polynesian and Melanesian groups identified three very distinct lineages, with a point of coalescence at 85 ka (Lum *et al.*, 1994). The three lineages group (1) remote Oceania (Hawaii – 95%; Samoa – 90%; Tonga – 100%) with Indonesians, Native Americans, Micronesians, Malaysians, Japanese and Chinese; (2) less than 10% of Hawaiians, Samoans, the population of the Cook Islands and Papuan Melanesia; and (3) Samoans and Indonesians. These results are interpreted as demonstrating several sources for present Polynesian islanders, including Melanesian populations (Lum *et al.*, 1994). In terms of the model discussed here, the inclusion of Melanesian genes in the make-up of Polynesians would explain the very distant point of coalescence found by Lum and co-workers.

Americas

The timing and character of the colonisation of the Americas is a controversial issue. Researchers disagree as to whether it was an early (40–30 ka) or late (15–10 ka) event, whether it was a single early source or several small groups over a long period of time, and whether the colonisation occurred through an inland route or along the western North American coast. These issues have been much discussed. Archaeologically, there are those researchers who see in the chronologically secure and typologically patterned archaeological record after 12 ka, the earliest evidence of occupation (Grayson, 1988; Haynes, 1987; Jelinek, 1992; Lynch, 1990; Meltzer, 1993; Taylor *et al.*, 1985). Others accept the scant and archaeologically unpatterned data from few South American sites as evidence of an early occupation from about 35 ka (Dillehay & Collins, 1988; Gruhn & Bryan, 1991; Guidon & Delibrias, 1986). The hypothesis of a relatively late entrance to the Americas has also received biological and linguistic support. Greenberg *et al.* (1986) proposed a three migration model of colonisation, in which the first wave of migrants would have entered the continent around 12 ka, speaking a language ancestral to the Amerind superfamily of languages, and given rise to all the southern, central and most of the northern Native American populations. The ancestral population of the Amerinds is proposed to have been an Asiatic sinodont group. The second migratory wave would have been of Na-Dene speakers who occupied the interior of Alaska and the North American Pacific coast, while the third wave would have been that of the Eskimo, speakers of Aleut-Eskimo languages, who colonised the Arctic and sub-Arctic regions. However, this model has been criticised, mainly for the lumping of all non-Eskimo and non-Na-Dene languages and peoples into a single hyper-variable group 'Amerind' (Campbell, 1988; Gruhn, 1987, 1988; Morell, 1990; Nichols, 1990; Ossenberg, 1992; Szathmary, 1993; Torroni *et al.*, 1994).

In evolutionary terms, the Asian origin of Amerindian groups is not contested (Gilbert & Gill, 1990; Howells, 1973; Stewart, 1973; Turner, 1971, 1979, 1985a,b, 1986b, 1989). However, it is also generally accepted that there is more morphological variation in the Americas than among typical Mongoloids (Bryan, 1978; Davis *et al.*, 1980; Givens, 1968a,b; Hrdlička, 1937; Neumann, 1952; Stewart & Newman, 1951; Sciulli, 1990). All Amerindian fossil remains are Holocene, so that their temporo-spatial distribution does not contribute to the debate. However, the morphology of all fossil North American remains and the Lagoa Santa fossils from Brasil has been recently studied by Steele & Powell (1992) and Neves & Pucciarelli (1991) respectively, who find that these fossils do not conform to a typical Mongoloid morphological pattern. Research on the morphology of Fueguian and Patagonian populations reinforces this view (Lahr, 1995), indicating the possibility that less specialised Mongoloid groups belonging to the sundadont population that inhabited most of southern and eastern

Asia during the late Pleistocene (and of which the Ainu are a relic) also entered the American continent. Recent research into the dental morphology of Central and South American groups would suggest that sundadont frequencies may be found within the Americas (Haydenblit, unpublished data; Powell, 1993).

In the past few years a large number of studies tackled the problem of mtDNA diversity within native Americans. Schurr *et al.* (1990) proposed that most Amerindian mtDNA variation could be accommodated within four major haplotypes (named A to D). These findings were subsequently supported by other studies, and the four haploytpes have come to be considered the four founding maternal lineages to enter the Americas (Ginther *et al.*, 1993; Horai *et al.*, 1993; Shields *et al.*, 1993; Torroni *et al.*, 1993a,b; Ward *et al.*, 1991). The absence of haplotype B (associated with the 9bp deletion) in certain South American groups has been interpreted as reflecting an early wave of occupation lacking this mtDNA lineage (Torroni *et al.*, 1994), and therefore refuting the phylogenetic integrity of all 'Amerind' populations. Haplotybe B has not been found in either the 10–9.5 ka Acha mummy from Chile (Matheney & Woodward, cited in Turbońn *et al.*, 1994), the 9 ka remains from Florida (Pääbo *et al.*, 1988) and Fueguian individuals (Turbón *et al.*, 1994), supporting the view that early South American groups (of which the Fueguians are hypothesised as relics – Lahr & Haydenblit, 1995) lacked this haplotype. However, the geographical distribution of haplotype B in South America proposed by Torroni *et al.* (1994) has been questioned by other studies (Bailliet *et al.*, 1994), as has indeed been the uniqueness of the four lineages (Bailliet *et al.*, 1994; Merriwether *et al.*, 1993). Furthermore, the genetic models proposed by Rogers & Harpending (1992) and Harpending *et al.* (1993) clearly indicate that small isolated populations (such as the Onas of Tierra del Fuego) would have been more prone to lineage extinction than groups of faster demographic growth, and therefore the absence of the 9bp deletion in these groups may be just a factor of drift. As clearly argued by Cann (1994), the genetic studies of Native American groups are still developing a clear view of variation in the continent.

Immigrations and wanderings

It is clear that the initial colonisation of Australia, the Americas and Polynesia were not dispersals but immigrations and wanderings into unoccupied lands. However, while Australian morphological patterns of diversity suggest a founding effect at the root of the occupation of the continent, the variability between Amerindian groups can be interpreted in various ways. It is possible that indeed the Amerind mega-cluster of Greenberg *et al.* (1986) is the result of a founder group which developed great biological diversity as it expanded into the vast ecological diversity of the Americas. On the other hand, it is possible that several groups dispersed into North America, for a period of time. In this case, part of Amerindian diversity may reflect the diversity of Northeast Asian populations in the late Pleistocene, where sundadont (Minatogawa-Jomon-Ainu), sinodont (typical Mongoloids) and western (Upper Palaeolithic Siberia) populations existed.

Patterns, processes and mechanisms

The previous sections address a number of issues relating to modern human prehistory, describe some of the data relevant to these questions, and offer an interpretation within a framework of a single origin of modern humans in Africa, followed by multiple dispersals in Africa, out of Africa and outside Africa.

These interpretations suggest that the history of each population is the consequence of the specific chain of events within each region of the world, so that the processes that created modern regional diversity are unique to each case. Therefore, taking the patterns of variability as an indicator of the history and composition of the population, the high levels of variation, as seen in the sub-Saharan African and to a lesser degree the Southeast Asian samples, can be taken to indicate high levels of population differentiation within these groups. Indeed, on a regional scale as used in this study, sub-Saharan African crania actually represent several markedly different populations. If African groups have their origin in early modern dispersals within Africa, the differences between San-Khoi, Bantu and Pygmy, would be phylogenetically comparable to those between large groupings, like Europeans, East Asians or Australians. This differentiation within Africa may be a function of sub-Saharan African ecology promoting isolation and diversification, but genetic data suggest that it is also a function of the extent of time that modern people have inhabited the continent. Therefore, Africa is a continent where the populations have deep ancestral roots, and express this long-standing existence in high levels of differentiation, both genetic and morphological. In Southeast Asia, the high levels of variability most probably also reflect the sampling of different populations, like the Negritos and the 'southern Mongoloids', but in this case the high level of admixture between groups also plays an important role. More than the effects of time, the demographic history of this region must have been

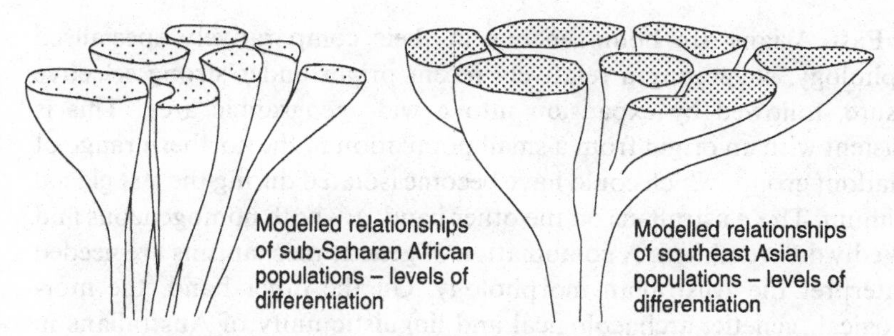

Figure 11.9 Schematic representation of the differences in origin of the morphological variabitity observed in sub-Saharan African and Southeast Asian crania.

strongly influenced by the formation and break-up of the Sunda landmass. Furthermore, the specialisation of peoples into exploiting different tropical environments within the area, like coastal, riverine or highland forest, and different subsistence strategies, like foraging, fishing, horticulture or agriculture, allowed the maintenance of a mosaic of populations, with very many levels of admixture. Genetic data suggest that indeed rainforest habitats allow the formation of genetic isolates (Stine *et al.*, 1992). Therefore, in terms of dispersals, African diversity may be interpreted as the result of early population expansions which maintained relatively independent geographical limits, while Southeast Asian diversity could be the consequence of the superimposition of the ranges of several groups and their various levels of admixture (Figure 11.9). Such a mosaic of overlapping old and new archaeological traditions, rather than a succession of cultures, is typical of island Southeast Asia (Coutts, 1984; Hutterer, 1976).

On the one hand, the low levels of variability may be the result of different processes, like a genetic bottleneck, followed by drift and founding, or the effects of very strong selection, or yet the results of very high levels of intra-regional gene flow. All of these processes may have acted to create the comparative homogeneity observed in Europeans, East Asians and Australians. The fact that Europeans are relatively homogeneous, but without a very derived morphology in comparison to other regional groups, would exclude both founding effects and strong selection. The levels of population movement within Europe, between Europe and western-central Asia, and between Europe and the Middle East, suggest that strong gene flow throughout the period maintained the morphological characteristics at a regional scale. This gene flow is expressed differently in different features. Some show an east–west cline, probably reflecting the last major genetic event with the introduction of agriculture (Cavalli-Sforza, 1991; Cavalli-Sforza *et al.*, 1994), while others reflect a north–south cline.

The East Asians are homogeneous in their comparatively specialised morphology, suggesting a relatively recent origin under strong selective pressure, followed by expansion into a wide geographic area. This is consistent with an origin from a small population in the northern range of sundadont groups which could have become isolated during the last glacial maximum. The Australians on the other hand, are both homogeneous and markedly differentiated. A combination of genetic mechanisms are needed to interpret the Australian morphology. On the other hand, the morphological, genetic, archaeological and linguistic unity of Australians in relation to other Southeast Asian populations strongly suggest that the population of that continent is the result of a founding effect. Sometime after the original occupation, the population expanded and a period of morphological variability is observed. At a certain point in its history, this population found itself under very strong selective pressures that favoured a number of morphological features, which may account for the intensification of certain characteristics in this population (and possibly in the Fueguian-Patagonian remains as well). Demographic reconstructions of the occupation of Australia show that there was an important decrease in population size or, at least, in the geographical area occupied by population groups around the last glacial maximum. Such a period of either small population size or higher population densities, followed by subsequent geographic and/or demographic expansion, would provide the conditions for a process of either drift or selection creating the morphological patterns observed in Holocene Australian Aboriginals.

Reconstructions of past and present diversity

Howells' (1989:66) multivariate analysis of worldwide cranial shape shows some important differences from the lineages reconstructed here (Figure 11.10a,b). The main difference is the cluster of African with Australian and Negrito crania. The pattern obtained by Howells could reflect important gene flow across Asia, creating a Eurasian mega-cluster. As mentioned before, such gene flow could have occurred in at least two independent situations. In the first case, it is possible that after 40 ka, when Upper Palaeolithic populations expanded from the northern source in North Africa/Middle East, the population that dispersed eastwards towards the Indian subcontinent reached parts of present mainland Southeast Asia mixing with modern populations already in the area. In the second case, gene flow between western Siberian populations deriving from the northern route and expanding sinodont populations in the late Pleistocene could

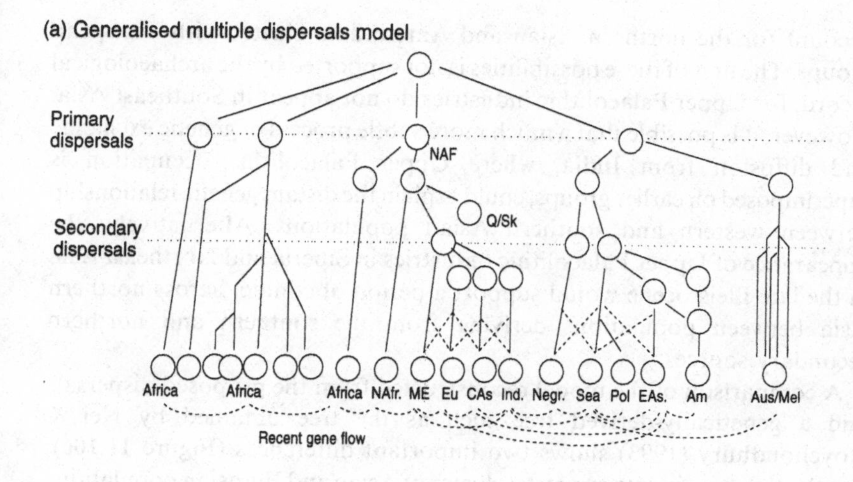

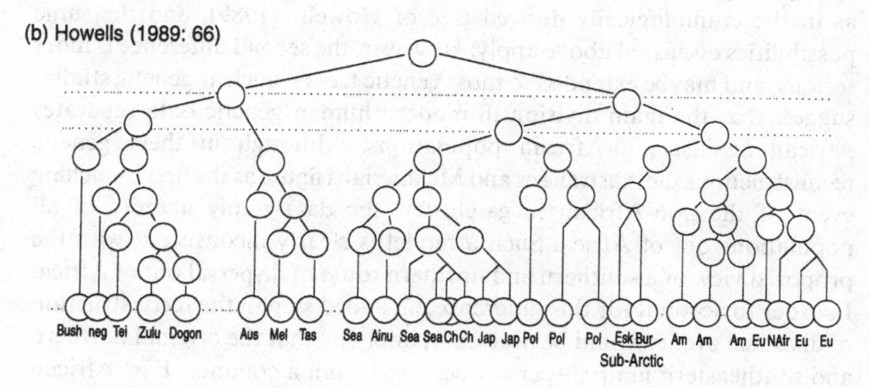

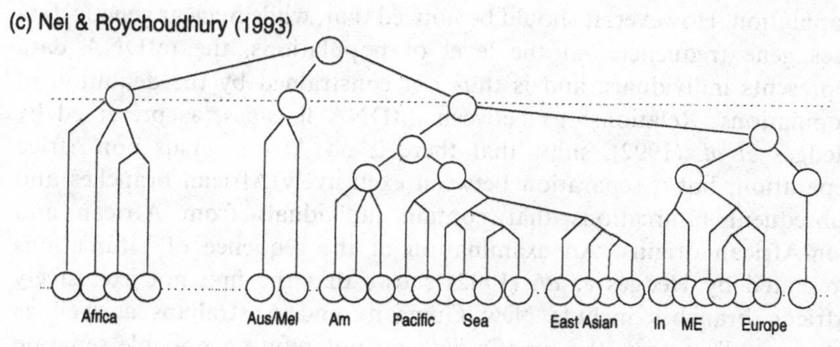

Figure 11.10 Comparison of the models of relationships and processes of differentiation of modern populations proposed by Howells' analyses of cranial shape (1989) and Nei & Roychoudhury's (1993) genetic data with those proposed by the Multiple Dispersals Model.

account for the northern Asian and Amerindian cluster with European groups. The first of these possibilities is not supported by the archaeological record, for Upper Palaeolithic industries do not appear in Southeast Asia. However, it is possible that a much more subtle process of genetic exchange and diffusion from India, where Upper Palaeolithic occupation is superimposed on earlier groups, could explain the distant genetic relationship between western and southern Asian populations. Alternatively, the appearance of Upper Palaeolithic industries in Siberia and Northeast Asia in the late Pleistocene would support a period of contact across northern Asia between populations deriving from the southern and northern secondary sources.

A comparison of the model reconstructed from the proposed dispersals and a genetically-derived tree such as the tree obtained by Nei & Roychoudhury (1993) shows two important differences (Figure 11.10c). Firstly, the genetic data suggest a cluster of Asian and Eurasian populations as in the craniologically derived tree of Howells (1989), and the same possibilities discussed above apply. However, the second difference is more serious, and maybe extended to most genetic trees – nuclear genetic studies suggest that the main division in modern human genetic data separates African against non-African populations. Although in these genetic reconstructions the Australians and Melanesians figure as the first branching event of the non-African mega-cluster, the data imply a unity of all populations out of Africa. Such a model is clearly inconsistent with the proposed view of a southern and northern route of dispersal out of Africa. In order to account for this difference, an added step in the reconstruction of dispersal events would be necessary, implying that the original northern and southeastern jump dispersals occurred from a common East African population. However, it should be noticed that, while nuclear genetic data uses gene frequencies at the level of populations, the mtDNA data represents individuals, and is thus not constrained by the definition of populations. Relationships between mtDNA lineages, as presented by Hedges *et al.* (1992), show that there is no Africa versus non-Africa separation, but a separation between exclusively African branches and subsequent bifurcations that contain individuals from African and non-African origins. An examination of the sequence of bifurcations presented by Hedges *et al.* (1992) shows that the first not exclusively African branches include New Guineans and Australians as well as Africans. Therefore, the genetic data do not refute a possible separate dispersal leading to Australo-Asian populations.

It could also be argued that the southeastern dispersal never occurred. However, two important points have to be taken into consideration.

Phylogenetic reconstructions of the evolution of modern diversity based on genetic and recent cranial data are strongly affected by the levels of gene flow resulting from recent population expansions, which the record shows were a fact of post-glacial populations almost worldwide. In contrast, the model of multiple dispersals takes the actual fossil and archaeological record into account. In this sense, it remains true that Australia was not part of the Upper Palaeolithic expansion, that the northern Middle Eastern route was not used by modern humans between 80 and 45 ka when Australia was colonised, and that the Upper Palaeolithic population that expanded out of North Africa and the Middle East at 40 ka is morphologically very different from fossil and recent Australian crania. All these points would indicate that a group of modern humans, independent of the Upper Palaeolithic expansion into Eurasia at 40 ka, dispersed out of Africa during the first half of the Upper Pleistocene to give rise to early Southeast Asian and Australo-Melanesian populations. This could have been achieved by two means. Either a southeastern route from East Africa towards southern Asia as suggested here, or by a dispersal of the early modern population in the Middle East (Skhūl and Qafzeh) towards India and Southeast Asia during the last Interglacial. Current biogeographical evidence does not support such a dispersal at this time, but until the Indian and Southeast Asian climatic, archaeological and fossil records are better known, this issue will not be resolved.

By identifying primary and secondary sources of dispersal, I have grouped populations into major lineages, originating from the first dispersals within Africa and out of Africa (Figure 11.11). Such a reconstruction implies that differences between some African groups are of similar magnitude as all Eurasian or all Mongoloid populations, although admixture resulting from recent expansions of post-glacial and of agriculturally related peoples may have erased some of the patterning. I have hypothesised that each of the two secondary sources of dispersal represent the source of another human lineage. Therefore, the Southeast Asian secondary source would have given rise to a lineage containing the present Australians, Melanesians, Negritos, certain peoples of India, southern Mongoloids, typical Mongoloids, Polynesians, Micronesians and Amerindians, while the northern African secondary source would have given rise to a lineage containing present North Africans, Middle Eastern, European, Central Asian, Indian and western Siberian populations. However, it is clear that the ranges of these lineages are not discrete today, nor were they in the past. As discussed above, during the late Pleistocene there was major overlap of the two lineages in the Indian subcontinent, across Siberia, across the Sahara, and possibly in Southeast Asia during the

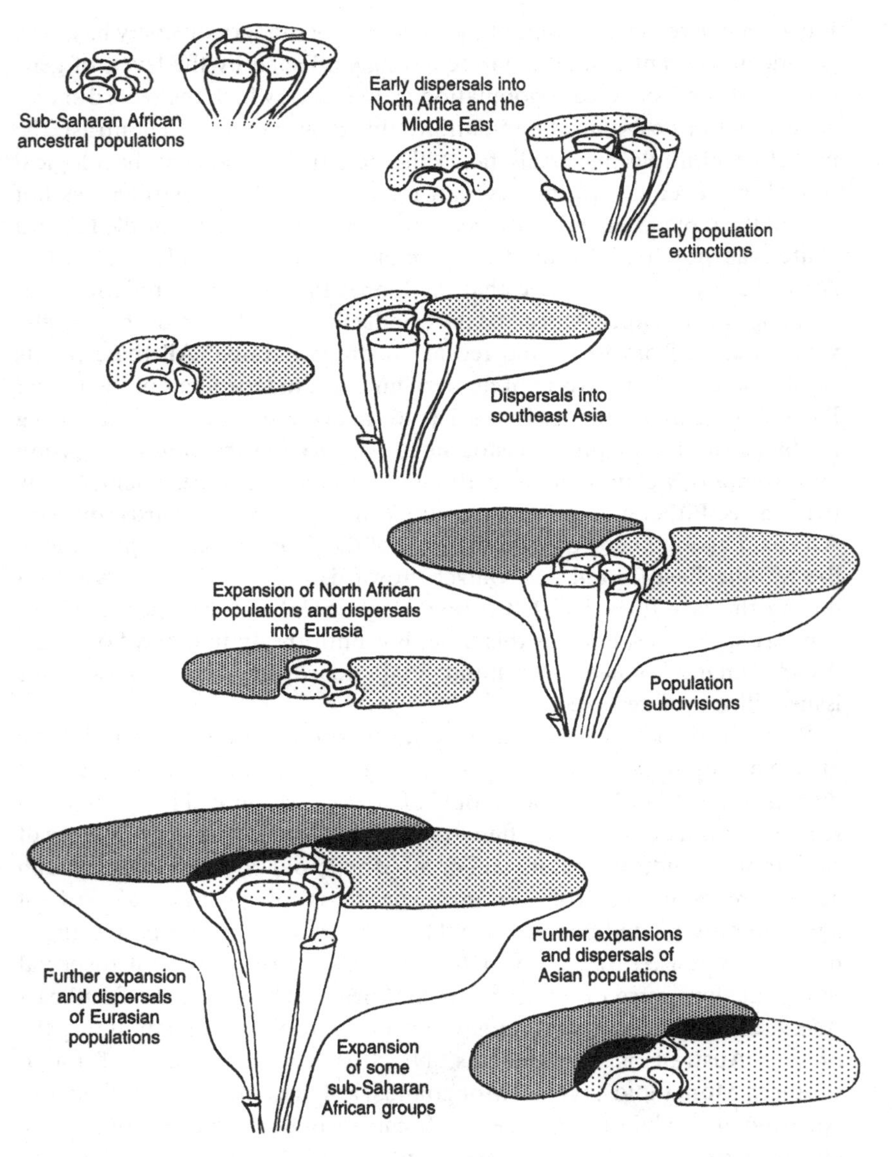

Figure 11.11 Schematic representation of the relationships of the typically defined three 'racial' clusters (Negroid, Caucasoid, Mongoloid), showing the difficulty in applying such divisions and definitions when studying modern diversity from an evolutionary perspective.

expansion of moderns from the Middle East at around 40 ka. If these reconstructions are correct, the only area outside Africa which was not influenced by the major expansion of modern humans from North Africa around 40 ka was the Sahul landmass. In recent times, the expansion of post-glacial, agriculturalist and later populations changed the density at which most of the world was occupied, and partly masked earlier Pleistocene population patterns.

The origins of modern human cranial diversity

As the modern human origins debate moves closer to accepting a single origin of modern humans in the late Middle Pleistocene of Africa, there is the need for a formal hypothesis to account for the process of modern differentiation from this ancestral source. The model for the evolution of modern cranial diversity involving multiple dispersal events proposed by Lahr & Foley (1994) and expanded here stands as such a hypothesis. As any other working hypothesis, this model can be tested with more detailed regional studies, new fossil, genetic and chronological information, a thorough integration with the archaeological record and eventually be supported or refuted, partly or wholly, by the new data.

12 *Final conclusions*

The evolutionary origin of modern humans has always been a controversial subject of study. Different theoretical assumptions and interpretations of the fossil record exist, and in recent years research has become polarised into two main opposing hypotheses. Both these hypotheses are concerned with explaining the origins of modern humans from an archaic hominid ancestor, either from a single or multiple geographical and biological source of ancestry. However, the origin of modern human diversity, an issue that is intimately linked with the process of origins of modern humans as a whole has received much less attention.

In evolutionary terms, the Multiregional Model offers an explanation for the evolution of modern human diversity. By proposing that a modern morphology developed from the regional archaic populations in several regions of the world, this model implies that the modern cranial form was superimposed on existing regional variations, and the differences between recent populations would be as old as the establishment of the archaic regional pattern. Therefore, this model sees regional differences as very ancient and very stable, and the process of modern diversification could be de coupled from that of modern origins. In contrast, the theoretical basis of the opposing view, i.e. a single and recent origin of all modern humans, implies that the origin of the species and that of the groups within the species were the result of different processes occurring at different times. Therefore, it is impossible to address the problem of the evolution of modern diversity without addressing the issue of modern human origins first, since the model of human origins accepted will determine the mode of diversification.

To explore the patterns and processes of the evolution of modern human diversity, either the multiregional or the single origin model has to be assumed with confidence. This book has thus begun with a review of the two opposing views on the origins of modern humans and the existing supporting evidence. The literature indicates that most lines of research, whether palaeontological, chronological, biogeographical, demographical or genetic, indicate that a single and recent origin of all modern humans took place. This evidence rests on the early dates of some African and

Middle Eastern *Homo sapiens* fossils that antedate the appearance of modern people elsewhere, on the apparent contemporaneity of supposedly ancestral-descendant populations in some areas, on the patterns of faunal movements and associations with the hominid fossils suggesting the dispersal of early modern humans out of Africa, on reconstructions of past demographic parameters showing that gene flow could not have been maintained until a critical world population size was achieved late in the Pleistocene, and on extensive nuclear and mtDNA genetic studies which indicate the high probability of an African ancestry of all modern people in the recent past. In spite of this impressive body of evidence indicating a single source of ancestry of all modern people, the controversy is not resolved, for there are a number of cranial features that seem to link regional archaic populations with the recent modern inhabitants in those regions, and thus prove a multiregional pattern of descent.

Since the hypothesis of a single origin is supported by several lines of evidence, the solution to the problem lies in confirming or refuting the multiple archaic–modern regional lines of descent. The first section of this book examined these lines of descent in space and time, as well as the assumption that the particular features that reflect the multiregional evolutionary process (the 'regional continuity traits') can be taken to represent stable and independent phylogenetic markers. Therefore, the work presented in the first section of this book studied the problem of modern human origins from a different perspective from most other studies. It is argued that a re-evaluation of the character, temporo-spatial expression and phylogenetic validity of those traits proposed by the Multiregional Model as showing regional continuity through time would either substantiate or discard the morphological evidence for this model. A study of the regional distribution and anatomical relationships of these traits using a worldwide sample of modern human crania was carried out. Furthermore, using a worldwide cranial sample as representative of modern morphological variability, a standardised scoring technique was developed. This scoring system is based on the definition of the observed degrees of expression of a number of features into categorical grades, which in turn allowed quantification and statistical testing of their regional incidence.

The re-evaluation of the morphological evidence for multiregional evolution clearly shows that the particular features that have been proposed as characterising the evolution of Chinese and Javanese *H. erectus* into modern Chinese and Australian aboriginal populations fail to do so in several accounts. Firstly, in a number of cases the recent spatial distribution does not confirm certain features as typical of either Chinese or Australian Aborigines, either because their expression is not exclusive to those regions, or because they do not occur in these two populations at all. Furthermore, it was found that a number of supposedly Javanese–Australian features were not typical of the Ngandong sample, while in other cases these archaic hominids clearly show anatomical structures non-homologous with the modern condition. In terms of the temporal distribution of the 'regional continuity traits', it was not possible to identify a clear morphological lineage in the late Pleistocene modern humans of China, strongly rejecting a view of morphological stability over a period of a million years. Fossil modern humans in Australia share with recent inhabitants of that continent a morphological pattern of robusticity that supports the view of an evolutionary lineage in the late Pleistocene, although this particular pattern does not extend to Javanese *H. erectus* remains. The variable early modern sample from Africa and the Middle East was found to be a better model of the ancestral morphology of all recent populations based on the very features that have been argued to show multiregional evolution throughout the Pleistocene. In terms of the validity of this particular set of traits proposed by the Multiregional Model to represent stable phylogenetic markers, it was found that the high level of correlation with craniometrical parameters of the skull, as well as high inter-trait correlation, precludes their use as independent sources of data.

On the basis that the morphological evidence for regional lineages of descent from the early Pleistocene is partly incorrect and partly wrongly applied to evolutionary studies, the results obtained strongly refute a multiregional evolution of modern humans. In the absence of contradicting morphological evidence, a single and recent origin of all modern humans, as indicated by all other lines of evidence, can be assumed.

Accepting a single and recent origin for the modern morphology implies that all the differentiation between modern populations occurred subsequent to this evolutionary event. The second section of this book addressed the problem of understanding this process of diversification. The first step taken was to examine the levels of diversity within and between recent regional populations in order to establish the cranio-metrical parameters that define modern humans as a whole and their expression through time. This study led to the conclusion that very few cranial traits characterise all modern humans (a few features related to parietal and basicranial dimensions), the recent morphologies having developed in a stepwise manner in which features were acquired and lost at each node of diversification between groups. Furthermore, the observed temporal and spatial variation in primitive and specialised cranial traits, together with the patterns of regional differentiation among recent populations, indicates that the parameters used to describe modern humans should be widened to encompass a number of features that are currently considered archaic. This in turn, suggests that a number of fossil specimens in the African late Middle Pleistocene and early Upper Pleistocene which are usually excluded from the early 'modern' sample for exhibiting certain archaic features, are actually very important in explaining the variability observed in later crania, and should be included within a variable ancestral population.

These conclusions have strong implications for studies of late Pleistocene human evolution. Firstly, by recognising that there are few cranial apomorphies that characterise modern humans; secondly, by widening the range of variation accepted as characterising modern peoples, both in terms of specialised and primitive features; and thirdly, by recognising that this range of variation varied both spatially and temporally between modern populations. Human diversity increased and decreased in the past in relation to the level of diversification and extinction of different groups.

The process of human diversification in the Upper Pleistocene was examined by establishing spatial and temporal patterns of variation in cranial form. The time and rate at which different regional populations seem to have diversified during the late Upper Pleistocene suggest that this process was not synchronous throughout the world, showing varying levels of specialisation in relation to an ancestral morphology at any one time. Such differences are consistent with the view that different evolutionary mechanisms were acting on each population at any one time, reflecting the historical component in the determination of regional variation. In this context, the different levels of gene flow, selection and drift acting upon populations to create diversity would have been closely associated with the different geographical parameters of each group. Therefore, the specific circumstances affecting each population at different times had a historical component (in terms of the character and time of origin of the population) and a geographical component (in terms of the level of isolation in which that population is responding to the genetic mechanisms involved). These different components had a major role in determining the level of distinctiveness of populations. Groups like the San-Khoi and Australo-Melanesians are all examples of relatively early diversifications without great amounts of gene flow from other groups as reflected by the existence of clear evolutionary lineages of descent in the late Pleistocene and their levels of genetic diversity. This pattern probably applies to the Negritos and Pygmies, but the lack of any supporting fossil evidence precludes a similar conclusion. Other populations such as the Europeans also show a regional line of descent, but in this case it is postulated that large amounts of intra-regional gene flow have had a homogenising effect in the population of the area. Superimposed on these spatially and temporally specific evolutionary pathways, a common trend towards gracilisation is observed. The association between robusticity and size is the main formal structure constraining modern cranial form, and the variation in its expression the main trend of modern human cranial evolution.

The evidence of modern human diversification suggests that geographical differentiation was not only an integral part of the process, but that early geographical divisions account for the major divisions among humans today. However, the data further indicate that early geographical separation was not followed by gradual change at a regional level, but rather by a very dynamic process of population expansions and contractions, constantly altering the balance between stability and change. These patterns are best explained by a model of multiple dispersals from the ancestral source in Africa and from the secondary sources outside Africa, giving rise to more than one hierarchical chain of events. Depending on the level of ecological fragmentation determining the level of gene flow between groups, the process of multiple dispersals superimposed on the range of previous groups within each region will either create a mosaic of distinctive populations or relative homogeneous morphologies with some clinal variations. It is suggested that the tropical area of Southeast Asia, with its variable and numerous habitats, promoted such a mosaic of relatively distinct groups like the Negritos, southern Mongoloids and typical Mongoloids, while the Eurasian plains promoted long distance population contact and a level of biological and cultural homogeneity not observed in other parts of the world. In Africa, although populations expanded and contracted as in other continents, the biological data (both morphological and genetic) strongly suggest long-standing differentiation rather than several layers of populations. However, the fossil record of sub-Saharan Africa for most of the Pleistocene is very poor and lacking chronological control, so that substantiating fossil evidence for such a temporal pattern does not exist.

An important implication from these studies is the need to disassociate events in the history of the evolution of modern human populations. It is apparent that the genetic bottleneck that gave rise to modern humans as a group occurred before we were able to identify the anatomical changes associated with a modern form. Furthermore, the time of coalescence of mtDNA lineages probably precedes this event. In terms of behavioural changes, the appearance and expansion of modern humans is not associated to a significant behavioural event, and it is possible to hypothesise that whichever changes in behaviour that allowed a small population to survive the stringent conditions that drove it to reduce its population size dramatically, and later allowed it to expand over wide geographical areas, had already occurred when a modern morphology appears in the fossil record. This implies that such a behavioural change was not reflected in the type of stone tool technology, for late archaics, early modern humans and Neanderthals all share a technological complex. The distinctive Upper Palaeolithic of the late Pleistocene was a localised event associated with the expansion of modern humans into Eurasia, and not with modern human origins *per se*. This implies that particular tool technologies are not associated with biological levels of organisation, and therefore, their small stylistic variations most likely reflect the identity of the groups producing them. And lastly, it is clear that the origins of a modern morphology have to be disassociated from the process of skeletal gracilisation, which although characterises most modern populations at certain stages, never got to be a universal trend and was not part of the process of modern differentiation from an archaic source.

These last two points are very important in understanding the problems of interpreting the fossil record. Because certain aspects of either behaviour or morphology came to be widely distributed among recent modern humans (in this case the Upper Palaeolithic and skeletal gracilisation) they came to be associated with modern humans as a whole, and thus with the origins of the group. According to such a view, it is the few discrepancies, like the robusticity in early modern, Australian and Fueguian skulls, or the absence of Upper Palaeolithic technologies in these same groups, that need explaining in some special way. The Fueguian peculiarities are usually explained as extreme drift under extreme conditions, the Australian pattern as a local evolution from archaic forms, and the early moderns by excluding or including one or other specimen within an ancestral hypodigm. However, once the dynamic demographic processes of the past, reflected in the expansion and contraction of lithic traditions during the last glacial cycle are considered, it becomes clear that a large proportion of the patterning observed today is related to the marked expansion of groups after the last glacial maximum, during the Neolithic, and in more recent times. Therefore, it is the patterns preceding these expansions that should provide the most useful information on the original relationship of populations.

The spatial and temporal distribution of populations before the last glacial maximum shows clearly that these two features usually generalised as 'modern human features', i.e. cranial gracility and the Upper Palaeolithic, had a much more restricted distribution. All the Asian fossil remains at that time are relatively robust, showing that levels of robusticity comparable to those observed in Australia today were more widely dispersed, and that the Australians have retained (and possibly intensified) this ancestral condition.

Furthermore, the absence of Upper Palaeolithic tools becomes a reality of most of eastern and southern Asia and parts of Africa at that time, showing that the appearance of the Upper Palaeolithic was a late process into some of these areas. In evolutionary terms, these temporal comparisons show that what has been thought to be an apomorphy of all modern humans (and therefore excludes certain fossil and recent populations from the common ancestry), are instead particular synapomorphies of certain modern groups, disclosing the patterns of diversification within the group as a whole. This confirms the view that the Upper Palaeolithic and gracile skeletons are derived features of certain populations acquired after the origin of modern humans as a group. In the same way that modern populations like the Mongoloids are wrongly characterised by the apomorphic character of the most differentiated form within the group, while the remaining populations are somehow explained as resulting from admixture with other forms, modern humans cannot be characterised by the few apomorphies of a once small and presently large number of groups. The features shared by all modern peoples are those that differentiate the group from its ancestral and sister taxa, and those are neither the Upper Palaeolithic nor skeletal gracility.

Taking the features that characterise different modern populations, the question arises as to whether taxonomical distinctions at the subspecific level exist. Subspecies are defined as geographical variants of a species, and many times the terms subspecies and race are used interchangeably. However, the actual definition of subspecies requires that 75% of the individuals of one subspecies be distinguishable from 100% of individuals belonging to other subspecies of the same species (Mayr *et al.*, 1953), which statistically equates with 90% mutual non-overlap (Groves, 1989a). Modern human regional populations may be very discrete in some characters which could comply to the 75% rule, but not in their combined biological parameters. Furthermore, there is so much variation within these populations, that the thresholds of racial units are virtually impossible to determine. It could be argued that typical Mongoloids form a race defined on the basis of a number of features, like epicanthic fold, facial flatness, hair form and a blue patch on the base of the spine (Coon, 1962; Howells, 1959). These features are visible in the vast majority of typical Mongoloids, and virtually absent in Australians, Europeans or sub-Saharan Africans (although the San and Khoi may have epicanthic folds and certainly show facial flatness). Southern Mongoloids have, for the most part, most of these features, although in a much less distinct expression, and would thus have to be included in a Mongoloid race. Since southern Mongoloids have markedly interbred with Negritos, with whom they share a number of features and probably a remote common ancestor, the Negritos could be included within Mongoloids. However, Negritos share a number of features with Melanesians, who themselves received extensive gene flow from Polynesia, so it could be argued that all these groups should be included in a Mongoloid race. And clearly, whichever classification applies to Melanesian peoples should also encompass Australian aborigines. This grouping equates with the population lineage resulting from the southern dispersal. However, at this stage, the point of classifying the Mongoloids into a 'race' has been defeated, since Chinese and Australians show a number of morphological and genetic differences.

Any attempt at classifying modern human populations into a number of distinct races is doomed to face the same problems as briefly exemplified above. Taxonomical classifications are an attempt to order the variability of biological organisms into meaningful units. Most biologists today recognise that the most meaningful criterion for such classifications is phylogeny, whereby organisms are classified into hierarchical units that reflect evolutionary relationships. Classifications of human races have traditionally by-passed this principle, and classify very small and relatively homogeneous units in which clear distinctive features may be recognised as a 'race', while other less discrete groups of people are taken as the result of extensive admixture. This line of thought led to the belief that pure races existed, and that these pure races once had pure racial ancestors. Anthropological research has shown how variable modern peoples are, and that while populations like Mongols, Siberians, Chinese, Koreans, Japanese, mainland Indonesian, insular Indonesian, Negrito, Micronesian, Polynesian, Melanesian and Australian Aborigines may be characterised individually on the basis of a small number of features (although there are any number of morphological clines between some of them), these groups maybe joined into hierarchical groupings reflecting the actual evolutionary bifurcations. The ancestral form at the node of each bifurcation was less derived than the present descendant, so that the classification of those populations within the clade would have to include the ancestral relative generalised condition. A possibility is to separate clusters within a clade based on some important shared derived features. Accordingly, in the examples used above, it is possible and indeed common to separate the Australo-Asian clade into Australo-Melanesians and Mongoloids, on the basis that the latter group shares a number of traits that allow its characterisation. By this method one may reach some level of consensus in identifying population lineages that can be characterised morphologically as the result of an early differentiation later conserved in all populations deriving from it. However, this characterisation will not necessarily equate with the external features that allow the identification of recent populations, and because of the level of later differentiation and admixture, these lineages would not classify as races in the biological sense of subspecies. Furthermore, palaeoanthropological research has shown that given the morphological variability in early modern fossils, morphological unity did not exist even in the ancestral population of us all. This is consistent with the mechanics of evolutionary change, whereby the behavioural, genetic and morphological novelties of speciation events are not necessarily expressed at the same time.

This book has examined the problems of the origins of modern humans and modern human diversity by establishing spatial and temporal patterns of diversity in different sets of data and interpreting such patterns through the evolutionary process of biological differentiation. The data presented in this study support a single origin of modern humans in Africa in the late Middle to early Upper Pleistocene, followed by a process of multiple dispersals and early spatial differentiation throughout the Upper Pleistocene. This hypothesis rests on the integration of levels of morphological variation and distribution of archaeological remains, and is broadly consistent with statistical reconstructions of population relationships based on recent cranial and genetic data. However, the integration of different sources of prehistoric data is still very limited and faced with the problem of interpreting the patterns of cultural versus biological diversity, and those of recent versus past diversity. Discovering the specific history of each modern population, including all those whose existence we can only infer from the fossil record, remains the greatest challenge.

Appendix I Grading of the morphological 'regional continuity traits': Trait description and grade scores

Sagittal keeling: mid-sagittal crest, with parasagittal depression (Figure AI.1)

Definition: Sagittal keeling refers to the orientation of the parietal bones in the coronal plane, which may define a central keel or blunt ridge along the sagittal suture. This keeling may extend posteriorly to the lambda and anteriorly into the frontal bone, disappearing completely before reaching the supraorbital area, but the most usual form it takes is as a ridge or elevation from around bregma (where it is usually broadest) to the obelion area. This description differs from that of Weidenreich's (1943a) in two aspects: first, the keel or ridge is not necessarily highest within the bregma region, and second, parasagittal depressions are not always found in association with sagittal keeling. This latter point is supported by the works of Larnach & Macintosh (1966, 1974) and Larnach (1978), who found that while sagittal keeling was present in 93.0% of New South Wales crania and 94.8% of Queensland crania, associated parasagittal depressions were found in only 2.6% and 0.9% respectively. Cases in which only keeling of the frontal bone (along the fused metopic line) was observed were not recorded as showing sagittal keeling. This kind of keel is common in sub-Saharan African crania, and it was also observed to certain a degree in the Fueguian and Patagonian sample.

Grades:

SK 1 – Rounded calvarium in the coronal dimension (Figure AI.1a).
SK 2 – Moderate angling of the parietal bones towards the sagittal suture, but always lacking a distinct ridge or 'crest' (Figure AI.1b).
SK 3 – Pronounced angling of the parietal bones towards the sagittal suture, with the formation of a distinct ridge or 'crest' along part or all of the sagittal suture (Figure AI.1c).

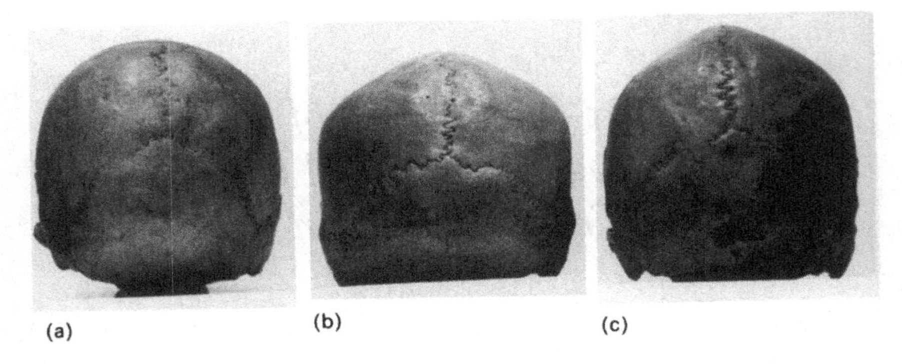

Figure AI.1 Sagittal keeling: (a) SK 1; (b) SK 2; (c) SK 3. For full explanation of grades see text.

Profile of the nasal saddle and nasal roof (Figure AI.2) and infraglabellar notch (Figure AI.3)

In studying the variability in shape of the nasal saddle, it was found that this is composed of two elements: (1) the curve or angle formed by the nasal bones with each other; and (2) the curve or angle between the nasal bones and the frontal bone in profile. Weidenreich also discussed these two components, coronal and sagittal, of the profile of the nasal bridge. The first one is considered to represent East Asian trait 3, profile of nasal saddle and nasal roof. The second one was found to co-vary so much with the degree of development of glabella, that it was impossible to treat them independently. Therefore, a grade score was made that combined degree of projection of glabella and sagittal angle of the nasal bones at around the naso-frontal suture, thus representing Southeast Asian trait 4, profile of the infraglabellar notch.

Definition: 'The bridge itself is formed by the two nasal bones and the two bridge portions of the maxilla which are bound by the naso-maxillar margin, on one side, and the crista lacrimalis anterior, on the other. Height and lowness of the saddle depend upon two factors: first on the breadth of the individual bones and second on the position in which they are placed to one another . . . The profile of the *Sinanthropus* nasal bridge shows a continuous curvature . . .' (Weidenreich, 1943a:pp.73–74).

Grades:

Profile of the nasal saddle and nasal roof

NS 1 – Flat nasal bones (Figure AI.2a).
NS 2 – Slightly raised nasals, forming a small curve (Figure AI.2b).

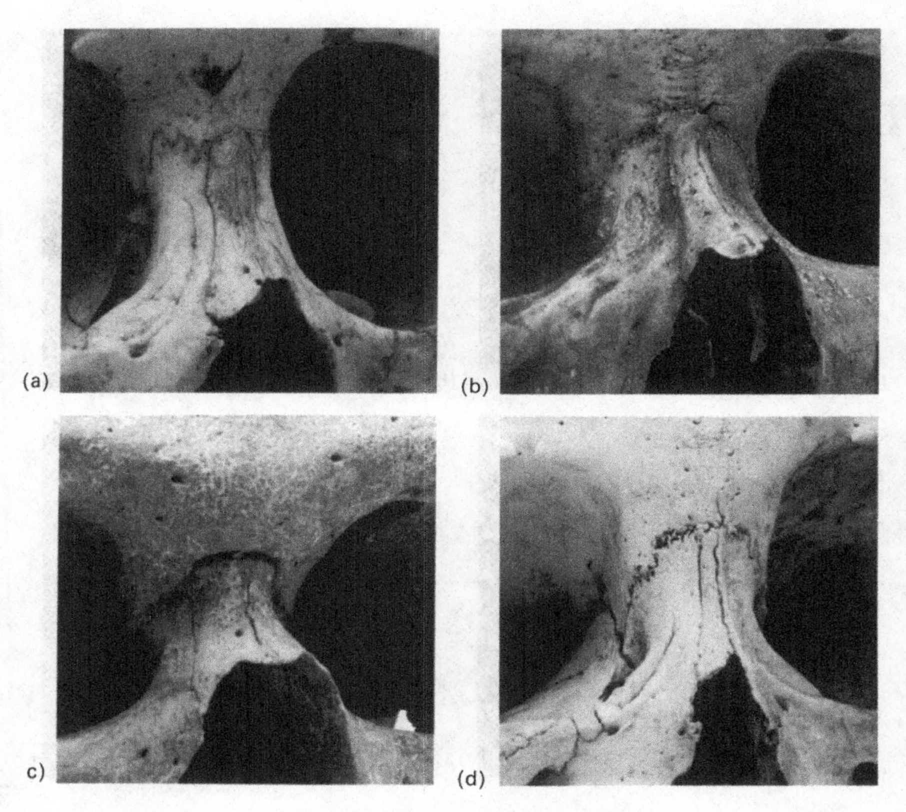

Figure AI.2 Nasal saddle and nasal roof: (a) NS 1; (b) NS 2; (c) NS 3; (d) NS 4. For full explanation of grades see text.

NS 3 – Nasals forming a well-defined curve, ranging in size from medium to large (Figure AI.2c).

NS 4 – Deep angled nasal bones, forming a 'pinched nose' (Figure AI.2d).

Profile of the infraglabellar notch

IN 1 – Non-projecting glabella and flat nasion; both landmarks on approximately the same level (Figure AI.3a).

IN 2 – Slightly or non-projecting glabella, but angled nasals, forming a slight curve in profile (Figure AI.3b).

IN 3 – Prominent glabella and relatively deep and wide nasion angle (Figure AI.3c).

IN 4 – Very prominent glabella, with very deep and narrow nasion angle (Figure AI.3d).

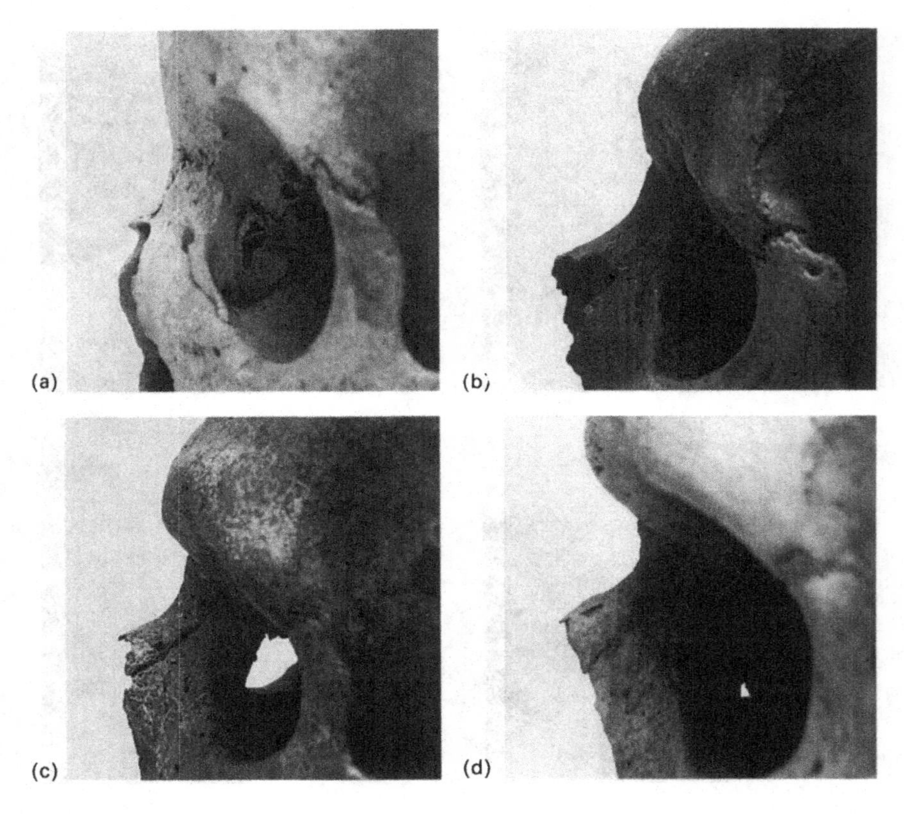

Figure AI.3 Infraglabellar notch: (a) IN 1; (b) IN 2; (c) IN3; (d) IN 4. For full explanation of grades see text.

Supraorbital ridges/torus (Figure AI.4)

Definition: The supraorbital area in modern humans can be either flat or projecting, with extreme cases showing pronounced ridges over the orbits. In the latter case, it is possible to recognise three elements within the supraorbital structure, the glabellar portion, the superciliary ridges and the lateral portion or trigone. The degree of development of these three elements does not always correspond, cases of very pronounced glabella and non-projecting superciliary ridges and vice-versa being observed. However, even in the case of the three elements being very pronounced, they do not, in most cases, form a true supraorbital torus, for a strong supraglabellar depression and a groove ascending obliquely upwards and outwards from the supraorbital notch separate the lateral trigones from the medial portion. Cases of a complete supraorbital torus in modern humans

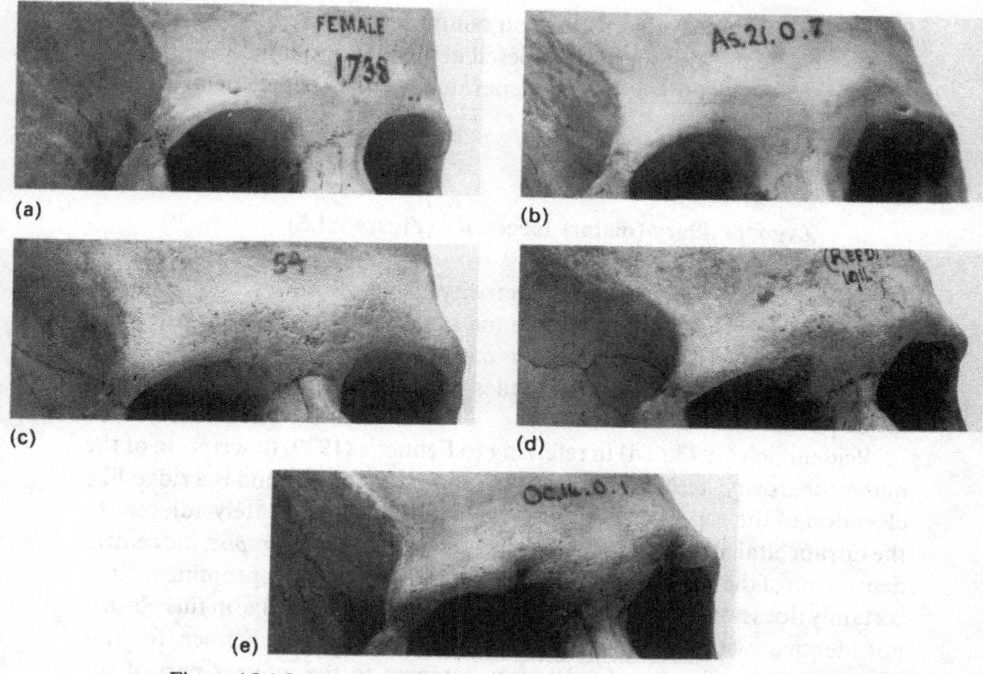

Figure AI.4 Supraorbital ridges/torus: (a) ST 1; (b) ST 2; (c) ST 3; (d) ST 4; (e) ST 5. For full explanation of grades see text.

have been observed (Burkitt & Hunter, 1922; Lahr, 1992), but these seem to be extremely rare.

Grades:

ST 1 – Flat or very slightly projecting superciliary ridges and glabella (Figure AI.4a).

ST 2 – Supercilliary ridges well-defined as two distinct units, separated medially by a flat glabella (Figure AI.4b).

ST 3 – Supercilliary ridges ranging from visible to well-developed, prolonged medially to form a prominent glabella[1] (Figure AI.4c).

ST 4 – Supercilliary ridges very pronounced, joined medially by a very pronounced and projecting glabella, clearly separated from the rest of the frontal bone (Figure AI.4d).

[1] The odd cases of a prominent glabella, but with little development of the superciliary arches, were classified as grades 2 or 3, according to the degree of projection of the arches. This combination was relatively frequent among the fossil skulls from the sites of Afalou and Taforalt.

ST 5 – Continuous ridge formation composed of the three superciliary elements, although the superciliary ridges and lateral trigone can still be identified, but not separated, as different elements (Figure AI.4e).

Zygomaxillary (malar) tuberosity (Figure AI.5)

Definition: The zygomaxillary tuberosity is an elevation on the malar surface between its orbital and free margins. This elevation may develop into a pronounced ridge, running parallel to the inferior free border (Fenner, 1939), which virtually divides the malar surface in to upper and lower portions.

Weidenreich (1943a:84) in referring to Fenner's (1939) description of the malar tuberosity states 'What Fenner apparently has in mind is a ridge-like elevation of the superior part of the malar surface immediately adjacent to the infraorbital margin. Such a feature is absent in *Sinanthropus*; the central depression of the surface... makes the superior part slightly prominent but it certainly does not deserve to be called a ridge. Besides, a ridge in this place is not identical with the malar tuberosity as suggested by Fenner, for this formation is confined under all circumstances to the inferior part of the surface of the inferior border.' Although the uncertainty about the occurrence of malar tuberosities among the *Sinanthropus* crania and the lack of facial elements among the Javanese fossils (the data on Sangiran 17 is contradictory – Habgood, 1989), this trait has continued to be used in the Multiregional Model. Furthermore, work by Larnach & Macintosh (1966) suggests that this trait may be sexually dimorphic: 'In 78.1% of the males (Australians), the malar tuberosity was moderately to markedly developed, while only 2.1% of the females showed this degree of development. In 97.9% of the females it showed little or no development against 21.9% of the males.'

Grades:

ZT 1 – The surface of the malar bone is smooth and flat (Figure AI.5a).
ZT 2 – Presence of a tubercle of small dimensions (Figure AI.5b).
ZT 3 – Tubercle present, more pronounced and horizontally extended (Figure AI.5c).
ZT 4 – Very large tubercle, forming a ridge along the surface of the malar, parallel to the lower free margin of the bone (Figure AI.5d).

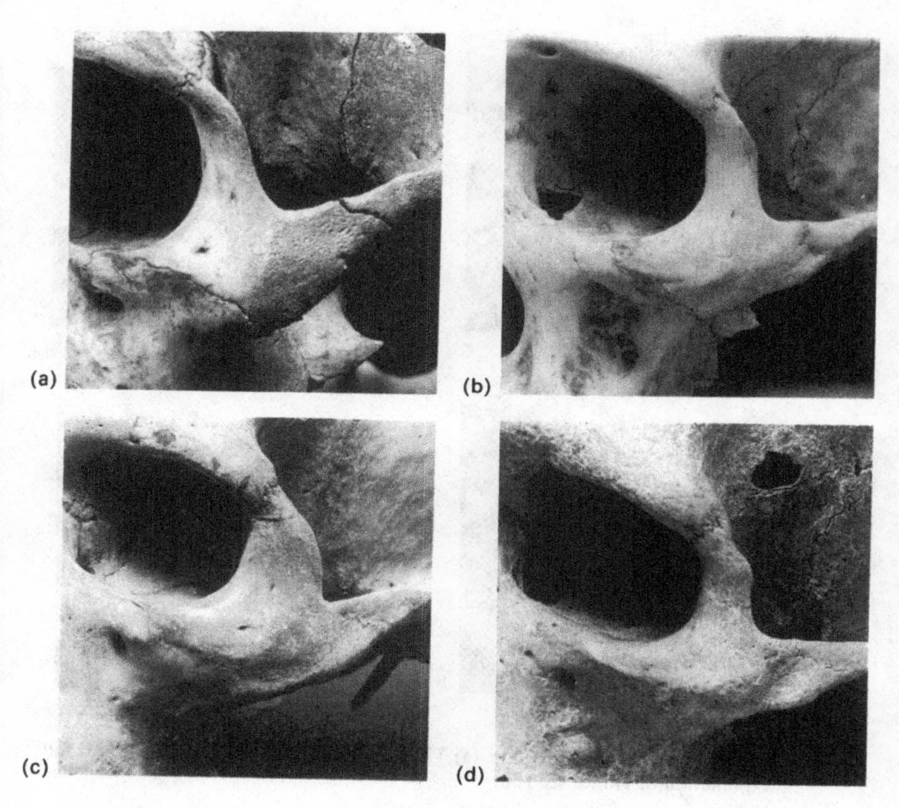

Figure AI.5 Zygomaxillary (malar) tuberosity: (a) ZT 1; (b) ZT 2; (c) ZT 3; (d) ZT 4. For full explanation of grades see text.

Zygomatic trigone (Figure AI.6)

Definition: The zygomatic trigone is the lateral portion of the supraorbital area formed by the frontal bone and the immediately adjacent portion of the frontal process of the zygomatic. In modern populations, this area has typically reduced in all dimensions (including its antero-posterior width) and it is a separate element in relation to the superciliary ridges. However, in some cases the zygomatic trigone can be rounded, prominent and well-developed.

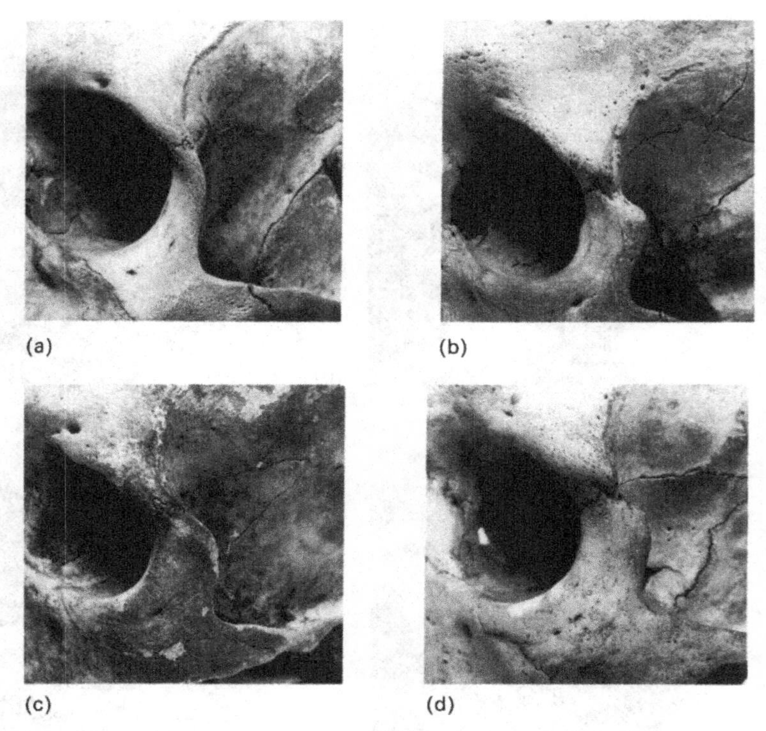

Figure AI.6 Zygomatic trigone: (a) TR 1; (b) TR 2; (c) TR 3; (d) TR 4. For full explanation of grades see text.

Grades:

TR 1 – Completely smooth trigone or very slightly salient, if at all. Generally very thin (Figure AI.6a).

TR 2 – Formation of a raised surface just adjacent to the fronto-malar suture, and extending anteriorly or posteriorly along the orbital margin or the temporal line, for up to 5 mm (Figure AI.6b).

TR 3 – The whole area of the trigone is inflated and widened, but retains a smooth surface (Figure AI.6c).

TR 4 – Pronounced development of the trigone area, very salient, with or without a rugged surface. The frontal region adjacent to the zygomatico-malar suture may be considerably larger than its relative malar portion (Figure AI.6d).

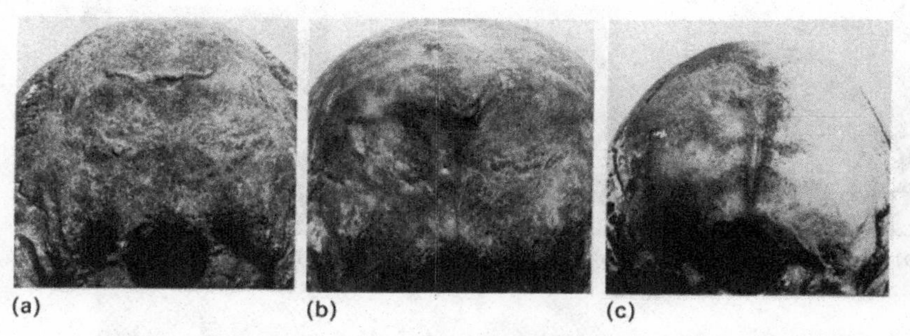

Figure AI.7 Occipital crest: (a) OCR 1; (b) OCR 2; (c) OCR 3. For full explanation of grades see text.

Occipital crest (Figure AI.7)

Definition: The external occipital crest is a bony ridge that separates left and right masses of nuchal muscles (*complexus superior* and *rectus capitis inferior*). It runs sagittally on the nuchal surface of the occipital bone, from the posterior border of the foramen magnum to the superior nuchal lines[2] (that mark the limit of the attachment of the nuchal muscles) (Hublin, 1978a). The occipital crest can be divided into superior and inferior portions as above and below the inferior nuchal lines (Waldeyer, 1880). The degree of expression of the external occipital crest is very variable. It may just represent a reflexion of lateral reliefs, with no protrusion, it may be a wide blunt area, or a line, or a pronounced ridge.

Grades:

OCR 1 – The occipital crest is only visible as a line or very slightly raised surface (Figure AI.7a).
OCR 2 – Either the superior or the inferior occipital crest visible as a sagittal ridge (Figure AI.7.b).
OCR 3 – Clearly raised ridge or crest along the whole course between the tuberculum linearum and the foramen magnum (Figure AI.7c).

[2] Contrary to *Gray's Anatomy* and other textbooks, Hublin (1978a) shows that the superior nuchal lines join medially into the tuberculum linearum and not the external occipital protuberance (found at the medial junction of the supreme nuchal lines). It is therefore, the tuberculum linearum together with the external occipital crest that give attachment to the Ligamentum Nuchae. In European crania it is common to see the two points joined, but in other cases where there is a distance between the superior and supreme nuchal lines (with or without formation of an occipital torus), these two points are apart and Hublin's observations apparent.

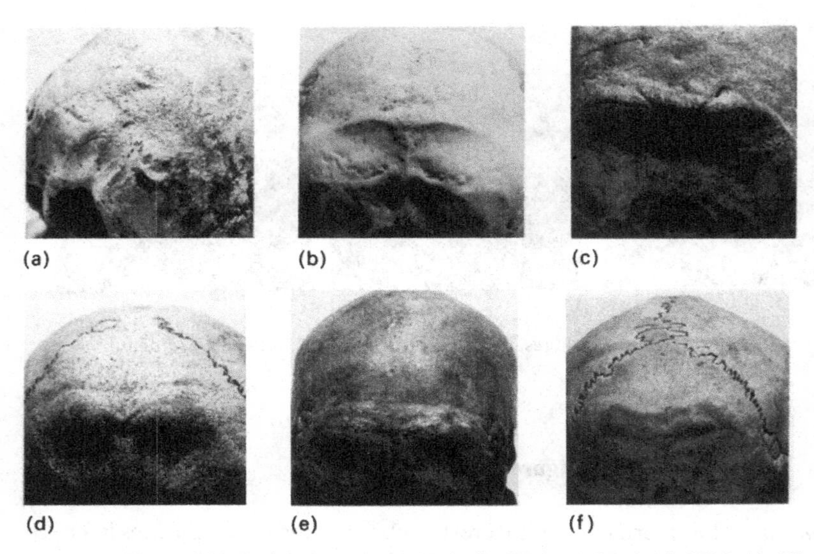

Figure AI.8 Occipital torus: (a) OT 2; (b) OT 3; (c) OT 4; (d) OT 5; (e) OT 6; (f) OT 7. For full explanation of grades see text.

Occipital torus (Figure AI.8)

Definition: An occipital torus is formed between the superior and supreme nuchal lines, which separate the area of attachment of the nuchal muscles from the rest of the occipital squama. It is characterised by having a concave superior margin called the supratoral sulcus.

Grades:

On the basis that the occipital torus is formed by the superior and supreme nuchal lines, being present in its complete projecting form only in few individuals, both the variability in configuration of the nuchal lines and formation of a torus were recorded. Therefore, the following categories were scored, being that they do not represent just a directional grading scale as other scores in this system.

OT 1 – Supreme nuchal lines not visible.[3]

OT 2 – All that is visible of the supreme nuchal lines is the external occipital protuberance (Figure AI.8a).

[3] This grade was not observed in any of the samples used in this study.

OT 3 – Supreme nuchal lines visible, separated medially from the superior nuchal lines, i.e. the external occipital protuberance and tuberculum linearum are separate (Figure AI.8b).

OT 4 – Supreme and superior nuchal lines visible and joined medially by the external occipital protuberance (Figure AI.8c).

OT 5 – Same configuration as Grade 4, but superimposed on a torus, forming a 'Supranuchal Tubercle' of Hasebe (1935) (Figure AI.8d).

OT 6 – Occipital torus, with a supratoral sulcus visible, but with a medial 'indent' caused by the presence of an external occipital protuberance along the superior toral margin (Figure AI.8e).

OT 7 – Occipital torus present, with no visible external occipital protuberance (Figure AI.8f).

Rounding of the inferolateral margin of the orbits (Figure AI.9)

Definition: The inferolateral margin of the orbital rim can be either sharp or rounded, and the floor of the orbit can be depressed or levelled in relation to this margin. In the case of a rounded orbital margin, in which the floor of the orbit is at the same level, no distinct boundary can be observed, just a gradual curve between the surface of the malar and the orbital cavity.

Grades:

RO 1 – Sharp, high line dividing the floor of the orbit from the facial portion of the malar (Figure AI.9a).

RO 2 – Relatively rounded orbital margin, but raised in relation to the floor of the orbit. This is the most common condition, present in the majority of skulls[4] (Figure AI.9b).

RO 3 – Pronounced rounding of the inferior lateral border, which is levelled with the floor of the orbit. In extreme cases, the description of Larnach & Macintosh (1966) applies : 'When the margin is flattened there are three distinct surfaces at the margin; (1) the facial surface of the malar; (2) the wide flattened area along the top of the margin; and (3) the orbital surface of the malar' (Figure AI.9c).

[4] Those cases that show a pronounced rounding of the inferolateral margin of the orbit but not levelled with the floor of the orbit are considered grade 2 (RO 2).

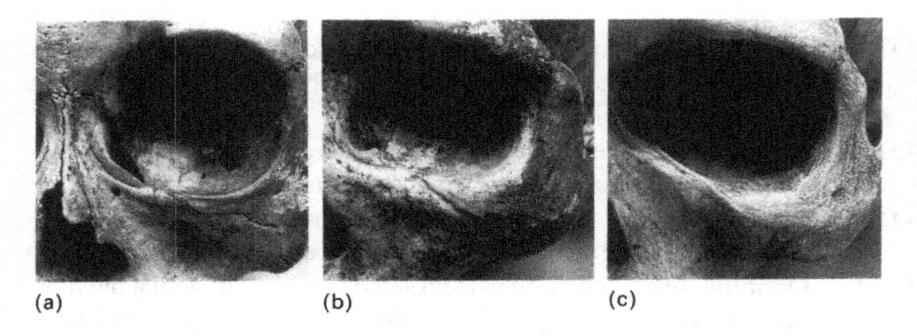

Figure AI.9 Rounding of the infero-lateral margin of the orbits: (a) RO 1; (b) RO 2; (c) RO 3. For full explanation of grades see text.

Bregmatic eminence ('eminentia cruciata')

Definition: Described by Weidenreich (1943a) in the *Sinanthropus* skulls as an eminence coinciding with the cranial vertex, at or slightly behind bregma, which together with a slight elevation along the coronal suture decreasing from bregma towards the pterion, turns the eminence into a cross-like elevation (especially pronounced in *Sinanthropus* skull XI).

Grades:

GRADE 0 – absent
GRADE 1 – present

This trait was observed in only two skulls of the grading series, one of which was casted as representing the character (American Indian DC1842, California). The other skull (Australian Oc.0.0.1), shows the trait in a much less well-defined form, and because of the lack of comparison with other skulls showing the feature, it was not considered as an intermediate score. In several Australian skulls the cranial vertex is found at or near the bregma, creating an elevated area that may give the impression of representing a bregmatic eminence. However, these cases do not show any of the other descriptive characteristics of this trait, such as the cross-like extension along the coronal suture, and were thus not considered as showing a true bregmatic eminence.

Appendix II Complementary measurements

Standard measurements

A list of standard cranial measurements was taken from Howells (1973), and definition, method of measuring and calipers employed for the following measurements follow the instructions in Appendix A of his work.

GOL – Glabello-occipital length
BNL – Basion-nasion length
XPB – Maximum parietal breadth
AUB – Biauricular breadth
STB – Bistephanic breadth
XFB – Maximum frontal breadth
NPH – Nasion-prosthion length
NLB – Nasal breadth
OBB – Orbital breadth
MAB – Palate breadth, external
MDB – Mastoid width
SSS – Bimaxillary subtense
NAS – Nasio-frontal subtense
EKB – Biorbital breadth
IML – Malar length, inferior
WMH – Cheek height
GLS – Glabella projection
FOL – Foramen magnum length
RS – Nasion-bregma subtense
PAC – Bregma-lambda chord
PAF – Bregma-subtense fraction
OCS – Lambda-opisthion subtense
VRR – Vertex radius
SSR – Subspinale radius
ZOR – Zygoorbitale radius
EKR – Ectoconchion radius
AVR – Molar alveolar radius
NOL – Naso-occipital length
BBH – Basion-bregma height
ASB – Biasterionic breadth
WCB – Minimum cranial breadth
ZYB – Bizygomatic breadth
WFB – Minimum frontal breadth
NLH – Nasal height
OBH – Orbital height
JUB – Bijugal breadth
MDH – Mastoid height
ZMB – Bimaxillary breadth
FMB – Bifrontal breadth
NBS – Biorbital-nasion subtense
WNB – Simotic chord
XML – Malar length, maximum
SOS – Supraorbital projection
BPL – Basion-prosthion length
FRC – Nasion-bregma chord
FRF – Nasion-subtense fraction
PAS – Bregma-lambda subtense
OCC – Lambda-opisthion chord
OCF – Lambda-subtense fraction
NAR – Nasion radius
PRR – Prosthion radius
FMR – Frontomalare radius
ZMR – Zygomaxillare radius
XCB – Maximum cranial breadth
FRA – Frontal angle
PAA – Parietal angle
OCA – Occipital angle

Other measurements

Cranial

Palate length (PALL): measured from prosthion to a line transversely joining the two third molars.
Palate depth (PALD): taken as a subtense from inter-molar alveolar breadth, at the level of M1. Because this measurement was always taken at the level of M1, it does not necessarily represent maximum palatal depth, but if not it approximates this measure very closely.
Minimum interorbital breadth (WIOB): taken from dacryon to dacryon.
Naso-frontal breadth (NFRB): taken from one end of the naso-frontal suture to another. The comparison of this value with minimum frontal breadth shows whether the latter is placed at the suture or further down along the nasals.
Minimum basilar breadth (WBAB): measures the minimum breadth of the basilar portion of the occipital bone, before or at the basisphenoidal synchondrosis (hormion).
Foramen magnum breadth (FOB)
Frontal arch (FARCH)
Parietal arch (PARCH)
Occipital arch (OARCH)
Intermeatal length (EAMEAM)

In order to obtain a measure of basal flexion, the following measurements were taken:
Basion-hormion length (BAH)
Palate-basion length (ALVOC): the palate point used as landmark is the end of the medial palatal suture, at a point where it forms a small triangular depression. This point does not always coincide with a transverse line joining the distal end of both M3s, neither represents the most distal point of the palate, which may sometimes continue for a few millimetres. However, it was found to be the only constant sagittal point that could be used to measure the basicranial triangle.
Palate-hormion height (HPAL)
Palate-basion subtense (D1)
Palate-opisthion subtense (D2)

Dimensions of the parietal bone (the measurements involving pterion were always taken from the anteriormost parietal point):
Pterion-bregma length (PTEBR)
Pterion-asterion length (PTEAST)
Asterion-lambda length (ASTLAM)

Asterion-bregma length (ASTBR)
Pterion-lambda length (PTELAM)

Relative position of inion:
Lambda-inion length (LAMINI): The definition of inion used here is that by Hublin (1978a), i.e. the point formed by the junction of the superior nuchal lines, and as such coincides with the tuberculum linearum. In the case of an occipital torus being present and inion as a landmark not obvious, the medial sagittal point of the inferior margin of the torus (formed by the superior nuchal lines) was used.
Lambda-opistocranium length (LAMOPIS): Opistocranium was determined, and marked by pencil, while taking glabello-occipital length.
Inion-opisthion length (INOP)
Inion-opistocranium lenth (INIOPIS)

Cranial measurements of temporal muscle size:

The landmarks used for the following measurements were: for the height – (1) a point above porion on the surface of the temporal bone, at the root of the zygomatic arch anteriorly and supramastoid crest posteriorly, to mark the lowermost point of the temporal muscle; and (2) the highestmost point on the parietal bone of either the inferior or superior temporal lines; and for the length – (1) the anterior-most point along the temporal crest on the frontal bone, usually along the posterior margin of the zygomatic trigone; and (2) the most distal point on the parietal bone of either the inferior or superior temporal lines[1]. All measurements were taken on the left side, and only if this was not possible due to damage, was the right side used.
Inferior temporal line height chord (ITLHCH)
Inferior temporal line height arch (ITLHTA)
Superior temporal line height chord (STLHCH)
Superior temporal line height arch (STLHTA)
Inferior temporal line length chord (ITLLCH)
Inferior temporal line length arch (ITLLTA)

[1] Maximum length of the temporal muscle proved to be a difficult measurement to take, since on several occasions there appeared to be two temporal lines along its superior course that then descended towards the base of the skull and divided in three. This was seen both in Australian and Inuit crania. In these cases, the third line observed reached distally the margins of the occipital bone, crossing the lambdoid suture in some cases, and the determination of its exact limits was difficult due to the other nuchal markings and reliefs. In these cases both superior temporal lines were measured, but to maintain comparative levels, only the first one was used in the analysis. The exact meaning of this third distal temporal line in terms of whether it represents further sub-divisions of muscle fibres and a middle fascia is not clear.

Superior temporal line length chord (STLLCH)
Superior temporal line length arch (STLLTA)
Minimum distance between left and right inferior temporal lines across the parietals (SILLPA)
Minimum distance between left and rigth superior temporal lines across the parietals (IILLPA)
Distance between left and right inferior temporal lines across the occipital bone (IILLOC)
Distance between left and right superior temporal lines across the occipital bone (SILLOC)

Dimensions of the temporal fossa:
Temporal fossa length (TFL)
Temporal fossa breadth (TFB)
Thickness of the malar bone at the zygomaticomaxillary suture (MALSU)
Thickness of the malar bone at its free border (MALBO)

Relative size of anterior and posterior dentitions (measured at the alveolus):
Left medial incisor-canine length (LI-C)
Rigth medial incisor-canine length (RI-C)
Left P3/M3 length (LP-M)
Right P3/M3 length (RP-M)

Dental

Size of the anterior teeth
Bucco-lingual and mesiodistal dimensions of the posterior teeth

References

Abbie, A.A. (1966) Physical characteristics. In: B.C. Cotton (ed.) *Aboriginal Man in South and Central Australia*, pp. 9–45. Adelaide: Govern. Print.

Abbie, A.A. (1968) The homogeneity of Aborigines. *Archaeology and Physical Anthropology in Oceania*, **3**: 221–31.

Abbie, A.A. (1975) *Studies in Physical Anthropology II*. Canberra: Australian Institute of Aboriginal Studies.

Adam, W. (1940) *The Jaws and Dentition of the Tasmanian Aborigines*. DDSc.Th., University of Melbourne.

Aiello, L.C. (1992) The origins of modern humans: the fossil evidence. *American Anthropologist*, **95**: 73–96.

Aiello, L.C. & Dean, M.C. (1990) *An Introduction to Human Evolutionary Anatomy*. London: Academic Press.

Aigner, J.S. (1976) Chinese Pleistocene cultural and hominid remains: a consideration of their significance in reconstructing the pattern of human bio-cultural development. In: A.K. Ghosh (ed.) *Le Paléolithique Inferieur et Moyen en Inde, en Asie Centrale, en Chine et dans le sud-Est Asiatique*, pp. 65–90. Paris: UISPP Coll. VII, C.N.R.S.

Aigner, J.S. (1978) Important archaeological remains from North China. In: Ikawa-Smith (ed.) *Early Paleolithic in South and East Asia*, pp. 163–232. The Hague: Mouton.

Aitken, M.J. & Valladas, H. (1992) Luminescence dating relevant to human origins. *Philosophical Transactions of the Royal Society of London, Series B*, **337**: 139–44.

Alimen, H. (1975) Les 'isthmes' hispano-marocain et sicilo-tunisien aux temps acheuléens. *L'Anthropologie*, **19**: 399–436.

Allchin, B., Goudie, A. & Hedge, K.T.M. (1978) *The Prehistory and Palaeogeography of the Great Indian Desert*. London: Academic Press.

Allen, J. (1989) Excavations at Bone Cave, south central Tasmania, January-February 1989. *Australian Archaeology*, **28**: 105–6.

Allen, J., Gosden, C., Jones, R. & White, J.P. (1988) Pleistocene dates for the human occupation of New Ireland, Melanesia. *Nature*, **331**: 707–9.

Allsworth-Jones, P. (1986) Middle Stone Age and Middle Palaeolithic: The evidence from Nigeria and Cameroon. In: G.N. Bailey & P. Callow (eds.) *Stone Age Prehistory*, pp. 153–68. Cambridge: Cambridge University Press.

Allsworth-Jones, P. (1993) The archaeology of archaic and early modern *Homo sapiens*: an African perspective. *Cambridge Archaeological Journal*, **3**: 21–39.

Ambergen, A.W. & Schaafsma, W. (1984) Interval estimates for posterior probabilities, applications to Border Cave. In: G.N. van Vark & W.W. Howells (eds.) *Multivariate Statistical Methods in Physical Anthropology*, pp. 115–34. Dordrecht: Reidel.

Anderson, D.D. (1987) A Pleistocene-Early Holocene rock shelter in peninsular Thailand. *National Geographic Research*, **3**: 184–98.
Anderson, J.E. (1968) Late Paleolithic skeletal remains from Nubia. In: F. Wendorf (ed.) *The Prehistory of Nubia*, pp. 996–1040. Dallas: Southern Methodist University Press.
Andrews, P. (1984) On the characters that define *Homo erectus*. *Courier Forschungsinstitut Senckenberg*, **69**: 167–75.
Anoutchine, D.N. (1878) Sur la conformation du pterion chez diverses races humaines et les primates. *Bulletin Societé Anthropologie*, **1**: 330–3.
Anton, A.C. (1989) Facial structure: the influence of intentional cranial vault deformation. *American Journal of Physical Anthropololgy*, **78**: 184.
Arambourg, C. & Balout, L. (1955) L'ancien lac de Tihodaine et ses gisements préhistoriques. In: L. Balout (ed.) *Congrés Panafrican de Préhistoire, Algiers, 1952*, pp. 281–92. Paris: Arts et Métiers Graphiques.
Arensburg, B. (1977) New Upper Palaeolithic human remains from Israel. *Eretz Israel*, **13**: 208–15.
Arensburg, B. (1981) Recent evolution in Israel. In: D. Ferembach (ed.) *Les Processus de l'Hominisation*, pp. 195–201. Paris: C.N.R.S.
Arensburg, B. & Bar-Yosef, O. (1973) Human remains from Ein Gev I, Jordan Valley, Israel. *Paléorient*, **1**(2): 201–6.
Arkell, A.J. (1964) *Wanyanga and an archaeological reconnaisance of the South West Libyan Desert: The British Ennedi expedition, 1957*. London: Oxford University Press.
Ascenzi, A. & Balistreri, P. (1975) Aural exostoses in a roman skull excavated at the 'baths of the swimmer' in the ancient town of Ostia. *Journal of Human Evolution*, **4**: 579–84.
Avise, J.C. (1986) Mitochondrial DNA and the evolutionary genetics of higher animals. *Philosophical Transactions of the Royal Society of London, Series B*, **312**: 325–42.
Avise, J.C., Ball, R.M. & Arnold, J. (1988) Current versus historical population sizes in vertebrate species with high gene flow: a comparison based on mitochondrial DNA lineages and imbreeding theory for neutral mutations. *Molecular Biology and Evolution*, **5**: 331–44.
Baba, H. & Aziz, F. (1993) Reconstruction of Sangiran 17 skull. Paper presented at the 13th International Congress of Anthropological and Ethnological Sciences, 29 July–4 August, Mexico City.
Bahn, P.G. & Vertut, J. (1988) *Images of the Ice Age*. New York: Facts on File.
Bailliet, G., Rothammer, F., Carnese, F.R., Bravi, C.M. & Bianchi, N.O. (1994) Founder mitochondrial haplotypes in amerindian populations. *American Journal of Human Genetics*, **54**: 27–33.
Balakrishnan, V. (1978) A preliminary study of genetic distances among some populations of the Indian subcontinent. *Journal of Human Evolution*, **7**: 67–75.
Ballinger, S.W., Schurr, T.G., Torroni, A., Gan, Y.Y., Hodge, J.A., Hassan, K., Chen, K.H. & Wallace, D.C. (1992) Southeast Asian mitochodrial DNA analysis reveals genetic continuity of ancient mongoloid migrations. *Genetics*, **130**: 139–52.
Bar-Yosef, O. (1988) The date of south-west Asian Neanderthals. In: E. Trinkaus (ed.) *L'homme de Neanderthal 3: L'anatomie*, pp. 31–8. Liège: Études et

Recherches Arch., Université de Liège.
Bar-Yosef, O. (1989a) Geochronology of the Levantine Middle Palaeolithic. In: P. Mellars & C.B. Stringer (eds.) *The Human Revolution*, pp. 589–610. Edinburgh: Edinburgh University Press.
Bar-Yosef, O. (1989b) Upper Pleistocene cultural stratigraphy in southwest Asia. In: E.Trinkaus (ed.) *The Emergence of Modern Humans*, pp. 154–80. Cambridge: Cambridge University Press.
Bar-Yosef, O. (1992a) Middle Palaeolithic chronology and the transition to the Upper Palaeolithic in southwest Asia. In: G. Bräuer & F.H. Smith (eds) *Continuity or Replacement? Controversies in* Homo sapiens *evolution*, pp. 261–72. Rotterdam: Balkema.
Bar-Yosef, O. (1992b) The role of western Asia in the origins of modern humans. *Philosophical Transactions of the Royal Society of London, Series B*, **337**: 193–200.
Bar-Yosef, O. & Cohen, A. (1988) The early Upper Palaeolithic in Levantine caves. *BAR, International Series*, **437**: 23–41.
Bar-Yosef, O. & Valla, F.R. (1991) *The Natufian Culture in the Levant*. Ann Arbor: International Monographs in Prehistory, Archaeological Series 1.
Barbujani, G., Nasidze, I.S. & Whitehead, G.N. (1994) Genetic diversity in the Caucasus. *Human Biology*, **66**: 639–68.
Barstra, G.J. (1982) *Homo erectus erectus*: the search for his artefacts. *Current Anthropology*, **23**: 318–20.
Barstra, G.J., Soegondho, S. & Wijk, A. van der (1988) Ngandong man: age and artefacts. *Journal of Human Evolution*, **17**: 325–37.
Bass, W.M. (1987) *Human Osteology*. Special Publ. No. 2, Miss. Arch. Soc.
Beadle, L.C. (1974) *The Inland Waters of Tropical Africa*. London: Longman.
Beaton, J.M. (1983) Comment. Does intensification account for changes in the Australian Holocene archaeological record? *Archaeology in Oceania*, **18**: 94–7.
Beaumont, P.B., De Villiers, H. & Vogel, J.C. (1978) Modern man in sub-Saharan Africa prior to 49,000 years B.P.: A review and evaluation with particular reference to Border Cave. *South African Journal of Science*, **74**: 409–19.
Belgraver, P. (1938) *Over exostoses van de uitwendige gehoorgang*. Diss. University of Leiden.
Bellwood, P.S. (1975) The prehistory of Oceania. *Current Anthropology*, **16**: 9–28.
Bellwood, P.S. (1978) *Man's Conquest of the Pacific*. Auckland: Collins.
Bellwood, P.S. (1985) *The Prehistory of Indo-Malaysian Archipelago*. Sydney: Academic Press.
Bellwood, P.S. (1986) A hypothesis for Austronesian origins. *Asian Perspectives*, **26**.
Bellwood, P.S. (1987) The prehistory of island Southeast Asia: a multidisciplinary review of recent research. *Journal of World Prehistory*, **1**: 171–224.
Bellwood, P.S. (1990) From late Pleistocene to early Holocene in Sundaland. In: C.S. Gamble and O. Soffer (eds.) *The World at 18,000 BP*, vol 2, pp. 155–63. London: Unwin Hyman.
Bellwood, P. S & Koon, P. (1988) The Tingkayu sites. In: P. Bellwood (ed.) *Archaeological Research in South-eastern Sabah*, pp. 38–50. Sabah Museum and State Activities. *Sabah Museum Monograph 1988*, vol. 2.
Bergman, C.A. & Stringer, C.B. (1989) Fifty years after: Egbert, an early Upper Palaeolithic juvenile from Ksar Akil, Lebanon. *Paléorient*, **15**(2): 99–111.
Berry, A.C. (1975) Factors affecting the incidence of non-metrical skeletal variants.

Journal of Anatomy, **120**: 519–35.
Bhasin, M.K.; Walter, H. & Danker-Hopfe, H. (1994) *People of India: An Investigation of Biological Variability in Ecological, Ethno-economic and Linguistic Groups*. Delhi: Kamla – Raj Enterprises.
Biberson, P. (1961) Le cadre paleogeographique de la Prehistoire du Maroc Atlantique. *Publications du Service des Antiquites du Maroc*, **1**: 37–92.
Biegert, J. (1963) The evaluation of characteristics of the skull, hands, and feet for primate taxonomy. In: S. Washburn & F.C. Howell (eds.) *Classification and Human Evolution*, pp. 116–45. London: Methuen.
Billy, G. (1975) Étude anthropologique des restes humains de l'Abri Pataud. In: H.L. Movius (ed.) *Excavation of the Abri Pataud, Les Eyzies (Dordogne)*, pp. 99–201. Harvard: Peabody Museum Archaeology and Ethnology.
Billy, G. (1986) L'*Homo sapiens sapiens* fossilis en Europe. In: D. Ferembach, C. Susanne & M.C. Chamla (eds.) *L'Homme, son Évolution, sa Diversité*, pp. 217–24. Paris: C.N.R.S.
Binford, L.R. (1981) *Bones: Ancient Men and Modern Myths*. New York: Academic Press.
Binford, L.R. (1982) Comment on R. White: Rethinking the Middle/Upper Palaeolithic transition. *Current Anthropology*, **23**: 177–81.
Binford, L.R. (1984) *Faunal Remains from Klasies River Mouth*. New York: Academic Press.
Binford, L.R. (1989) Isolating the transition to cultural adaptations: an organizational approach. In: E. Trinkaus (ed.) *The Emergence of Modern Humans*, pp. 18–41. Cambridge: Cambridge University Press.
Birdsell, J.B. (1949) The racial origin of the extinct Tasmanians. *Records of the Queensland Vict. Museum II*, **3**: 105–22.
Birdsell, J.B. (1967) Preliminary data on the trihybrid origin of the Australian Aborigines. *Archaeology and Physical Anthropology in Oceania*, **2**: 100–55.
Birdsell, J.B. (1972) *Human Evolution: An Introduction to the New Physical Anthropology*. Chicago: Rand McNally & Co.
Birdsell, J.B. (1977) The recalibration of a paradigm for the first peopling of Greater Australia. In: J. Allen, J.Golson & R. Jones (eds.) *Sunda and Sahul: Prehistoric Studies in Southeast Asia, Melanesia and Australia*, pp. 113–67. London: Academic Press.
Birdsell, J.B. (1993) *Microevolutionary Patterns in Aboriginal Australia: A Gradient Analysis of Clines*. Oxford: Oxford University Press.
Bischoff, J.L., Soler, N., Maroto, J. & Julia, R. (1989) Abrupt Mousterian/Aurignacian boundary at c. 40 ka bp: accelerator 14C dates from l'Abreda Cave (Catalunia, Spain). *Journal of Archaeological Sciences*, **16**: 563–76.
Bloom, W.M. (1988) Late Quaternary sediments and sea-levels in Bass Basin, south-eastern Australia: A preliminary report. *Search*, **19**: 94–6.
Bolk, L. (1920) Uber metopismus. *Zeitschrift für Morphologie und Anthropologie*, **21**: 200–26.
Bonin, G. (1912) Zur morphologie der fossa praenasalis. *Archiv für Anthropologie*, **11**, S. 185.
Bonnefille, R. & Hamilton, A. (1986) Quaternary and Late Tertiary history of Ethiopian vegetation. *Symbolae Botanicae Uppsalienses*, **26**: 48–63.
Boule, M. (1911) L'homme fossile de la Chapelle-aux-Saints. *Annals de Paléontologie*,

6: 109–72; **7**: 105–92, **8**: 1–62.
Bowcock, A.M., Bucci, C., Hebert, J.M., Kidd, K.K., Friedlaender, J.A. & Cavalli-Sforza, L.L. (1987) Study of 47 DNA markers in five populations from four continents. *Gene Geography*, **1**: 47–64.
Bowcock, A.M., Kidd, J.R., Mountain, J.L., Hebert, J.M., Carotenuto, L., Kidd, K.K. & Cavalli-Sforza, L.L. (1991a) Drift, admixture and selection in human evolution: A study with DNA polymorphisms. *Proceedings of the National Academy of Sciences, USA*, **88**: 839–43.
Bowcock, A.M., Hebert, J.M., Mountain, J.L., Kidd, J.R., Kidd, K.K. & Cavalli-Sforza, L.L. (1991b) Study of an additional 58 DNA markers in five human populations from four continents. *Gene Geography*, **5**: 151–73.
Bowcock, A.M., Ruiz-Linares, A., Tomfohrde, J., Minch, E., Kidd, J.R. & Cavalli-Sforza, L.L. (1994) High resolution of human evolutionary trees with polymorphic microsatellites. *Nature*, **368**: 455.
Bowdler, S. (1988) Tasmanian Aborigines in the Hunter Islands in the Holocene: Island use and seasonality. In: G. Bailey & J. Parkington (eds.) *The Archaeology of Prehistoric Coastlines*, pp. 42–52. Cambridge: Cambridge University Press.
Bowler, J.M. (1976) Recent developments in reconstructing late Quaternary environments in Australia. In: R.L. Kirk and A.G. Thorne (eds.) *The Origins of the Australians*, pp. 55–77. Canberra: Australian Institute of Aboriginal Studies.
Bowler, J.M., Hope, G.S., Jennings, J.N., Singh, G. & Walker, D. (1976) Late Quaternary climates of Australia and New Guinea. *Quaternary Research*, **6**: 359–94.
Boyce, A.J., Harrison, G.A., Platt, C.M., Hornabrook, R.W., Serjeantson, S., Kirk, R.L. & Both, P.B. (1978) Migration and genetic diversity in an island population: Karkar, Papua New Guinea. *Proceedings of the Royal Society of London, Series B*, **202**: 269–95.
Boyd, W.C. (1963) Four achievements of the genetical method in physical anthropology. *American Anthropologist*, **65**: 243–52.
Brace, C.L. (1967) *The Stages of Human Evolution*. Englewood Cliffs, NJ: Prentice-Hall.
Brace, C.L. & Hunt, K.D. (1990) A non-racial craniofacial perspective on human variation: A(ustralia) to Z(uni). *American Journal of Physical Anthropology*, **82**: 341–60.
Brace, C.L. & Nagai, M. (1982) Japanese tooth size: past and present. *American Journal of Physical Anthropology*, **59**: 399–411.
Brace, C.L. & Tracer, D.P. (1992) Craniofacial continuity and change: a comparison of Late Pleistocene and recent Europe and Asia. In: T. Akazawa, K. Aoki & T. Kimura (eds.) *The Evolution and Dispersal of Modern Humans in Asia*, pp. 439–71. Tokyo: Hokusen-sha.
Brace, C.L., Shao, X. & Zhang, Z. (1984) Prehistoric and modern tooth size in China. In: F.H. Smith & F. Spencer (eds.) *The Origin of Modern Humans: A World Survey of the Fossil Evidence*, pp. 485–516. New York: Alan R. Liss.
Brace, C.L., Brace, M.L. & Leonard, W.R. (1989) Reflections on the face of Japan: A multivariate craniofacial and odontometric perspective. *American Journal of Physical Anthropology*, **78**: 93–113.
Brace, C.L., Brace, M.L., Dodo, Y., Hunt, K.D., Leonard, W.R., Yongyi, L., Sangvichien, S., Xiang-Qing, S. and Zhenbiao, Z. (1990) Micronesians,

Asians, Thais and relations: A craniofacial and odontometric perspective. *Micronesia (Suppl.)*, **2**: 323–48.

Brace, C.L., Tracer, D.P., Allen Yaroch, L., Robb, J., Brandt, K. & Nelson, A.R. (1992) Clines and clusters versus 'race': A test in ancient Egypt and the case of a death on the Nile. *Yearbook of Physical Anthropology*, **36**: 1–31.

Bradley, R.S. (1985) *Quaternary Paleoclimatology: Methods of Paleoclimatic Reconstruction*. Boston: Allen & Unwin.

Bräuer, G. (1978) The morphological differentiation of anatomically modern man in Africa, with special regard to recent finds from East Africa. *Zeitschrift für Morphologie und Anthropologie*, **69**: 266–92.

Bräuer, G. (1982) Early anatomically modern man in Africa and the replacement of the Mediterranean and European Neanderthals. *1st International Congress of Human Palaeontology*, Nice, 1982, Resumés: 112.

Bräuer, G. (1984a) The 'Afro-European sapiens hypothesis' and hominid evolution in East Asia during the late Middle and Upper Pleistocene. *Courier Forschungsinstitut Senckenberg*, **69**: 145–65.

Bräuer, G. (1984b) A craniological approach to the origin of anatomically modern *Homo sapiens* in Africa and implications for the appearence of modern Europeans. In: F.H. Smith & F. Spencer (eds) *The Origin of Modern Humans: A World Survey of the Fossil Evidence*, pp. 327–410. New York: Alan R. Liss.

Bräuer, G. (1989) The evolution of modern humans: a comparison of the African and non-African evidence. In: P. Mellars & C.B. Stringer (eds.) *The Human Revolution: Behavioural and Biological Perspectives on the Origins of Modern Humans*, pp. 123–54. Edinburgh: Edinburgh University Press.

Bräuer, G. (1992a) Africa's place in the evolution of *Homo sapiens*. In: G. Bräuer & F.H. Smith (eds.) *Continuity or Replacement? Controversies in* Homo sapiens *Evolution*, pp. 83–98. Rotterdam: Balkema.

Bräuer, G. (1992b) L'hypothese Africaine de l'origine des hommes modernes. In: J.J. Hublin & A.M. Tillier (eds.) *Aux origines d'Homo sapiens*, pp. 181–215. Paris: Presses Universitaires de France.

Bräuer, G. & Mehlman, M.J. (1988) Hominid molars from a Middle Stone Age level at the Mumba Rock Shelter, Tanzania. *American Journal of Physical Anthropology*, **75**: 69–76.

Bräuer, G. & Smith, F.H. (eds.) (1992) *Continuity or Replacement? Controversies in* Homo sapiens *evolution*. Rotterdam: Balkema.

Brooks, A.S. & Yellen, J.E. (1989) An archaeological perspective on the African origins of modern humans. *American Journal of Physical Anthropology*, **78**: 197.

Broste, K., Fischer-Moller, K. & Pederson, P. (1944) The Mediaeval Norseman at Gardar. *Meddelelser om Grönland*, **89**: 48–51.

Brothwell, D. (1960) Upper Pleistocene human skull from Niah caves, Sarawak. *Sarawak Museum Journal*, **9**: 323–49.

Brothwell, D.R. (1975) Possible evidence of a cultural practice affecting head growth in some Late Pleistocene east Asian and Australasian populations. *Journal of Archaeological Sciences*, **2**: 75–7.

Brothwell, D. R. & Krzabowski, W. (1974) Evidence of biological differences between early British populations from Neolithic to Medieval times, as revealed by eleven commonly available cranial vault measurements. *Journal of Archaeological Sciences*, **1**: 249–60.

Brown, P. (1981) Artificial cranial deformation: a component in the variation in

Pleistocene Australian Aboriginal crania. *Archaeology and Physical Anthropology in Oceania*, **16**: 156–67.
Brown, P. (1989) *Coobool Creek*. Terra Australis 13, Canberra: The Australian National University.
Brown, P. (1990) Osteological definitions of 'Anatomically modern' *Homo sapiens*: A test using modern and terminal Pleistocene *Homo sapiens*. In: L. Freedman (ed.) *Is Our Future Limited by Our Past?* pp. 51–74. Proceedings of the Australasian Society for Human Biology, No. 3. Perth: The University of Western Australia.
Brown, T. (1967) *Skull of the Australian Aboriginal. A Multivariate Analysis of Craniofacail Associations*. Adelaide: Dept. of Dental Science, University of Adelaide.
Brown, W.L. (1957) Centrifugal speciation. *Quarterly Review of Biology*, **32**: 247–77.
Brown, W.M. (1985) The mitochondrial genome of animals. In: R.J. MacIntyre (ed.) *Molecular Evolutionary Genetics*, pp. 85–130. New York: Plenum Press.
Brown, W.M., George, M. & Wilson, A.C. (1979) Rapid evolution of animal mitochondrial DNA. *Proceedings of the National Academy of Sciences, USA*, **76**: 1967–71.
Brown, W.M., Prager, E.M., Wang, A. & Wilson, A.C. (1982) Mitochondrial DNA sequences of primates: tempo and mode of evolution. *Journal of Molecular Evolution*, **18**: 225–39.
Brues, A.M. (1977) *People and Races*. New York: Macmillan.
Bryan, A.L. (1978) *Early Man in America from a Circum-Pacific Perspective*. Edmonton: University of Alberta Press.
Burkitt, A.N. & Hunter, J. (1922) The description of a neanderthaloid Australian skull. *Journal of Anatomy*, **57**: 31–54.
Burkitt, A.N. & Lightoller, G.H.S. (1923) Preliminary observations on the nose of the Australian aboriginal, with a table of aboriginal head measurements. *Journal of Anatomy*, **57**: 295–312.
Bush, G.L. (1969) Sympatric host race formation and speciation in frugivorous flies of the genus *Rhagoletis* (Diptera, Tephritidae). *Evolution*, **23**: 237–51.
Cabrera, V. & Bischoff, J.L. (1989) Accelerator 14C dates for early Upper Palaeolithic (Basal Aurignacian) at El Castillo Cave (Spain). *Journal of Archaeological Sciences*, **16**: 577–84.
Campbell, L. (1988) Review of Language in the Americas, by J.H. Greenberg. *Language*, **64**: 591–615.
Campbell, N.A. (1984) Some aspects of allocation and discrimination. In: G.N. van Vark & W.W. Howells (eds.) *Multivariate Statistical Methods in Physical Anthropology*, pp. 177–92. Dordrecht: Reidel.
Camps, G., Delibrias, G. & Thommeret, J. (1973) Chronologie des civilisations préhistoriques du Nord de l'Afrique d'après le radiocarbone. *Libyca*, **21**: 65–89.
Cann, R.L. (1988) DNA and human origins. *Annals Revue of Anthropology*, **17**: 127–43.
Cann, R.L. (1994) Invited editorial. mtDNA and native Americans: a Southern Perspective. *American Journal of Human Genetics*, **55**: 7–11.
Cann, R.L., Stoneking, M. & Wilson, A.C. (1987) Reply. *Nature*, **329**: 111–12.
Carson, H.L. (1975) The genetics of speciation at the diploid level. *American Naturalist*, **109**: 83–92.
Caspari, R. & Wolpoff, M.H. (1990) The morphlogical affinities of the Klasies

River Mouth skeletal remains. *American Journal of Physical Anthropology*, **81**: 203.

Cavalli-Sforza, L.L. (1965) Population structure and human evolution. *Philosophical Transactions of the Royal Society of London, Series B*, **164**: 362–79.

Cavalli-Sforza, L.L. (1974) The genetics of human populations. *Scientific American*, **231**: 81–9.

Cavalli-Sforza, L.L. (1991) Genes, people and languages. *Scientific American, November*, **265** (5): 72–8.

Cavalli-Sforza, L.L. & Bodmer, W.F. (1971) *The Genetics of Human Populations*. San Francisco: Freeman.

Cavalli-Sforza, L.L. & Edwards, A.W. (1967) Phylogenetic analysis: models and estimation procedures. *American Journal of Human Genetics*, **23**: 235–52.

Cavalli-Sforza, L.L., Piazza, A., Menozzi, P. & Mountain, J. (1988) Reconstruction of human evolution: bringing together genetic, archaeological and linguistic data. *Proceedings of the National Academy of Sciences, USA*, **85**: 6002–6.

Cavalli-Sforza, L.L., Piazza, A. & Menozzi, P. (1994) *History and Geography of Genes*. Princeton: Princeton University Press.

Cesnys, G. & Konduktorova, T.S. (1982) Non-metric features of the skull in people of the Chernyakhovskaya culture. *Vopr. Anthrop.*, **70**: 62–76.

Chamla, M.C. (1968) Les populations anciennes du Sahara et des régions limitrophes. Étude des rests osseux humains Néolithiques et protohistoriques. *Memoires du Centre de Recherches Anthropologiques, Préhistoriques et Ethnographiques*, **9**: 1–249.

Chamla, M.C. (1978) Le peuplement de l'Afrique du Nord de l'Epipaleolithique a l'epoque actuelle. *L'Anthropologie*, **82**: 385–430.

Chappell, J.M.A. (1976) Aspects of late Quaternary palaeogeography of the Australian-east Indonesian region. In: R.L. Kirk and A.G. Thorne (eds.) *The Origins of the Australians*, pp. 11–22. Canberra: Institute of Aboriginal Studies.

Chappell, J. & Shackleton, N.J. (1986) Oxygen isotopes and sea-level. *Nature*, **324**: 137–40.

Chase, P.G. (1989) How different was Middle Palaeolithic subsistence? A zooarchaeological perspective on the Middle to Upper Palaeolithic transition. In: P. Mellars & C.B. Stringer (eds.) *The Human Revolution*, pp. 321–37. Edinburgh: Edinburgh University Press.

Chen, C. and Olsen, J.W. (1990) China at the last glacial maximum. In : O. Soffer & C. Gamble (eds.) *The World at 18,000 BP*, vol 1, pp. 276–95. London: Unwin Hyman.

Cheverud, J.M. (1988) A comparison of genetic and phenotypic correlations. *Evolution*, **42**: 958–68.

Cheverud, J.M. & Buikstra, J.A. (1981) Quantitative genetics of skeletal non-metric traits in the rhesus macaques on Cayo Santiago II. Phenotypic genetic and environmental correlations between traits. *American Journal of Physical Anthropology*, **54**: 51–8.

Clark, G. (1987) From the Mousterian to the Metal Ages: long-term change in the human diet of Northern Spain. In: O. Soffer (ed.) *The Pleistocene Old World: Regional Perspectives*, pp. 293–316. New York: Plenum Press.

Clark, J.D. (1982) The transition from Lower to Middle Paleolithic in the African continent. In: A. Ronen (ed.) *The Transition from Lower to Middle Paleolithic*

and the Origin of Modern Man. BAR International Series, **151**: 235–55.
Clark, J.D. (1993) The Aterian of the Central Sahara. In: L. Krzyzaniak, M. Kobusiewicz & J. Alexander (eds.) *Environmental Change and Human Culture in the Nile Basin and Northern Africa Until the Second Millennium B.C.*, pp. 49–68. Poznan: Museum Archeologiczne W Poznaniu.
Clark, J.D., Williamson, K.W., Michels, M.J. & Marean, C.A. (1984) A Middle Stone Age occupation site at Porc Epic Cave, Dire Dawa (East Central Ethiopia). *African Archaeological Revue*, **2**: 37–71.
CLIMAP Project Members (1976) The surface of the ice-age earth. *Science*, **191**: 1131–7.
Close, A. (1993) BT-14, a stratified Middle Palaeolithic site at Bir Tarfawi, Western Desert of Egypt. In: L. Krzyzaniak, M. Kobusiewicz & Alexander, J. (eds.) *Environmental Change and Human Culture in the Nile Basin and Northern Africa Until the Second Millennium B.C.*, pp. 113–27. Poznan: Museum Archeologiczne W Poznaniu.
Close, A.E. & Wendorf, F. (1990) North Africa at 18,000 BP. In: G. Gamble & O. Soffer (eds.) *The World at 18,000* BP, vol. 2., pp. 41–57. London: Unwin Hyman.
Collins, H.B. (1925) The pterion in Primates. *American Journal of Physical Anthropology*, **8**: 261–74.
Collins, H.B. (1926) The temporo-frontal articulation in man. *American Journal of Physical Anthropology*, **9**: 343–8.
Comas, J. (1942) Contribution a l'étude du metopism. *Archives Suisses d'Anthropologie et Genetique*, **10**: 273–412.
Condemi, S. (1985) *Les hommes fossiles de Saccopastore (Italie) et leurs relations phylogenetiques*. PhD Th., Univ. of Bordeaux.
Conrad, G. (1969) *L'evolution continentale post-hercynienne du Sahara Algerien (Saoura, Erg Chech-Tarezrouft-Ahnet-Mouydir)*. Paris: Centre Rech. Zones Arides, C.N.R.S.
Cook, J., Stringer, C.B., Currant, A.P., Schwarcz, H.P. & Wintle, A.G. (1982) A review of the chronology of the European Middle Pleistocene hominid record. *Yearbook of Physical Anthropology*, **25**: 19–65.
Coon, C.S. (1962) *The Origin of Races*. NY: Knopf.
Coon, C.S., Garn, S.M. and Birdsell, J.B. (1950) *Races: A Study of the Problems of Race Formation in Man*. Springfield, Ill.: Charles C. Thomas.
Coon, C.S. & Ralph, E.K. (1955) Radiocarbon dates for Kara Kamar, Afghanistan. *Science*, **122**: 921–2.
Corruccini, R.S. (1974) An examination of the meaning of cranial discrete traits for human skeletal biological studies. *American Journal of Physical Anthropology*, **40**: 425–46.
Corruccini, R.S. (1992) Metrical reconsidertion of the Skhūl IV and IX and Border Cave 1 crania in the context of modern human origins. *American Journal of Physical Anthropology*, **87**: 433–45.
Cosgrove, R. (1989) Thirty thousand years of human colonization in Tasmania: New Pleistocene dates. *Science*, **243**: 1706–8.
Cosseddu, G.G., Floris, G. & Vona, G. (1979) Sex and size differences in the minor non-metrical cranial variants. *Journal of Human Evolution*, **8**: 685–92.
Coutts, P.J.F. (1984) A hunter-gatherer/agriculturalist interface on Panay Island, Philippines. In: D. Bayard (ed.) *Southeast Asian Archaeology at the XV Pacific*

Science Congress, pp. 218–34. Dunedin: Dept. of Anthropology, Otago University.

Curtis, G.H. (1981) Man's immediate forerunners – establishing a relevant time scale in anthropological and archaeological research. *Philosophical Transactions of the Royal Society of London, Series B*, **292**: 7–20.

Czarnetzki, A. (1975) On the question of correlation between the size of the epigenetic distance and the degree of allopatry in different populations. *Journal of Human Evolution*, **4**: 483–9.

Davida, E. (1914) Beitrage zur persistenz der transitorischen Nahte. *Anatomischer Anzeiger Jena*, **46**: 399–412.

Davis, E.L., Brown, K.H. & Nichols, J. (1980) *Evaluation of Early Human Activities and Remains in the California Desert*. Riverside, CA: Great Basin Foundation.

Davis, R.S. (1990) Central Asia hunter-gatherers at the Last Glacial Maximum. In: O. Soffer & C.S. Gamble (eds.) *The World at 18,000* BP: High Latitudes, pp. 267–75. London: Unwin Hyman.

Day, M.H. & Stringer, C.B. (1982) A reconsideration of the Omo Kibish remains and the *erectus-sapiens* transition. *1er. Congres de Paleontologie Humaine*, Nice, pp. 814–46.

Day, M.H., Twist, M.H.C. & Ward, S. (1991) Les vestiges post-craniens d'Omo I (Kibish). *L'Anthropologie*, **94**: 595–610.

De Villiers, H. (1968) *The Skull of the African Negro*. Johannesburg: Witwatesrand University Press.

De Villiers, H. (1973) Human skeletal remains from Border Cave, Ingwaruma district, Kwa Zulu, South Africa. *Annals of the Transvaal Museum*, **28**: 229–56.

De Villiers, H. & Fatti, L.P. (1982) The antiquity of the Negro. *South African Journal of Science*, **78**: 321–32.

De Vos, J. (1985) Faunal stratigraphy and correlation of the Indonesian hominid sites. In: E. Delson (ed.) *Ancestors: The Hard Evidence*, pp. 215–20. New York: Alan R. Liss.

Deacon, H.J. (1988) The origins of anatomically modern people and the South African evidence. *Palaeoecology of Africa*, **19**: 193–200.

Deacon, H.J. (1989) Late Pleistocene palaeoecology and archaeology in the southern Cape, South Africa. In: P. Mellars & C.B. Stringer (eds.) *The Human Revolution*, pp. 547–64. Edinburgh: Edinburgh University Press.

Deacon, H.J. (1992) Southern Africa and modern human origins. *Philosophical Transactions of the Royal Society of London, Series B*, **337**: 177–83.

Deacon, H.J. & Shuurman, R. (1992) The origins of modern people: The evidence from Klasies River. In: G.Bräuer & F.H. Smith (eds.) *Continuity or Replacement? Controverisies in* Homo sapiens *evolution*, pp. 121–9. Rotterdam: Balkema.

Debenath, A., Raynal, J.P., Roche, J., Texier, J.P. & Ferembach, D. (1986) Stratigraphie, habitat, typologie et devenir de l'Aterian marocain: donnes recentes. *L'Anthropologie*, **90**: 233–46.

Delson, E. (1981) Palaeoanthropology: Pliocene and Pleistocene human evolution. *Paleobiology*, **7**: 298–305.

Demes, B. (1987) Another look at an old face: biomechanics of the Neandertal facial skeleton reconsidered. *Journal of Human Evolution*, **16**: 297–303.

Demes, B. & Creel, N. (1988) Bite force, diet and cranial morphology of fossil hominids. *Journal of Human Evolution*, **17**: 657–70.

Deol, M.S. & Truslove, G.M. (1957) Genetic studies on the skeleton of the mouse. XX Maternal physiology and variation in the skeleton of C57BL mice. *Journal of Genetics*, **55**: 288–312.

Deuser, W.G., Ross, E.H. & Waterman, L.S. (1976) Glacial and pluvial periods: their relationship revealed in Pleistocene sediments of the Red Sea and Gulf of Aden. *Science*, **191**: 1168–70.

DeVries, H.L. & Waterbolk , H.T. (1958) Groningen readiocarbon dates 3. *Science*, **128**: 1550–6.

Diester-Haas, L. (1976) Late Quaternary climatic variation in northwest Africa deduced from east Atlantic sedimentary cores. *Quaternary Research*, **6**: 299–314.

Dillehay, T.D. & Collins, M.B. (1988) Early cultural evidence from Monte Verde in Chile. *Nature*, **32**: 150–2.

Dixon, R. (1980) *The Languages of Australia*. Cambridge: Cambridge University Press.

Dodo, Y. (1974) Non-metrical cranial traits in the Hokkaido Ainu and the Northern Japanese of recent times. *Journal of the Anthropological Society of Nippon*, **82**: 31–51.

Dodo, Y. (1986) Metrical and non-metrical analyses of Jomon crania from eastern Japan. In: T. Akazawa & C.M. Aikens (eds.) *Prehistoric Hunter-Gatherers in Japan*, pp. 137–61. The University Museum, The University of Tokyo, Bulletin No. 27.

Dodo, Y. & Ishida, H. (1990) Population history of Japan as viewed from cranial nonmetric variation. *Journal of Anthropological Society of Nippon*, **98**: 269–87.

Dodo, Y., Ishida, H. & Saitou, N. (1992) Population history of Japan: A cranial nonmetric approach. In: T. Akazawa, K. Aoki & T. Kimura (eds.) *The Evolution and Dispersal of Modern Humans in Asia*, pp. 479–92. Tokyo: Hokusen-sha.

Dorit, R.L., Akashi, H. & Gilbert, W. (1995) Absence of polymorphism at the ZAY locus on the human Y chromosome. *Science*, **268**: 1183–5.

Drennan, M.R. (1937) The torus mandibularis in the Bushman. *Journal of Anatomy*, **72**: 66–70.

Dreyer, T.F. (1936) The Florisbad skull in the light of the Steinheim discovery. *Zeitschrift für Rassenkunde*, **4**: 320–2.

Dunnett, G. (1993) Diving for dinner: Some implications from Holocene middens for the role of coasts in the late Pleistocene of Tasmania. In: M.A. Smith, Spriggs, M. & Fankhauser, B. (eds.) *Sahul in Review: Pleistocene Archaeology in Australia, New Guinea and Island Melanesia*, pp. 247–57. Canberra: Department of Prehistory, Research School of Pacific Studies, ANU. *Occasional Papers in Prehistory*, No. 24.

El-Najjar, M. Y. & Dawson, G.L. (1977) The effect of artificial cranial deformation on the incidence of Wormian bones in the lambdoid suture. *American Journal of Physical Anthropology*, **46**: 155–60.

Eldredge, N. & Gould, S.J. (1972) Punctuated equilibria: an alternative to phyletic gradualism. In: T. Schopf (ed.) *Models in Paleobiology*, pp. 82–115. San Francisco: Freeman and Cooper.

Endo, B. (1965) Distribution of stress and strain produced in the human face by masticatory forces. *Journal of the Anthropological Society of Nippon*, **73**: 123–36.

Endo, B. (1966) Experimental studies on the mechanical significance of the form of the human facial skeleton. *Journal of the Faculty of Science of the University of*

Tokyo (Section V, Anthropology), **3**: 1–106.
Endo, B. (1970) Analysis of stress around the orbit due to masseter and temporalis muscles respectively. *Journal of the Anthropological Society of Nippon*, **78**: 251–66.
Enlow, D.H. (1966) A comparative study of facial growth in *Homo* and *Macaca*. *American Journal of Physical Anthropology*, **24**: 293–307.
Enlow, D.H. (1982) *Handbook of Facial Growth*. 2nd edn. Philadelphia: Saunders.
Evans, I.H.N. (1937) *The Negritos of Malaya*. Cambridge: Cambridge University Press.
Fagnart, J.P. (1984) Le Paléolithique supérieur dans le Nord de la France : un état de la question. *Bulletin de la Société Préhistorique Française*, **81**: 291–301.
Falconer, D.S. (1965) The inheritance of liability to certain diseases, estimated from the incidence among relatives. *Annals of Human Genetics*, **29**: 51–76.
Falconer, D.S. (1967) The inheritance of liability to disease with variable age of onset, with particular reference to diabetes mellitus. *Annals of Human Genetics*, **31**: 1–20.
Farrand, W.R. (1971) Late Quaternary paleoclimates of the eastern Mediterranean area. In: K.K. Turekian (ed.) *The Late Cenozoic Glacial Ages*, pp. 529–64. New Haven: Yale University Press.
Fatti, L.P. (1986) Discriminant analysis in prehistoric physical anthropology. In: R. Singer & J.K. Lundy (eds.) *Variation, Culture and Evolution in African Populations: Papers in Honor of Dr. Hertha de Villiers*, pp. 27–34. Johannesburg: Witwatesrand University Press.
Fenner, F.J. (1939) The Australian aboriginal skull: its non-metrical morphological characters. *Transactions of the Royal Society of S. Australia*, **63**: 248–306.
Ferembach, D. (1976) Les hommes du Mesolithique. In: H. de Lumley (ed.) *La Préhistoire Française*, pp. 604–11. Paris: C.N.R.S.
Ferembach, D. (1979) Filiation paléolithique des hommes du Mesolithique. In: D. De Sonneville-Bordes (ed.) *La fin des temps glaciares en Europe*, pp. 179–88. Paris: C.N.R.S.
Ferembach, D. (1986a) Les Hommes de l'Holocène. In: D. Ferembach, C. Susanne & M.C. Chamla (eds.) *L'Homme, son Évolution, sa Diversité*, pp. 224–38. Paris: C.N.R.S.
Ferembach, D. (1986b) *Homo sapiens sapiens* en Asie jusqu'au néolitique. In: D. Ferembach, C. Susanne & M.C. Chamla (eds.) *L'Homme, son Évolution, sa Diversité*, pp. 239–44. Paris: C.N.R.S.
Ferembach, D. (1986c) *Homo sapiens sapiens* en Afrique: des origines au néolithique. In: D. Ferembach, C. Susanne & M.C. Chamla (eds.) *L'Homme, son Évolution, sa Diversité*, pp. 245–56. Paris: C.N.R.S.
Ferring, C.R. (1975) The Aterian in North African prehistory. In: F. Wendorf & A.E. Marks (eds.) *Problems in prehistory: North Africa and the Levant*, pp. 11–126. Dallas: Southern Methodist University Press.
Finnegan, M. (1972) *Population Definition on the Northwest Coast by Analysis of Discrete Character Variation*. PhD. Thesis, Univ. of Colorado.
Fischer-Moller, K. (1942) The Mediaeval Norse settlements in Greenland. *Meddelelser om Grönland*, **89**: 61–3.
Flenly, J.R. (1972) Evidence of Quaternary vegetational change in New Guinea. In: P. Ashton & K. Ashton (eds.) *The Quaternary Era in Malaysia*, pp. 98–108. Hull: University of Hull Press.

Flohn, H. & Nicholoson, S. (1980) Climatic fluctuations in the arid belt of the Old World since the last glacial maximum; possible causes and future implications. *Palaeoecology of Africa*, **12**: 3–21.

Flood, J. (1983) *Archaeology of the Dreamtime*. Sydney: Collins.

Florschutz, F., Menedez Armor, J. & Wijmstra, T.A. (1971) Palynology of a thick Quaternary succession in southern Spain. *Palaeogeography, Palaeoclimatology, Palaeoecology*, **10**: 23–6.

Foley, R.A. (1989) The ecological conditions of speciation: a comparative approach to the origins of anatomically modern humans. In: P. Mellars & C.B. Stringer (eds) *The Human Revolution*, pp. 298–320. Edinburgh: Edinburgh University Press.

Foley, R. & Lahr, M.M. (1992) Beyond 'Out of Africa': reassessing the origins of *Homo sapiens*. *Journal of Human Evolution*, **22**: 523–9.

Foley, W.A. (1986) *The Papuan languages of New Guinea*. Cambridge: Cambridge University Press.

Fontes, J.C. & Gasse, F. (1989) On the ages of humid and late Pleistocene phases in North Africa – remarks on 'Late Quaternary climatic reconstruction for the Maghreb (North Africa)' by P. Rognon. *Palaeogeography, Palaeoclimatology, Palaeoecology*, **70**: 393–8.

Fowler, E. & Osmum, P. (1942) New bone growth due to cold water in the ear. *Archives of Otolaryngology*, **36**: 455–66.

Fox, R.B. (1970) *The Tabon Caves*. Manila: National Museum Monographs No. 1.

Fox, R.B. (1978) The Philippine Palaeolithic. In: F. Ikawa-Smith (ed.) *Early Palaeolithic in South and East Asia*, pp. 59–85. The Hague: Mouton.

Franciscus, R.G. (1989) Neanderthal mesosterna and noses: implications for activity and biogeographical patterning. Paper presented at the 58th Ann. Meet. Am. Assoc. Phys. Anthrop., San Diego, 1989.

Frayer, D.W. (1978) Evolution of the dentition in Upper Paleolithic and Mesolithic Europe. *University of Kansas Publications in Anthropology*, **10**.

Frayer, D.W. (1984) Biological and cultural change in the European late Pleistocene and early Holocene. In: F.H. Smith & F. Spencer (eds) *The Origin of Modern Humans: a World Survey of the Fossil Evidence*, pp. 211–50. New York: Alan R. Liss.

Frayer, D.W. (1986) Cranial variation at Mladeč and the relationship between Mousterian and Upper Palaeolithic hominids. *Anthropos*, **23**: 243–56.

Frayer, D.W., Jelinek, J., Minugh-Purvis, N., Oliva, M., Seitl, L., Smith, F.H. & Wolpoff, M.H. (1991) Late Pleistocene human remains from Mladeč Cave, Moravia. (Unpublished data).

Frayer, D.W., Wolpoff, M.H., Thorne, A.G., Smith, F.H. & Pope, G.G. (1993) Theories of modern human origins: the palaeontological test. *American Anthropology*, **95**: 14–50.

Freedman, L. & Lofgren, M. (1979a) The Cossak skull and a dihybrid origin of the Australian aborigines. *Nature*, **282**: 298–300.

Freedman, L. & Lofgren, M. (1979b) Human skeletal remains from Cossak, western Australia. *Journal of Human Evolution*, **8**: 283–99.

Freedman, L. & Lofgren, M. (1983) Human skeletal remains from Lake Tandou. *Archaeology in Oceania*, **18**: 98–105.

Friedlaender, J.S. & Merriwether, D.A. (1995) New indications of Bougainville genetic distinctiveness. Paper presented at the 1995 AAPA annual meeting, March 38–April 1, Oakland, California.

Furst, C.M. & Hansen, F. (1915) *Crania Groenlandica*. Copenhagen: Reitzel.

Gaboni-Csank, V. (1970) C14 dates of the Hungarian paleolithic. *Acta Archaeologica Hungara*, **22**: 1–11.

Gabriel, B., Voigt, B. & Ghod, M.M. (1993) Quaternary Ecology in Northern Somalia: a preliminary field report. In: L. Krzyzaniak, M. Kobusiewicz & Alexander, J. (eds.) *Environmental change and Human Culture in the Nile Basin and Northern Africa until the second millennium* BC, pp. 449–57. Poznan: Museum Archeologiczne W Poznaniu.

Gamble, C.S. (1993) *Timewalkers: The Prehistory of Global Colonization*. Cambridge, Ma.: Harvard University Press.

Garn, S.M. (1971) *Human Races*. Springfield, Ill.: Charles C. Thomas.

Gasse, F., Rognon, P. & Street, F.A. (1980) Quaternary history of the Afar and Ethiopian Rift lakes. In: M.A.J. Williams & H. Faure (eds.) *The Sahara and the Nile. Quaternary Environments and Prehistoric Occupations in Northern Africa*, pp. 361–400. Rotterdam: A.A. Balkema.

Gates, R.R. (1956) A study of Ainu and early Japanese skulls. *Zeitschrift für Morphologie und Anthropologie*, **48**: 55–70.

Gaven, C., Hillaire-Marcel, C. & Petit-Marie, N. (1981) A Pleistocene lacustrine episode in southeastern Libya. *Nature*, **290**: 131–3.

Gilbert, R. & Gill, G.W. (1990) A metric technique for identifying American Indian femora. In: G.W. Gill & S. Rhine (eds) *Skeletal Attribution of Race: Methods for Forensic Anthropology*, pp. 97–9. Maxwell Museum of Anthropology, Anthropology Paper 4. Albuquerque: Maxwell Museum.

Gilead, I. (1989) The Upper Palaeolithic in the southern Levant: periodization and terminology. *BAR International Series*, **497**: 231–54.

Giles, E. (1976) Cranial variation in Australia and neighbouring areas. In: R.L. Kirk & A.G. Thorne (eds.) *The Origin of the Australians*. Human Biol. Series 6, Canberra: Australian Institute of Aboriginal Studies.

Giles, R.E., Blanc, H., Cann. H.M. & Wallace, D.C. (1980) Maternal inheritance of human mitochondrial DNA. *Proceedings of the National Academy of Sciences, USA*, **77**: 6715–19.

Gillieson, D. & Mountain, M.-J. (1983) The environmental history of Nombe rockshelter in the Highlands of Papua New Guinea. *Archaeology in Oceania*, **18**: 53–62.

Ginther, C., Corach, D., Penacino, G.A., Rey, J.A., Carnese, F.R., Hutz, M.H. & Anderson, A. (1993) Genetic variation among the Mapuche indians from the Patagonian region of Argentina: mitochondrial DNA sequence, nuclear variation and allele frequencies of several nuclear genes. In: S.D.J. Pena, R. Chakraborty, J.T. Epplen & A.J. Jeffrey (eds) *DNA Fingerprinting: State of the Science*, pp. 211–19. Basel: Birkhauser.

Givens, D.H. (1968a) On the peopling of America. *Current Anthropology*, **9**: 120.

Givens, D.H. (1968b) A preliminary report on excavations at Hitzfelder Cave. *Bulletin of the Texas Archeological Society*, **38**: 47–50.

Gladilin, V.N. (1989) The Korolevo Paleolithic site: Research methods, stratigraphy. *Anthropologie*, **27**: 93–103.

Glinka, J. (1981) Racial history of Indonesia. In: H. Suzuki (ed.) *Rassengeschichte der Mensheit 8, Lieferung Asien I: Japan, Indonesien, Ozeanien*, pp. 79–113. Munich: Oldenbourg.

Glover, I. (1981) Leang Burung 2: An Upper Palaeolithic rock shelter in South Sulawesi, Indonesia. *Modern Quaternary Research in Southeast Asia*, **6**: 1–38.

Glover, I. (1987) *Archaeology in eastern Timor 1966–67*. Terra Australis, No. 11. Canberra: Department of Prehistory, Australia National University.

Glover, I. & Presland, G. (1985) Microliths in Indonesian stone industries. In: P. Bellwood & V.N. Misra (eds.) *Recent Advances in Indo-Pacific Prehistory*. New Dehli.

Golson, J. (1977) No room at the top: Agricultural intensification in the New Guinea Highlands. In: J. Allen, J. Golson & R. Jones (eds) *Sunda and Sahul*, pp. 601–38. London: Academic Press.

Golson, J. (1981) New Guinea agricultural history: A case study. In: D. Denoon & C. Snowden (eds) *A Time to Plant and a Time to Uproot*, pp. 55–64. Port Moresby: Institute of Papua New Guinea Studies.

Goodfriend, G.A. & Magaritz, M. (1988) Paleosols and late Pleistocene rainfall fluctuations in the Negev Desert. *Nature*, **332**: 144–6.

Gosden, C. & Robertson, N. (1991) Models for Matenkupkum: Interpreting a late Pleistocene site from southern New Ireland, Papua New Guinea. In: J. Allen & C. Gosden (eds) *Report of the Lapita Homeland Project*, pp. 20–45. Canberra: Department of Prehistory, Research School of Pacific Studies, ANU. *Occasional Papers in Prehistory*, No. 20.

Goudie, A.S. (1977) *Environmental Change*. Oxford: Clarendon Press.

Gould, S.J. & Vrba, E. (1982) Exaptation – a missing term in the science of form. *Paleobiology*, **8**: 4–15.

Gower, C. (1923) A contribution to the morphology of the apertura piriformis. *American Journal of Physical Anthropology*, **6**: 27–36.

Gowlett, J.A.J. (1987) The coming of modern man. *Antiquity*, **61**: 210–19.

Gramly, R.M. & Rightmire, G.P. (1973) A fragmentary cranium and dated Later Stone Age assemblage from Lukenya Hill, Kenya. *Man*, **8**: 571–9.

Grayson, D.K. (1988) Perspectives on the archaeology of the first Americans. In: R. Carlisle (ed.) *Americas Before Columbus: Ice Age Origins*, pp. 107–23. Ethnology Monographs 12. University of Pittsburgh.

Green, R.C. (1979) Lapita. In: J.D. Jennings (ed.) *The Prehistory of Polynesia*, pp. 27–60. Cambridge: Harvard University Press.

Greenberg, J.H. (1953) Historical linguistics and unwritten languages. In: A.L. Kroeber (ed.) *Anthropology Today*, pp. 265–86. Chicago: University of Chicago Press.

Greenberg, J.H. (1963) *The Languages of Africa*. The Hague: Mouton.

Greenberg, J.H. (1971) The Indo-Pacific Hypothesis. In: T.A. Sebeok (ed.) Linguistics in Oceania. *Current Trends in Linguistics*, **8**: 807–71.

Greenberg, J.H. (1987) *Language in the Americas*. Stanford: Stanford University Press.

Greenberg, J.H., Turner, C.G. II & Zegura, S.L. (1986) The settlement of the Americas: a comparison of linguistic, dental and genetic evidence. *Current Anthropology*, **27**: 477–97.

Greene, D.L. & Armelagos, G. (1972) *The Wadi Halfa Mesolithic Population*. Amherst: University of Massachusetts, Department of Anthropology Research Report No. 11.

Gregg, J.B. & Bass, W. (1970) Exostoses in the external auditory canals. *Annals Otology Rhinology and Laryngology*, **79**: 834–9.
Griffin, B. (1985) A contemporary view of the shift from hunting to horticulture: The Agua Case. In: V.N. Misra & P.S. Bellwood (eds.) *Recent Advances in Indo-Pacific Prehistory*, pp. 349–52. New Delhi: Oxford & IBH.
Grine, F.E. & Klein, R.G. (1985) Pleistocene and Holocene human remains from Equus Cave, South Africa. *Anthropology*, **8**: 55–98.
Grine, F.E., Klein, R.G. & Volman, T.P. (1991) Dating archaeology and human fossils from the Middle Stone Age levels of Die Kelders, South Africa. *Journal of Human Evolution*, **21**: 363–95.
Groube, L.M., Chappell, J., Muke, J. & Price, D. (1986) A 40,000-year-old human occupation site at Huon Peninsula, Papua New Guinea. *Nature*, **324**: 453–5.
Groves, C.P. (1989a) *A Theory of Human and Primate Evolution*. Oxford: Oxford Science Publications.
Groves, C.P. (1989b) A regional approach to the problem of the origin of modern humans in Australasia. In: P. Mellars & C.B. Stringer (eds) *The Human Revolution*, pp. 274–85. Princeton: Princeton University Press.
Groves, C.P. (1992) How old are subspecies? A tiger's eye-view of human evolution. *Perspectives in Human Biology*, **2**: 153–60.
Groves, C.P. & Lahr, M.M. (1994) A bush not a ladder: Speciation and replacement in human evolution. *Perspectives in Human Biology*, **4**: 1–11.
Gruhn, R. (1987) On the settlement of the Americas: South American evidence for an expanded time frame. *Current Anthropology*, **28**: 363–5.
Gruhn, R. (1988) Linguistic evidence in support of the coastal route of earliest entry into the New World. *Man*, **23**: 77–100.
Gruhn, R. & Bryan, A.L. (1991) A review of Lynch's description of South American Pleistocene sites. *American Antiquity*, **56**: 342–8.
Grün, R. & Stringer, C.B. (1991) Electron spin resonance dating and the evolution of modern humans. *Archaeometry*, **33**: 153–99.
Grün, R., Beaumont, P.B. & Stringer, C.B. (1990) ESR dating evidence for early modern humans at Border Cave in South Africa. *Nature*, **344**: 537–9.
Guidon, N. & Delibrias, G. (1986) Carbon–14 dates point to man in the Americas 32,000 years ago. *Nature*, **321**: 769–71.
Gyllensten, U., Wharton, D. Josefsson, A. & Wilson, A.C. (1991) Paternal inheritance of mitochodrial DNA in mice. *Nature*, **352**: 255–7.
Ha Van Tan (1980) Nouvelles recherches préhistoriques et protohistoriques au Vietnam. *Bulletin de l'École Française d'extrême-orient*, **68**: 113–54.
Habgood, P.J. (1985) The origin of the Australian Aborigines: an alternative approach and view. In: *Hominid Evolution: Past, Present and Future*, pp. 367–80. New York: Alan R. Liss.
Habgood, P.J. (1986) The origin of the Australians: a multivariate approach. *Archaeology in Oceania*, **21**: 130–7.
Habgood, P.J. (1987) Reestimation des variations observées sur le material osseux Australien. *L'Anthropologie*, **91**: 797–802.
Habgood, P.J. (1989) The origin of anatomically modern humans in Australasia. In: P. Mellars & C.B. Stringer (eds.) *The Human Revolution*, pp. 245–73. Edinburgh: Edinburgh University Press.
Habgood, P.J. (1992) The origin of anatomically modern humans in east Asia. In:

G. Bräuer & F.H. Smith (eds.) *Continuity or Replacement? Controversies in* Homo sapiens *Evolution*, pp. 273–88. Rotterdam: Balkema.
Hagelberg, E. & Clegg, J.B. (1993) Genetic polymorphisms in prehistoric Pacific islanders determined by analysis of ancient bone DNA. *Philosophical Transactions of the Royal Society of London, Series B*, **252**: 163–70.
Hahn, J. (1977) Aurignacian: Das altere Jungpalaolithikum in Mittle und Osteuropa. *Fundamenta*, **9**, ser. A: 1–355.
Hamilton, A.C. (1978) The significance of patterns of distribution shown by forest plants and animals in tropical Africa for the reconstruction of Upper Pleistocene environments. *Palaeoecology of Africa*, **9**: 63–97.
Han, K. & Pan, Q. (1984) The study of the racial elements of ancient China. *Acta Anchaeologica Sinica*, **4**: 245–63.
Hanihara, K. (1985) Geographic variation of modern Japanese crania and its relationship to the origin of Japanese. *Homo*, **36**: 1–10.
Hanihara, K. (1987) Estimation of the number of early migrants to Japan: A simulative study. *Journal of the Anthropological Society of Nippon*, **95**: 391–403.
Hanihara, K. (1991) Dual structure model for the population history of Japanese. *Japanese Review*, **21**: 1–33.
Hanihara, T. (1989a) Comparative studies of dental characteristics in the Aogashima islanders. *Journal of the Anthropological Society of Nippon*, **97**: 9–22.
Hanihara, T. (1989b) Comparative studies of geographically isolated populations in Japan based on dental measurements. *Journal of the Anthropological Society of Nippon*, **97**: 95–107.
Hanihara, T. (1989c) Affinities of the Philippine Negritos as viewed from dental characters: A preliminary report. *Journal of the Anthropological Society of Nippon*, **97**: 327–339.
Hanihara, T. (1990a) Affinities of the Philippine Negritos with modern Japanese and the Pacific populations based on dental measurements: the basic populations of East Asia I. *Journal of the Anthropological Society of Nippon*, **98**: 13–27.
Hanihara, T. (1990b) Dental anthropological evidence of affinities among the Oceania and the Pan-Pacific populations: The basic populations in East Asia II. *Journal of the Anthropological Society of Nippon*, **98**: 233–46.
Hanihara, T. (1990c) Studies on the affinities of Sakhalin Ainu based on dental characters: the basic populations in East Asia III. *Journal of the Anthropological Society of Nippon*, **98**: 425–37.
Hanihara T (1991a) *Population Prehistory of East Asia and the Pacific in the Late Pleistocene and Holocene: A Dental and Craniofacial Perspective*. Doctoral Diss., The University of Tokyo.
Hanihara, T. (1991b) The origin and microevolution of Ainu as viewed from dentition. *Journal of the Anthropological Society of Nippon*, **99**: 345–61.
Hanihara, T. (1991c) Dentition of Nansei islanders and peopling of the Japanese archipelago. *Journal of the Anthropological Society of Nippon*, **99**: 399–409.
Hanihara, T. (1992a) Negritos, Australian Aborigines and the 'proto-Sundadont' dental pattern: the basic populations in East Asia V. *American Journal of Physical Anthropology*, **88**: 183–96.
Hanihara, T. (1992b) Dental and cranial evidence on affinities of the East Asian and Pacific populations. In: K. Hanihara (ed.) *Japan in the World IV: Japanese as a Member of the Asian and Pacific Populations*, pp. 119–37. International

Research Center Symposium No. 4, Kyoto: IRCJS Press.
Hanihara, T. (1992c) Dental and cranial affinities among the populations in East Asia and the Pacific: the basic populations in East Asia IV. *American Journal of Physical Anthropology*, **88**: 163–82.
Hanihara, T. (1992d) Biological relationships among Southeast Asians, Jomonese, and the Pacific populations as viewed from dental characters. *Journal of the Anthropological Society of Nippon*, **100**: 53–67.
Hanihara, T. (1993a) Population prehistory of East Asia and the Pacific as viewed from craniofacial morphology: The basic populations in East Asia VII. *American Journal of Physical Anthropology*, **91**: 173–87.
Hanihara, T. (1993b) Dental affinities among Polynesian and Circum-Polynesian Populations. *Japan Review*, **4**: 59–82.
Harpending, H. (1994) Signature of ancient population growth in a low resolution mitochodrial DNA mismatch distribution. *Human Biology*, **66**: 591–600.
Harpending, H., Sherry, S.T., Rogers, A.R. & Stoneking, M. (1993) The genetic structure of ancient human populations. *Current Anthropology*, **34**: 483–96.
Harris, E.F., Turner, C.G. II & Underwood, J.H. (1975) Dental morphology of living Yap islanders, Micronesia. *Archaeology and Physical Anthropology in Oceania*, **10**: 218–34.
Harrison, D.F.N. (1962) The relationship of osteomata of the external auditory meatus to swimming. *Annals of the Royal College of Surgeons*, **3**: 187–201.
Harrison, T. (1959) New archaeological and ethnological results from the Niah caves, Sarawak. *Man*, **59**: 1–8.
Harrold, F.B. (1989) Mousterian, Chatelperronian and early Aurignacian in western Europe: continuity or discontinuity? In: P. Mellars & C.B. Stringer (eds) *The Human Revolution*, pp. 677–713. Edinburgh: Edinburgh University Press.
Hasebe, K. (1935) Knochenerbebungen in der Schläfen und Nackengegend der Schäel der Mikronesier. *Arbeiten aus dem Anatomie Institute Kaiserl Japanese University, Sendai*, **17**: 1–9.
Hauser, G. & De Stefano, G.F. (1989) *Epigenetic Variants of the Human Skull*. Stuttgart: Verlag.
Haydenblit, R. Heterogeneity in dental morphology among prehispanic Mexican populations. Unpublished data.
Haynes, C.V. (1987) Clovis origins update. *The Kiva*, **52**: 83–93.
Hedges, R.E.M. Hiously, R.A., Law, I.A., Perry, C. & Hendy, E. (1988) Radiocarbon dates from the Oxford AMS system. *Archaeometry*, **30**: 291–305.
Hedges, R.E.M., Houseley, R.A., Bronk, C.R. & Van Klinken, G.J. (1992) Radiocarbon dates from the Oxford AMS system; Archaeometry datelist 14. *Archaeometry*, **34**: 141–59.
Hedges, S.B., Kumar, S., Tamura, K. & Stoneking, M. (1991) Comment on 'Human origins and analysis of mitochondrial DNA sequences' by A. Templeton. *Science*, **255**: 737–9.
Henke, W. (1991) Biological distances in late Pleistocene and early Holocene human populations in Europe. *Variability and Evolution*, **1**: 39–64.
Herman, Y., Thonmeret, Y. & Vergnaud-Grazzini, C. (1969) Micropaleontology, paleotemperatures and radiocarbon dates of Quaternary Mediterranean deep-sea cores. *International Quaternary Association, 8th Congress*, Resumes, Paris. 174 bis.

Hilloowala, R.A. & Trent, R.B. (1988a) Supraorbital ridge and masticatory apparatus I: Primates. *Human Evolution*, **3**(5): 343–50.

Hilloowala, R.A. & Trent, R.B. (1988b) Supraorbital ridge and masticatory apparatus II: Humans (Eskimos). *Human Evolution*, **3**(5): 351–6.

Hiscock, P. (1981) Comments on the use of chipped stone artefacts as a measure of 'intensity of site usage'. *Australian Archaeology*, **13**: 30–4.

Hoáng Xuân Chinh (1991) Faunal and cultural changes from Pleistocene to Holocene in Vietnam. In: *Indo-Pacific Prehistory 1990: Proceedings of the 14th congress of the Indo-Pacific Prehistory Association*, Vol. 1: 74–78. Canberra and Jakarta: IPPA and Asosiasi Prehistorisi Indonesia. *Indo-Pacific Prehistory Association Bulletin* **10**.

Hofman, S. (1933) Radiografski prilog patologij Krapinskog pracovjeka. *Lijecnicki Vjesnik*, **10**: 1–3.

Holl, M. (1882) Uber die fossae praenasalis. *Wiener medizinsche Wochenschrift*, **32**: 721–53.

Holloway, R.L. (1981) Volumetric and asymmetry determinations on recent hominid endocasts: Spy I and II, Djebel Irhoud I, and the Salé *Homo erectus* specimens with some notes on Neanderthal brain size. *American Journal of Physical Anthropology*, **55**: 385–93.

Hooghiemstra, H. & Agwu, C.O.C. (1988) Changes in the vegetation and trade winds in equatorial northwest Africa 140,000–70,000 yr BP as deduced from two marine pollen records. *Palaeogeography, Palaeoclimatology, Palaeoecology*, **66**: 173–213.

Hooijer, D.A. (1983) Remarks upon the Dubois collection of fossil mammals from Trinil and Kedungbrubus in Java. *Proceedings of the Koninkijke Nederlande Akademie van Wetenschappen, Ser. B*, **55**: 436–43.

Hooton, E.A. (1918) On certain Eskimoid characters in Icelandic skulls. *American Journal of Physical Anthropology*, **1**: 53–76.

Hope, G.S. (1976) The vegetational history of Mount Wihelm, Papua New Guinea. *Journal of Ecology*, **64**: 627–64.

Horai, S., Kondo, R., Nakagawa-Hattori, Y., Hayashi, S., Sonoda, S. & Tajima, K. (1993) Peopling of the Americas, founded by four major lineages of mitochondrial DNA. *Molecular Biology and Evolution*, **10**: 23–47.

Horton, D.R. (1986) Dispersal and speciation: Pleistocene biogeography and the modern Australian biota. In: M. Archer & G. Clayton (eds) *Vertebrate Zoogeography and Evolution in Australia*, pp. 113–18. Perth: Hesperian Press.

Howell, F.C. (1978) Hominidae. In: V.J. Maglio & H.B.S. Cook (eds) *Evolution of African Mammals*, pp. 154–248. Cambridge, MA: Harvard University Press.

Howell, F.C. (1984) Introduction. In: F.H. Smith & F. Spencer (eds) *The Origin of Modern Humans: A World Survey of the Fossil Evidence*, pp. xiii–xxii. New York: Alan R.Liss.

Howells, W.W. (1959) *Mankind in the Making*. London: Mercury Books.

Howells, W.W. (1960) The distribution of man. *Scientific American*, **203**: 112–27.

Howells, W.W. (1966) The Jomon population of Japan: A study by discriminant analysis of Japanese and Ainu crania. *Peabody Museum Papers*, **57**(1): 1–43.

Howells, W.W. (1969) *Early Man in the Far East*. Anthropological Publications, Oosterhout, The Netherlands.

Howells, W.W. (1970) Mount Carmel man: morphological relationships. *Proceedings of the VIIIth International Congress of the Anthropological and Ethnological Sciences*, **1**: 269–72.

Howells, W.W. (1973) *Cranial Variation in Man. A Study by Multivariate Analysis of Patterns of Difference Among Recent Populations*. Papers of the Peabody Museum, Archaeolgy and Ethnology, vol. 67. Cambridge, MA: Harvard University Press.

Howells, W.W. (1974) Neanderthals: names, hypotheses, and scientific method. *American Anthropologist*, **76**: 24–38.

Howells, W.W. (1976) Explaining modern man: evolutionists versus migrationists. *Journal of Human Evolution*, **5**: 477–95.

Howells, W.W. (1980) *Homo erectus:* who, when and where, a survey. *Yearbook of Physical Anthropology*, **23**: 1–23.

Howells, W.W. (1983) Origins of the Chinese people: Interpretations of the recent evidence. In: D.N. Keightley (ed.) *The Origins of Chinese Civilization*, pp. 297–319. Berkeley, CA: University of California Press.

Howells, W.W. (1984) Introduction. In: G.N. van Vark & W.W. Howells (eds.) *Multivariate Statistical Methods in Physical Anthropology*, pp. 1–11. Dordrecht: D. Reidel.

Howells, W.W. (1986) Physical anthropology of the prehistoric Japanese. In: R. Pearson (ed.) *Windows on the Japanese Past*, pp. 85–99. Ann Arbor: Michigan Center for Japanese Studies.

Howells, W.W. (1989) *Skull Shapes and the Map*. Papers of the Peabody Museum, Archaeolgoy and Ethnology, vol. 79, Cambridge, MA: Harvard University Press.

Howells, W.W. (1990) Micronesia to Macromongolia: Micro-Polynesian craniometrics and the Mongoloid population complex. *Micronesia (Suppl.)*, **2**: 363–72.

Howells, W.W. (1993) *Getting Here: A Story of Human Evolution*. Washington: The Compass Press.

Hrdlička, A. (1910) Contributions to the anthropology of Central and Smith Sound Eskimo. *Anthropological Papers of the American Museum of Natural History*, **5**: 177–280.

Hrdlička, A. (1914) Anthropological work in Peru in 1913. *Smithsonian Miscellaneous Collections*, **61**(19): 69.

Hrdlička, A. (1928) Catalogue of the human crania in the United States National Museum. Australians, Tasmanians, South African Bushmen, Hottentots and Negro. *Proceedings of the U.S. National Museum*, **71**(24): 1–140.

Hrdlička, A. (1935) Ear exostoses. *Smithsonian Miscellaneous Collections*, **93**: 1–100.

Hrdlička, A. (1937) Early man in America: What have the bones to say? In: G.G. MacCurdy (ed.) *Early Man as Depicted by the Leading Authorities at the International Symposium at the Academy of Natural Sciences, Philadelphia, March 1937*, pp. 93–104. Philadelphia: J.B. Lippincott.

Hrdlička, A. (1940) Mandibular and maxillary hyperostoses. *American Journal of Physical Anthropology*, **27**: 1–68.

Hublin, J.J. (1978a) *Le torus occipital transverse et les structures associée: évolution dans le genre Homo*. PhD Thesis, University of Paris.

Hublin, J.J. (1978b) Quelque characteres apomorphes du crâne neandertalien et leur interpretation phylogenique. *Comptes Rendus, Academie des Sciences, Paris*. **287**: 923–6.

Hublin, J.J. (1985) Human fossils from North Africa Middle Pleistocene and the origin of *Homo sapiens*. In: E. Delson (ed.) *Ancestors: The Hard Evidence*, pp. 283–8. New York: Alan R. Liss

Hublin, J.J. (1992) Recent human evolution in Northern Africa. *Philosophical Transactions of the Royal Society of London, Series B*, **337**: 185–92.

Hublin, J.J.(1994) The Zafarraya Mousterian site: New evidence on the contemporaneity of modern humans and Neanderthals in Southwestern Europe. *Palaeoanthropology Society Abstracts*, **3**: 7–8.

Hublin, J.J. & Tillier, A.M. (eds.) (1992a) *Aux origines d'Homo sapiens*. Paris: Presses Universitaires de France.

Hublin, J.J. & Tillier, A.M. (1992b) L'*Homo sapiens* en Europe occidentale; gradualisme et rupture. In: J.J. Hublin & A.M. Tillier (eds.) *Aux origines d'Homo sapiens*. Paris: Presses Universitaires de France, pp. 291–327.

Hurni, H. (1981) Simen Mountains – Ethiopia: paleoclimate of the last cold period (Late Wurm). *Palaeoecology of Africa*, **13**: 127–37.

Hutterer, K.L. (1976) An evolutionary approach to the Southeast Asian cultural sequence. *Current Anthropology*, **17**: 221–42.

Hutterer, K.L. (1977) Reinterpreting the Southeast Asian Paleolithic. In: J. Allen, J. Golson & R. Jones (eds) *Sunda and Sahul: Prehistoric studies in Southeast Asia, Melanesia and Australia*, pp. 31–71. London: Academic Press.

Hylander, W.L. (1977) The adaptive significance of the Eskimo craniofacial morphology. In: Dahlberg & Graber (eds) *Orofacial Growth and Development*, pp. 129–69. The Hague: Mouton.

Hylander, W.L. (1979) Mandibular function in *Galago crassicaudatus* and *Macaca fascicularis*: An in vivo approach to stress analysis of the mandible. *Journal of Morphology*, **159**: 253–96.

Hylander, W.L. (1984) Stress and strain in the mandibular symphysis of primates: A test of competing hypotheses. *American Journal of Physical Anthropology*, **64**: 1–46.

Hylander, W.L., Picq, P.G. & Johnson, K.R. (1991) Masticatory-stress hypotheses and the supraorbital region of primates. *American Journal of Physical Anthropology*, **86**: 1–36.

Jacob, T. (1967) *Some Problems Pertaining to the Racial History of the Indonesian Region*. Utrecht: Drukkerij Neerlandia.

Jacob, T. (1976) Early populations in the Indonesian region. In: R.L. Kirk & A.G. Thorne (eds) *The Origins of the Australians*, pp. 81–93. Canberra: Australian Institute of Aboriginal Studies.

Jaeger, J.J. (1981) Les hommes fossiles du Pleistocene Moyen du maghreb dans leur cadre geologique, chronologique et paleoecologique. In: B.A. Sigmon & J.S. Cybulski (eds) Homo erectus: *Papers in Honor of Davidson Black*, pp. 159–264. Toronto: University of Toronto Press.

Jelinek, A. (1982a) The Tabun cave and paleolithic man in the Levant. *Science*, **216**: 1369–75.

Jelinek, A. (1982b) The Middle Paleolithic in the southern Levant, with comments on the appearence of modern *Homo sapiens*. In: A. Ronen (ed.) *The transition from Lower to Middle Paleolithic and the Origin of Modern Man. BAR International Series*, **151**: 57–101.

Jelinek, A.J. (1992) Perspectives from the Old World on the habitation of the New.

American Antiquity, **57**: 345–7.

Jelinek, J. (1969) Neanderthal man and *Homo sapiens* in central and eastern Europe. *Current Anthropology*, **10**: 475–503.

Jelinek, J. (1976) The *Homo sapiens neanderthalensis* and *Homo sapiens sapiens* relationship in central Europe. *Anthropologie (Brno)*, **14**: 79–81.

Jelinek, J. (1978) Earliest *Homo sapiens sapiens* from central Europe (Mladeč, Czechoslovakia). Paper presented at the Xth International Congress Anthropological and Ethnological Sciences, Delhi, India.

Jelinek, J. (1980) Neanderthal remains in Kulna Cave, Czechoslovakia. In: Schwidetzky, B. Chiarelli & O. Necrasov (eds) *Physical Anthropology of European Populations*. The Hague: Mouton.

Jelinek, J. (1985) The European, Near East, and North African finds after *Australopithecus* and the principal consequences for the picture of human evolution. In: P.V. Tobias (ed.) *Hominid Evolution: Past, Present and Future*, pp. 341–54. New York: Alan R. Liss.

Jennings, J.N. (1971) Sea level changes and land links. In: D. J. Mulvaney & J. Golson (eds) *Aboriginal Man and Environment in Australia*, pp. 1–13. Canberra: Australian National University.

Jia, L. & Huang, W. (1985) The late Palaeolithic in China. In: R.Wu & J.W. Olsen (eds.) *Palaeoanthropology and Palaeolithic Archaeology in the People's Republic of China*, pp. 211–23. Orlando: Academic Press.

Johnson, E.H. (1937) The narial margins in man. *Journal of Anatomy (Lond.)*, **71**: 356–61.

Johnson, M.J., Wallace, D.C., Ferris, S.D., Rattazzi, M.C. & Cavalli-Sforza, L.L. (1983) Radiation of human mitochondrial DNA types analysed by restriction endonuclease cleavage patterns. *Journal of Molecular Evolution*, **19**: 255–71.

Jones, R. (1968) The geographical background to the arrival of man in Australia and Tasmania. *Archaeology and Physical Anthropology in Oceania*, **3**: 185–215.

Jones, R. (1989) East of Wallace's Line: issues and problems in the colonization of the Australian continent. In: P. Mellars & C.B. Stringer (eds) *The Human Revolution*, pp. 743–82. Edinburgh: Edinburgh University Press.

Jones, R. (1992) The human colonisation of the Australian continent. In: G. Bräuer & F.H. Smith (eds) *Continuity or Replacement? Controversies in* Homo sapiens *Evolution*, pp. 289–301. Rotterdam: Balkema.

Kadanoff, D. & Mutafov, S. (1968) Uber die variationen der typisch lokalisierten uberzahligen Knochen und Knochenfortsatze des Hirnschadels beim Meanschen. *Anthropologischer Anzeiger*, **31**: 28–39.

Kajeva, Y. (1912) Die zahne der Lappen. Anthropologische Zahnstudie. *Proceedings of the Finnish Dental Society*, **10**: 1–64.

Kamminga, J. & Wright, R.V.S. (1988) The Upper Cave at Zhoukoudian and the origins of the Mongoloids. *Journal of Human Evolution*, **17**: 739–67.

Kanaseki, T. (1956) The question of the Yayoi people. In: S. Sugihara (ed.) *Symposium on Japanese Archaeology*, pp. 238–52. Tokyo: Kawade Shobo.

Kanaseki, T. (1966) People of the Yayoi period. In: S. Wajima (ed.) *Archaeology of Japan, vol. 3*, pp. 460–71. Tokyo: Kawade Shobo.

Katayama, K. (1988) A comparison of the incidence of non-metric cranial variants in several Polynesian populations. *Journal of the Anthropological Society of Nippon*, **96**(3): 357–69.

Katayama, K. (1990) A scenario on prehistoric Mongoloid dispersals into the South Pacific, with special reference to hypothetic proto-Oceanic connection. *Man and Culture in Oceania*, **6**: 151–9.
Keith, A. (1927) A report on the Galilee skull. In: Turville Petre (ed.) *Researches in Prehistoric Galilee*, pp. 53–106. British School of Archaeology in Jerusalem.
Kellock, W.L. & Parsons, P.A. (1970) Variation of minor non-metrical cranial variants in Australian aborigines. *American Journal of Physical Anthropology*, **32**: 409–22.
Kennedy, G.E. (1973) *The Anatomy of the Middle and Lower Pleistocene Hominid Femora*. PhD Th., Univ. of London.
Kennedy, G.E. (1984), Are the Kow Swamp hominids 'archaic'? *American Journal of Physical Anthropology*, **65**: 163–8.
Kennedy, G.E. (1985) Bone thickness in *Homo erectus*. *Journal of Human Evolution*, **14**: 699–708.
Kennedy, G.E. (1986) The relationship between auditory exostoses and cold water: an altitudinal analysis. *American Journal of Physical Anthropology*, **71**: 401–15.
Kennedy, G.E. (1992) The evolution of *Homo sapiens* as indicated by features of the postcranium. In: G. Bräuer & F.H. Smith (eds.) *Continuity or Replacement? Controversies in* Homo sapiens *Evolution*, pp. 209–18. Rotterdam: Balkema.
Kennedy, K.A.R. (1977) The deep skull of Niah. *Asian Perspectives*, **20**: 32–50.
Kenndey, K.A.R. & Deraniyagala, S.U. (1989) Fossil remains of 28,000 year old hominids from Sri Lanka. *Current Anthropology*, **30**: 394–9.
Kidder, J.H., Jantz, R.L. & Smith, F.H. (1992) Defining modern humans: a multivariate approach. In: G. Bräuer & F.H. Smith (eds) *Continuity or Replacement? Controveries in* Homo sapiens *Evolution*, pp. 157–77. Rotterdam: Balkema.
Kiernan, K., Jones, R. & Ranson, D. (1983) New evidence from Fraser Cave for glacial age man in southwest Tasmania. *Nature*, **301**: 28–32.
Kingdon, J. (1993) *Self-Made Man and His Undoing*. London, Simon & Schuster.
Kirch, P.V. (1986) Rethinking east Polynesian prehistory. *Journal of the Polynesian Society*, **95**: 9–40.
Kirch, P.V. & Green, R.C. (1987) History, phylogeny and evolution in Polynesia. *Current Anthropology*, **28**: 431–56.
Kirch, P.V., Swindler, D.S. & Turner, C.G. II (1989) Human skeletal and dental remains from Lapita sites (1,600–500 BC) in the Mussau Islands, Melanesia. *American Journal of Physical Anthropology*, **79**: 63–76.
Kirk, R.L. (1980) Population movements in the Southwest Pacific: The genetic evidence. In: J.J. Fox (ed.) *Indonesia: The Making of a Culture*, pp. 45–56. Canberra: Research School of Pacific Studies.
Kirk, R.L. (1981) *Aboriginal Man Adapting*. Oxford: Oxford University Press.
Klaatsch, H. (1902) Occipitalia und temporalia der schadel von Spy verglichen mit denen von Krapina. *Zeitschrscheift Ethnologie*, **34**: 392–409.
Klaatsch, H. (1908) The skull of the Australian aboriginal. *Report of the Pathology Laboratory of the Lunacy Deptartment, New South Wales Government*, **1**(3): 43–167.
Klein, R.G. (1992) The archaeology of modern human origins. *Evolutionary Anthropology*, **1**: 5–14.
Knip, A.S. (1971) The frequencies of non-metrical variants in Tellem and Nokara

skulls from the Mali Republic, I, II. *Proceedings of the Koninklijke Nederlandse Akademie van Weterschappen, serie C*, **74**: 422–43.

Kozlowski, J.K. (1985) Sur la contemporanéité des differénts faciès du Magdalénien. In: *Jagen und Sammeln, Festschriftr für H.G. Bandi zum 65. Geburstag*. pp. 211–16. Bern: Stämpfli + Cie. AG.

Kozlowski, J.K. (1990) Northern central Europe *c.* 18,000 BP. In: O. Soffer & C.S. Gamble (eds) *The World at 18,000* BP: High Latitudes, pp. 204–27. London: Unwin Hyman.

Lahr, M.M. (1992) *The Origins of Modern Humans: A Test of the Multiregional Hypothesis*. PhD Thesis., University of Cambridge.

Lahr, M.M. (1994) The Multiregional Model of modern human origins: a reassessment of its morphological basis. *Journal of Human Evolution*, **26**: 23–56.

Lahr, M.M. (1995) Patterns of modern human diversification: implications for Amerindian origins. *Yearbook of Physical Anthropolgy*.

Lahr, M.M. Difference in robusticity between two Mediterranean Epi-Palaeolithic populations. (Unpublished data.)

Lahr, M.M. & Bowman, J.E. (1989) Non-metrical variation in a series of forty-five Amerindian crania. *Anales del VI Congreso Español de Antropologia Biológica*, pp. 367–75. E. Rebato & R. Calderón (eds.), Bilbao: Universidad del País Vasco.

Lahr, M.M. & Foley, R.A. (1992) Discriminant analysis of five traits used in the Multiregional Model of modern human origins. *American Journal of Physical Anthropology, Supp.* **4**: 104.

Lahr, M.M. & Foley, R.A. (1994) Multiple dispersals and modern human origins. *Evolutionary Anthropology*, **3**(2): 48–60.

Lahr, M.M. & Haydenblit, R. (1995) Traces of ancestral morphology in Tierra del Feugo and Patagonia. *American Journal of Physical Anthropology, Supplement 20*: 128

Lahr, M.M. & Wright, R.V.S. The question of robusticity and the relationship between cranial size and shape. *Journal of Human Evolution*. (In press.)

Laitman, J.T., Heimbuch, R.C. & Crelin, E.S. (1992) The basicranium of fossil hominids as an indicator of their upper respiratory systems. *American Journal of Physical Anthropology*, **51**: 15–34.

Lampert, R.J. & Hughes, P.J. (1988) Early human occupation of the Flinders Ranges. *Records of the South Australian Museum*, **22**: 139–68.

Larnach, S.L. (1978) Australian aboriginal craniology. *Oceania Monographs, No. 21*. Sydney: Sydney University Press.

Larnach, S.L. & Freedman, L. (1964) Sex determination of aboriginal crania from coastal New South Wales, Australia. *Records of the Australian Museum*, **26**: 295–308.

Larnach, S.L. & Macintosh, N.W.G. (1966) The craniology of the aborigines of coastal New South Wales. *Oceania Monographs 13*, Sydney.

Larnach, S.L. & Macintosh, N.W.G. (1974) A comparative study of Solo and Australian aboriginal crania. In: A.P. Elkin & N.W.G. Macintosh (eds) *Grafton Elliot Smith: The Man and His Work*, pp. 95–102. Sidney: Sidney University Press.

Le Double, A.F. (1903) *Traité des variations des os du crâne de l'homme et leur signification au point de vue de l'anthropogie zoologique*. Paris: Vigot Freres.

Leveque, F. & Vandermeersch, B. (1981) Le neandertalien de Saint-Cesaire. *Recherche*. **12**: 242–4.

Lewontin, R.C. (1974) *The Genetic Basis of Evolutionary Change*. New York: Columbia University Press.

Lewontin, R.C. & Krakauer, J. (1973) Distribution of gene frequency as a test of the theory of the selective neutrality of polymorphisms. *Genetics*, **74**: 175–95.

Lieberman, P. (1989) The origins of some aspects of human language and cognition. In: F.H. Smith & F. Spencer (eds) *The Origins of Modern Humans: A World Survey of the Fossil Evidence*, pp. 391–414. New York: Alan Liss.

Loritz, J.B. (1916) Ueber Lagebeziehungen der Fissura petrotympanica zur Mediansagittalen. *Zeitschrift für Morphologie und Anthropologie*, **19**: 381–90.

Lourandos, H. (1987) Pleistocene Australia: Peopling a continent. In: O. Soffer (ed.) *The Pleistocene Old World: Regional Perspectives*, pp. 147–65. New York: Plenum Press.

Lucotte, G. (1992) African pygmies have the more ancestral gene pool when studied for Y-chromosome DNA haplotypes. In: G. Bräuer & F. H. Smith (eds) *Continuity or Replacement? Controversies in* Homo sapiens *Evolution*, pp. 75–81. Rotterdam: Balkema.

Lukacs, J.R. & Hemphill, B.E. (1993) Odontometry and biological affinity in South Asia: Analysis of three ethnic groups from Northwest India. *Human Biology*, **65**: 279–325.

Lum, J.K., Rickards, O., Ching, C. & Cann, R.L. (1994) Polynesian mitochondrial DNAs reveal three deep maternal lineage clusters. *Human Biology*, **66**: 567–90.

Lynch, T.F. (1990) Glacial age man in South America? A critical review. *American Antiquity*, **55**: 12–36.

Macalister, A. (1898) The apertura piriformis. *Journal of Anatomy and Physiology*, **32**: 223–30.

Macintosh, N.W.G. (1978) The Tabon Cave mandible. *Archaeology and Physical Anthropology in Oceania*, **13**: 143–59.

Macintosh, N.W.G. & Barker, B.C.W (1965) The osteology of aboriginal man in Tasmania. *Oceania Monographs, 12*. Sydney.

Macintosh, N.W.G. & Larnach, S.L. (1976) Aboriginal affinities looked at in world contest. In: R. L. Kirk & A.G. Thorne (eds) *The Origin of Australians*, pp. 113–26. Canberra: Australian Institute of Aboriginal Studies.

Maddison, D.R. (1991) African origin of human mitochondrial DNA reexamined. *Systematics Zoology*, **40**(3): 355–63.

Maddison, D.R., Ruvolo, M. & Swofford, D.L. (1992) Geographic origins of human mitochondrial DNA: Phylogenetic evidence from control region sequences. *Systematics Biology*, **41**: 111–24.

Majid, Z. (1982) The west mouth, Niah, in the prehistory of Southeast Asia. *Sarawak Museum Journal*, **31**(52), Special Monograph, No. 3. Kuching, Sarawak: The Museum.

Maley, J.J., Roset, R. & Servant, M. (1971) Nouveaux gisements préhistoriques au Niger oriental: localisation stratigraphique. *Bulletin de Liaison ASEQUA* **31**: 9–18.

Malhotra, K.C. (1978) Morphological composition of the people of India. *Journal of Human Evolution*, **7**: 45–63.

Mann, G.E. (1984) *The Torus Auditivus: A Re-appraisal and its Application to a Series of Ancient Egyptian Skulls*. MPhil Thesis, Cambridge University.

Marks, A.E. (1988) The Middle to Upper Palaeolithic transition in the southern Levant: Technological change as an adaptation to increasing mobility. In: M.

Otte (ed.) *L'Homme de Néanderthal*, vol. 8, pp. 109–24. Liège: Université de Liège.
Marks, A.E. (1990) The Middle and Upper Palaeolithic of the Near East and the Nile Valley: the problem of cultural transformations. In: P. Mellars (ed.) *The Emergence of Modern Humans*, pp. 56–80. Ithaca: Cornell University Press.
Martin, R. (1928) *Lehrbuch der Anthropologie in systematischer darstellung*. Jena: G.Fischer.
Masao, F.T. (1992) The Middle Stone Age with reference to Tanzania. In: G. Bräuer & F.H. Smith (eds) *Continuity or Replacement? Controversies in* Homo sapiens *Evolution*, pp. 99–109. Rotterdam: Balkema.
Maslovsky, V.V. (1927) O Metopizme. *Russkii Antropologicheskii Zhurnal*, **15**: 7–12.
Matsu'ura, S. (1982) A chronological framing for the Sangiran hominids. *Bulletin of the National Science Museum of Tokyo, Series D, Anthropology*, **8**: 1–53.
Mayhall, J. (1970) The effects of culture upon the Eskimo dentition. *Arctic Anthropology*, **7**: 117–21.
Maynard Smith, J. (1966) Sympatric speciation. *American Naturalist*, **100**: 637–50.
Mayr, E. (1942) *Systematics and the Origin of Species*. New York: Columbia University Press.
Mayr, E. (1963) *Animal Species and Evolution*. Cambridge, MA: Harvard University Press.
Mayr, E., Linsley, E.G. & Usinger, R.L. (1953) *Methods and Principles of Systematic Zoology*. New York: McGraw Hill.
McCown, T.D. & Keith, A. (1939) *The Stone Age of Mount Carmel*, vol. 2, *The Fossil Human Remains from the Levalloiso-Mousterian*. Oxford: Clarendon Press.
McBurney, C.B.M. (1967) *Haua Fteah and the Stone Age of the Southeast Mediterranean*. Cambridge: Cambridge University Press.
McNiven, I., Marshall, B., Allen, J., Stern, N. & Cosgrove, R. (1993) The Southern Forests Archaeological Project: An overview. In: M.A. Smith, Spriggs, M. & Fankhauser, B. (eds) *Sahul in Review: Pleistocene Archaeology in Australian, New Guinea and Island Melanesia*, pp. 213–24. Canberra: Department of Prehistory, Research School of Pacific Studies, ANU. *Occasional Papers in Prehistory*, No. 24.
Mellars, P. (1989) Technological changes at the Middle-Upper Palaeolithic transition: economic, social and cognitive perspectives. In: P. Mellars & C.B. Stringer (eds) *The Human Revolution*, pp. 338–65. Edinburgh: Edinburgh University Press.
Mellars, P. (1991) Cognitive changes and the emergence of modern humans in Europe. *Cambridge Archaeological Journal*, **1**: 63–76.
Mellars, P. (1992) Outlining the problems. *Philosophical Transactions of the Royal Society of London, Series B*, **337**: 225–34.
Mellars, P. & Stringer, C.B. (Eds) (1989) *The Human Revolution: Behavioural and Biological Perspectives on the Origins of Modern Humans*. Edinburgh: Edinburgh University Press.
Meltzer, D.J. (1993) Pleistocene peopling of the Americas. *Evolutionary Anthropology*, **1**: 157–69.
Mercier, N., Valladas, H., Joron, J.L., Reyss, J.L., Leveque, F. & Vandermeersch, B. (1991) Thermoluminscence dating of the late Neanderthal remains from Saint-Cesaire. *Nature*, **351**: 737–9.
Merkel, F. (1871) *Die linea nuchae suprema*. Leipzig: Engelmann.

Merriwether, D.A. ; Clark, A.G., Ballinger, S.W., Schurr, T.G., Soodyall, H., Jenkins, T., Sherry, S.T. & Wallace, D.W. (1991) The structure of human mitochondrial DNA variation. *Journal of Molecular Evolution*, **33**: 543–55.

Merriwether, D.A., Rothhammer, F. & Ferrel, R.E. (1993) Mitochondrial DNA D-loop sequence variation in Native South Americans. *American Journal of Human Genetics, Supplement* **53**: 833.

Meyer, K. (1949) Uber den einfluss des badens auf die exostosenbildung im ausseren Gehorgang. *Zeitschrift für Laryngologie, Rhinologie, Otologie und ihre Grenzgebiete*, **28**: 480–92.

Milne, N., Schmitt, L.H. & Freedman, L. (1983) Discrete trait variation in Western Australian Aboriginal skulls. *Journal of Human Evolution*, **12**: 157–68.

Mitchell, P. (1973) *Les basins des fleuves Senegal et Gambie: étude geomorphologique*. Mem. No. 63, Paris: Office du Recherches Scientific Techniques d'Outre-mer.

Mizoguchi, Y. (1988) Affinities of the protohistoric Kofun people of Japan with pre and protohistoric Asian populations. *Journal of the Anthropological Society of Nippon*, **96**: 71–109.

Moore, A.M.T. (1982) Agricultural origins in the Near East: a model for the 1980s. *World Archaeology*, **14**: 224–36.

Moorrees, D.F.A., Osborne, R.H. & Wilde, E. (1957) Torus mandibularis: Its occurrence in Aleut children and its genetic determinants. *American Journal of Physical Anthropology*, **10**: 319–29.

Morell, V. (1990) Confusion in earliest America. *Science*, **248**: 439–41.

Morlan, R.E. (1987) The Pleistocene archaeology of Beringia. In: M.H. Nitecki & D.V. Nitecki (eds) *The Evolution of Human Hunting*, pp. 267–307. New York, Plenum.

Morris, A.G. (1992) Biological relationships between Upper Pleistocene and Holocene populations in southern Africa. In: G. Bräuer & F.H. Smith (eds) *Contintuity or Replacement? Controversies in* Homo sapiens *Evolution*, pp. 131–43. Rotterdam: Balkema.

Moss, M.L. & Young, R.W. (1960). A functional approach to craniology. *Acta Physiologica Scandinavica*, **69**: 1–229.

Moss, M.L., Noback, C.R. & Robertson, G.G. (1956) Growth of certain human fetal cranial bones. *American Journal of Anatomy*, **48**: 191–204.

Mouri, T. (1986) *Geographical and Temporal Variation in Japanese Populations as Viewed from Nonmetric Traits of the Skull*. PhD Thesis., Kyoto University.

Mouri, T. (1988) Incidences of cranial nonmetric characters in five Jomon populations from West Japan. *Journal of the Anthropological Society of Nippon*, **96**: 319–37.

Murphy, T. (1956) The pterion in the Australian aborigine. *American Journal of Physical Anthropology*, **14**: 225–44.

Mussi, M. (1990) Continuity and change in Italy at the Last Glacial Maximum. In: O. Soffer & C.S. Gamble (eds) *The World at 18,000 BP: High Latitudes*, pp. 126–47. London: Unwin Hyman.

Mussi, M. & Zampetti, D. (1987) Frontiera e confini nel Gravettiano e nell'Epigravettiano dell'Italia. Prime considerazioni. *Scienze dell'Antichità*, **2**: 45–78.

Necrasov, O. & Cristescu, M. (1965) Données anthropologiques sur les populations de l'âge de pierre en Roumanie. *Homo*, **16**: 129–30.

Nei, M. & Roychoudhury, A.K. (1982) Genetic relationship and evolution of human races. In: M.K. Hecht, B. Wallace & G.T. Prance (eds) *Evolutionary Biology*, pp. 1–59. New York: Plenum Press.

Nei, M. & Roychoudhury, A.K. (1993) Evolutionary relationships of human populations on a global scale. *Molecular Biology and Evolution*, **10**: 927–43.

Neumann, G.K. (1952) Archeology and race in the American Indians. In: J.B. Griffin (ed.) *Archeology of the Eastern United States*, pp. 13–34. Chicago: University of Chicago Press.

Neves, W.A. & Pucciarelli H.M. (1991) Morphological affinities of the first Americans: an exploratory analysis based on early South American human remains. *Journal of Human Evolution*, **21**: 261–73.

Newell, R., Constandse-Westermann, T.S. & Meiklejohn, C. (1979) The skeletal remains of Mesolithic in Western Europe: An evaluative catalogue. *Journal of Human Evolution*, **8**: 1–228.

Nichols, J. (1990) Linguistic diversity and the first settlement of the New World. *Language*, **66**: 475–521.

O'Connor, S., Veth, P. & Hubbard, N. (1993) Changing interpretations of postglacial human subsistence and demography in Sahul. In: M.A. Smith, Spriggs, M. & Fankhauser, B. (eds) *Sahul in Review: Pleistocene Archaeology in Australian, New Guinea and Island Melanesia*, pp.95–105. Canberra: Department of Prehistory, Research School of Pacific Studies, ANU. *Occasional Papers in Prehistory*, No. 24.

Olivier, G. (1965) *Anatomie Anthropologique*. Paris: Vigot Freres.

Olsen, J.W. (1987) Recent developments in the Upper Pleistocene prehistory of China. In: O. Soffer (ed.) *The Pleistocene Old World: Regional Perspectives*, pp. 135–46. New York: Plenum Press.

Olsen, J.W. & Ciochon, R.L. (1990) A review of evidence for postulated Middle Pleistocene occupation of Viet Nam. *Journal of Human Evolution*, **19**: 761–88.

Omoto, K. (1984) The Negritos: Genetic origins and microevolution. *Acta Anthropogenetica*, **8**: 137–47.

Oschinsky, L. (1964) *The Most Ancient Eskimos*. Ottawa: University of Ottowa Press.

Ossenberg, N.S. (1969) *Discontinuous Morphological Variation in the Human Cranium*. PhD Thesis, University of Toronto.

Ossenberg, N.S. (1970) The influence of artificial cranial deformation on discontinuous morphological traits. *American Journal of Physical Anthropology*, **33**: 357–72.

Ossenberg, N.S. (1986) Isolate conservatism and hybridization in the population history of Japan: the evidence of non-metric cranial traits. *The University of Tokyo, Bulletin*, **27**: 198–214.

Ossenberg, N.S. (1992) Native people of the American Northwest: population history from the perspective of skull morphology. In: T. Akazawa, K. Aoki & T. Kimura (eds) *The Evolution and Dispersal of Modern Humans in Asia*, pp. 493–530. Tokyo: Hokusen-sha.

Ossenberg, N.S., Steele, S.S. & Howes, J. (1995) Occlusal load and temporomandibular reaction forces in prehistoric Eskimos vs. modern Eurasians: A comparison based on three dimensional analysis of static equilibrium from craniofacial measurements. Paper presented at the 1995 AAPA annual meeting, March 28–April 1, Oakland, California.

Otte, M. (1990) The northwestern European plain around 18,000 BP. In: O. Soffer &

C.S. Gamble (eds.) *The World at 18,000 BP: High Latitudes*, pp. 54–68. London: Unwin Hyman.
Pääbo, S., Gifford, J.A. & Wilson, A.C. (1988) Mitochondrial DNA sequences from a 7000-year old brain. *Nucleic Acid Research*, **16**(20): 9775–87.
Papillaut, G. (1896–1902) La suture metopique et ses rapports avec la morphologie cranienne. *Memoires de la Societé Anthropologique de Paris*, **3**: 1892–6.
Pardoe, C. (1984) *Prehistoric Human Morphological Variation in Australia.* PhD Thesis, National University of Australia.
Pardoe, C. (1991) Isolation and evolution in Tasmania. *Current Anthropology*, **32**: 1–21.
Parkin, D.W. & Shackleton, N.J. (1973) Trade winds and temperature correlations down a deep-sea core off the Saharan coast. *Nature*, **245**: 455–7.
Pearce, R.H. & Barbetti, M. (1981) A 38,000 year-old archaeological site at Upper Swan, Western Australia. *Archaeology in Oceania*, **16**: 173–8.
Perrot, J. (1968) La prehistoire palestinienne. In: Supplément au Dictionnaire de la Bible, **8**: 286–446.
Peterson, G.N., Webb, T., Kutzbach, J.E., van der Hammen, T., Wijmstra, T.A. & Street, F.A. (1979) The continental record of environmental conditions at 18,000 yr BP: An initial evaluation. *Quaternary Research*, **12**: 47–82.
Petit-Marie, N. (1989) Interglacial environments in presently hyperarid Sahara: Paleoclimatic implications. In: M. Leinen & M. Sarntheim (eds) *Paleoclimatology and Paleometeorology: Modern and past patterns of global atmospheric transport*, pp. 637–61. London: Kluwer Academic Press.
Phenice, T.W. (1969) A newly developed visual method of sexing the os pubis. *American Journal of Physical Anthropology*, **30**: 297–302.
Phillipson, D.W. (1985) *African Archaeology*. Cambridge: Cambridge University Press.
Picq, P.G. & Hylander, W.I. (1989). Endo's stress analysis of the primate skull and the functional significance of the supraorbital Region. *American Journal of Physical Anthropology*, **79**: 393–8.
Pielou, E.C. (1979) *Biogeography*. New York: John Wiley and Sons.
Pietrusewsky, M. (1973) A multivariate analysis of craniometric data from the Territory of Papua New Guinea. *Archaeology and Physical Anthropology in Oceania*, **8**: 12–23.
Pietrusewsky, M. (1976) *Prehistoric human skeletal remains from Papua New Guinea and the Marquesas*. Manoa: University of Hawaii.
Pietrusewsky, M. (1979) Craniometric variation in Pleistocene Australian and more recent Australian and New Guinea populations studies by multivariate procedures. *Occasional Papers in Human Biology*, **2**: 83–123. Canberra: Australian Institute of Aboriginal Studies.
Pietrusewsky, M. (1981) Cranial variation in early metal age Thailand and Southeast Asia studied by multivariate procedures. *Homo*, **32**: 1–26.
Pietrusewsky, M. (1983) Multivariate analysis of New Guinea and Melanesian skulls: a review. *Journal of Human Evolution*, **12**: 61–76.
Pietrusewsky, M. (1984) Metric and non-metric cranial variation in Australian Aboriginal populations compared with populations from the Pacific and Asia. *Occasional Papers in Human Biology*, **3**, Canberra: Australian Institute of Aboriginal Studies.

Pietrusewsky, M. (1985) The earliest Lapita skeleton from the Pacific: A multivariate analysis of a mandible fragment from Natunuku, Fiji. *Journal of the Polynesian Society*, **94**: 389–414.

Pietrusewsky, M. (1988) Multivariate comparisons of recently excavated Neolithic human crania from the Socialist Republic of Vietnam. *International Journal of Anthropology*, **3**: 267–83.

Pietrusewsky, M. (1989a) A study of skeletal and dental remains from Watom Island and comparisons with other Lapita people. *Records of the Australian Museum*, **41**: 235–92.

Pietrusewsky, M. (1989b) A Lapita-associated skeleton from Natunuku, Fiji. *Records of the Australian Museum*, **41**: 297–325.

Pietrusewsky, M. (1990a) Craniometric variation in Micronesia and the Pacific: A Multivariate study. *Micronesia (Suppl.)*, **2**: 373–402.

Pietrusewsky, M. (1990b) Craniofacial variation in Australasian and Pacific populations. *American Journal of Physical Anthropology*, **82**: 319–40.

Pietrusewsky, M. (1992) Japan, Asia and the Pacific: a multivariate craniometric investigation. In: K. Hanihara (ed.) *Japanese as a Member of the Asian and Pacific Populations*, pp. 9–52. International Symposium No. 4. Kyoto: International Research Center for Japanese Studies.

Pietrusewsky, M., Li, Y., Shao, Z. & Nguyen, Q.Q. (1992) Modern and near modern populations of Asia and the Pacific: A multivariate craniometric interpretation. In: K. Akazawa, T. Aoki & K. Kimura (eds) *The Evolution and Dispersal of Modern Humans in Asia*, pp. 531–58. Tokyo: Hokusen-sha.

Pope, G.G. (1983) Evidence on the age of the Asian Hominidae. *Proceedings of the National Academy of Sciences, USA*, **80**: 4988–92.

Pope, G.G. (1984) The antiquity and palaeoenvironments of the Asian Hominidae. In: R.O. Whyte (ed.) *The Evolution of the East Asian Environment*, pp. 822–47. Hong Kong: University of Hong Kong Press.

Pope, G.G. (1985) Taxonomy, dating and palaeoenvironment: the palaeoecology of the early Far Eastern hominids. *Modern Quaternary Research in Southeast Asia*, **9**: 65–81.

Pope, G.G. (1988) Recent advances in Far Eastern palaeoanthropology. *Annals Revue of Anthropology*, **17**: 43–77.

Pope, G.G. (1992) Craniofacial evidence for the origin of modern humans in China. *Yearbook of Physical Anthropology*, **35**: 243–98.

Powell, J.F. (1993) Dental evidence for the peopling of the New World: Some methodological considerations. *Human Biology*, **65**: 799–819.

Protsch, R. (1975) The absolute dating of Upper Pleistocene sub-Saharan fossil Hominids and their place in human evolution. *Journal of Human Evolution*, **4**: 297–322.

Rainer, F. & Simionescu, I. (1942) Sur le premier crâne d'homme paléolithique trouvé en Roumanie. *Analele Academiei Române*, **3**: 1–15.

Rak, Y. (1983) *The Australopithecine Face*. New York: Academic Press.

Rak, Y. (1985) Systematic and functional implications of the facial morphology of *Australopithecus* and early *Homo*. In: E. Delson (ed.) *Ancestors: The Hard Evidence*, pp. 168–70. New York: A. Liss.

Rak, Y. (1986) The Neanderthal: a new look at an old face. *Journal of Human Evolution*. **15**: 151–64.

Rak, Y. (1987) What is so robust about the hyperrobust *Australopithecus boisei*? *American Journal of Physical Anthropology*, **72**: 244.
Rak, Y. (1990) On the differences of two pelvises of Mousterian context from the Qafzeh and Kebara caves, Israel. *American Journal of Physical Anthropology*, **81**: 323–32.
Rak, Y. (1992) Morphological variation in *Homo neanderthalensis* and *Homo sapiens* in the Levant: a biogeographical model. In: Proc. 14th Ann. Spring Syst. Symp., 1991, Chicago.
Rak, Y. & Arensburg, B. (1987) Kebara 2 Neanderthal pelvis; first look at a complete inlet. *American Journal of Physical Anthropology*, **73**: 227–31.
Ranke, J. (1899) Die uberzahligen hautknochen des menschlichen Schadeldachs. *Abhandlungen der Königlich Bayrerischen Akademie der Wissenschaften*, **20**: 277–464.
Ravosa, M.J. (1988) Browridge development in Cercopithecidae: A test of two models. *American Journal of Physical Anthropology*, **76**: 535–55.
Ravosa, M.J. (1989) Browridge development in anthropoid primates. *American Journal of Physical Anthropology*, **78**: 287–8.
Ravosa, M.J. (1991a) Ontogenetic perspectives on mechanical and nonmechanical models of primate circumorbital morphology. *American Journal of Physical Anthropology*, **85**: 95–112.
Ravosa, M.J. (1991b) Interspecific perspective on mechanical and non-mechanical models of primate circumorbital morphology. *American Journal of Physical Anthropology*, **85**: 95–112.
Reiss, Z., Luz, B., Almogi-Labin, A., Halicz, E., Winter, A. & Wolf, M. (1980) Late Quaternary paleoceanography of the Gulf of Aqaba (Elat). *Red Sea Quaternary Research*, **14**: 294–308.
Renfrew, C. (1991) Before Babel: speculations on the origins of linguistic diversity. *Cambridge Archaeological Journal*, **1**(1): 3–23.
Reynolds, T.E.G. & Kaner, S.C. (1990) Japan and Korea at 18,000 BP. In: O. Soffer & C.S. Gamble (eds) *The world at 18,000 BP*, vol 1, pp. 296–311. London: Unwin Hyman.
Rigaud, J.P. (1976) Les civilizations du paléolithique supérieur en Périgord. In: H. de Lumley (ed.) *La Préhistoire Française*, pp. 1257–70. Paris: C.N.R.S.
Rigaud, J.P. (1989) From the Middle to the Upper Palaeolithic: transition or convergence? In: E.Trinkaus (ed.) *The Emergence of Modern Humans*, pp. 142–53. Cambridge: Cambridge University Press.
Rightmire, G.P. (1975a) Problems in the study of later Pleistocene man in Africa. *American Anthropologist*, **77**: 28–52.
Rightmire, G.P. (1975b) New studies of Post-Pleistocene human skeletal remains from the Rift Valley, Kenya. *American Journal of Physical Anthropology*, **42**: 351–70.
Rightmire, G.P. (1978) Human skeletal remains from the southern Cape Province and their bearing on the Stone Age prehistory of South Africa. *Quaternary Research*, **9**: 219–30.
Rightmire, G.P. (1981) More on the study of the Border Cave remains. *Current Anthropology*, **20**: 23–35.
Rightmire, G.P. (1984) *Homo sapiens* in sub-Saharan Africa. In: F.H. Smith & F. Spencer (eds) *The Origins of Modern Humans: A World Survey of the Fossil*

Evidence, pp. 295–325. New York: Alan R. Liss.
Rightmire, G.P. (1986) Species recognition and *Homo erectus*. *Journal of Human Evolution*, **15**: 823–6.
Rightmire, G.P. (1989) Middle stone age humans from eastern and southern Africa. In: P. Mellars & C.B. Stringer (eds) *The Human Revolution*, pp. 109–22. Edinburgh: Edinburgh University Press.
Rightmire, G.P. & Deacon, H.J. (1991) Comparative studies of late Pleistocene human remains from Klasies River Mouth, South Africa. *Journal of Human Evolution*, **20**: 131–56.
Riquet, R. (1982) Le crâne aurignacien de Cioclovina. *Bulletin de la Societé d'Anthropologie de Paris*, **17**: 75–9.
Ritchie, J.C. & Haynes, C.V. (1987) Holocene vegetational zonation in the eastern Sahara. *Nature*, **330**: 645–7.
Roberts, R.G. & Spooner, N. (1992) Luminescence dating of early occupation sites in northern Australia. Paper presented at the Australia Day, University of Cambridge.
Roberts, R.G., Jones, R. & Smith, M.A. (1990) Thermoluminescence dating of a 50,000 year old human occupation site in northern Australia. *Nature*, **345**: 153–6.
Roche, A.F. (1964) Aural exostoses in Australian aboriginal skulls. *Annals of Otology, Rhinology and Laryngology*, **73**: 1–10; 82–91.
Roche, J. (1976) Chronostratigraphie des restes atériens de la grotte des Contrebandiers à Temara (province de Rabat). *Bulletins et Memoires de la Societé d'Anthropologie de Paris, 3 (série XIII)*: 165–73.
Roche, J. & Texier, J.P. (1976) Decouverte des restes humains dans un niveau aterien superieur de la Grotte des Contrabandiers a Temara (Maroc). *Comptes rendus de l'Academie de Science de Paris*, **2821**: 45–7.
Rogers, A.R. Population structure and modern human origins. ((Unpublished data, a.)
Rogers, A.R. Genetic evidence for a Pleistocene population explosion. (Unpublished data, b.)
Rogers, A.R. & Harpending, H. (1992) Population growth makes waves in the distribution of pairwise genetic differences. *Molecular Biology and Evolution*, **9**: 552–69.
Rogers, A.R. & Jorde, L.B. (1995) Genetic evidence on modern human origins. *Human Biology*, **67**: 1–36.
Rogers, W.M. & Applebaum, E. (1941) Changes in the mandible following closure of the bite with particular reference to edentelous patients. *Journal of the American Dental Association*, **28**: 1573–86.
Rognon, P. (1976) Essai d'interprétation des variations climatiques au Sahara depuis 40.000 ans. *Revue de Géographie Physique et de Géologie Dynamique*, **18**: 251–82.
Rognon, P. (1987) Late Quaternary climatic reconstructions for the Maghreb (North Africa). *Palaeogeography, Palaeoclimatology, Palaeocology*, **58**: 11–34.
Rognon, P. & Williams, M.A.J. (1977) Late Quaternary climatic changes in Australia and North Africa: preliminary interpretation. *Palaeogeography, Palaeoclimatology, Palaeoecology*, **21**: 285–327.
Rossignol, M. (1961) Analyse pollinique de sediments marines quaternaires en Israel, I: sediments recents. *Pollen Spores*, **3**: 303.
Rossignol, M. (1962) Analyse pollinique de sediments marines quaternaires en

Israel, II: sediments Pleistocenes. *Pollen Spores*, **4**: 121.
Rubin, D.T., McLeod, K.J. & Brain, S.D. (1990) Functional strains and cortical bone adaptation: Epigenetic assurance of skeletal integrity. *Journal of Biomechanics*, **23**: 43–8.
Ruhlen, M. (1990) An overview of genetic classification. In: J.A. Hawkins & M. Gell-Mann (eds) *The Evolution of Human Languages*, pp. 1–27. Addison-Wesley.
Russell, M.D. (1983) Browridge development as a function of bending stress in the supraorbital region. *American Journal of Physical Anthropology*, **60**: 248.
Russell, M.D. (1985) The supraorbital torus: A most remarkable peculiarity. *Current Anthropology*, **26**: 337–50.
Ruttin, E. (1933) Uber exostosen und hyperostosen des ausseren gehorganges. *Acta Otolaryngology*, **18**: 381–445.
Saha, N., Mak, J.W., Tay, J.S.H., Liu, Y., Tan, J.A.M.A., Low, P.S. & Singh, M. (1995) Population genetic study among the Orang Asli (Semai Senoi) of Malaysia, Malayan Aborigines. *Human Biology*, **67**: 37–57.
Santa-Luca, A.P. (1980) *The Ngandong fossil hominids*. Yale University Publications in Anthropology, No. 78, New Haven.
Sarntheim, M. (1978) Sand deserts during glacial maximum and climatic optimum. *Nature*, **272**: 43–6.
Sarntheim, M., Stremme, H.E. & Mangini, A. (1986) The Holstein Interglaciation: time-stratigraphic position and correlation to stable-isotope stratigraphy of deep-sea sediments. *Quaternary Research*, **26**: 283–98.
Scarsini, C., Rossi, V. & Messeri, P. (1980) Caratteri cranici discontinui in Sardi di Nuoro. *Seminario di Sciencia e Antropologia*, **2**: 1–127.
Schebesta, P. (1927) *Among the Forest Dwarfs of Malaya*. London: Hutchinson & Co.
Schebesta, P. (1928) The jungle tribes of the Malay Peninsula. *Bulletin of the School of Oriental and African Studies*, **4**: 269–78.
Schnider,B. (1990) The last Pleiniglacial in the Paris Basin (22,500–17,000 BP). In: O. Soffer & C.S. Gamble (eds) *The World at 18,000* BP: *High Latitudes*, pp. 41–53. London: Unwin Hyman.
Schreiner, K.E. (1935) *Zur Osteologie der Lappen*, vol. 1. Cambridge, MA: Harvard University Press.
Schurr, T.G., Ballinger, S.W., Gan, Y.Y., Hodge, J.A., Merriwether, D.A., Lawrence, D.N., Knowler, W.C., Weiss, K.M. & Wallace, D.C. (1990) Amerindian mitochondrial DNAs have rare Asian mutations at high frequencies, suggesting they derive from four primary maternal lineages. *American Journal of Human Genetics*, **46**: 613–23.
Schwalbe, G. (1899) Studien uber *Pithecanthropus erectus Dubois*. *Zeitschrift für morphologie und Anthropologie*, **1**: 16–240.
Schwarcz, H.P. (1992) Uranium-series dating and the origins of modern man. *Philosophical Transactions of the Royal Society, London, Series B*, **337**: 131–7.
Schwarcz, H.P., Grün, R., Vandermeersch, B., Bar-Yosef, O., Vallois, H. & Tchernov, E. (1988) ESR dates for the hominid burial site of Qafzeh in Israel. *Journal of Human Evolution*, **17**: 733–7.
Schwarcz, H.P., Bietti, A., Buhay, W.M., Stiner, M.C., Grün, R. & Segre, A. (1991) U-series and ESR age data for the Neanderthal site of Monte Cicreo, Italy. *Current Anthropology*, **32**: 313–16.
Sciulli, P.W. (1990) Cranial metric and discrete trait variation and biological

differentiation in the terminal archaic of Ohio: the Duff site cemetery. *American Journal of Physical Anthropology*, **82**: 19–29.

Scott, J.H. (1957) Muscle growth and function in relation to skeletal morphology. *American Journal of Physical Anthropology* , **15**: 197–234.

Self, S.G. & Lamy, L. (1978) Heritability of quasi-continuous skeletal traits in random-bred populations of house mouse. *Genetics*, **88**: 109–20.

Sergi, S. (1930–32) Some comparison between the Gibraltar and Saccopastore skulls. *Proceedings of the 1st Congress of Prehistoric Sciences*, London.

Serjeantson, S.W. (1984) Migration and admixture in the Pacific. *Journal of Pacific History*, **19**: 160–71.

Seurat, L.G. (1934) *Études zoologiques sur le Sahara Central.* Memoires de la Societé d'Histoire Naturelle de l'Afrique du Nord, No. 4, Mission du Hoggar III.

Shea, B.T. (1986) On skull form and the supraorbital torus in primates.*Current Anthropology*, **27**: 257–60.

Sherry, S.T., Rogers, A.R., Harpending, H., Soodyall, H., Jenkins, T. & Stoneking, M. (1994) Mismatch distributions of mtDNA reveal recent human population expansions. *Human Biology*, **66**: 761–75.

Shields, G.F., Schmiechen, A.M., Frazier, B.L., Redd, A., Voevoda, M.I., Reed, J.K. & Ward, R.H. (1993) mtDNA sequences suggest a recent evolutionary divergence for Beringian and northern North American populations. *American Journal of Human Genetics*, **53**: 549–62.

Simek, J.F. (1992) Neanderthal cognition and the Middle to Upper Pleistocene transition. In: G. Bräuer & F.H. Smith (eds) *Continuity or Replacement? Controversies in* Homo sapiens *Evolution*, pp. 231–45. Rotterdam: Balkema.

Simek, J.F. & Snyder, L.M. (1988) Changing assemblage diversity in Perigord archaeofaunas. In: H. Dibble & A. Montet-White (eds) *Upper Pleistocene prehistory of western Eurasia*, pp. 321–32. Philadelphia: University of Pennsylvania Press.

Simmons, T., Falsetti, A.B. & Smith, F.H. (1990) Frontal bone morphometrics of southwest Asian Pleistocene hominids. *Journal of Human Evolution*, **20**: 249–69.

Singer, R. (1958) In: G. H. R. von Koenigswald (ed.) *Hundert Jahre Neanderthaler*, pp. 52–62. Utrecht: Kemink en Zoon.

Singer, R. & Wymer, J. (1982) *The Middle Stone Age at Klasies River Mouth in South Africa.* Chicago: University of Chicago Pres.

Sitliviy, V.I. (1989) The Mousterian complex Sorgeidy. *Anthropologie*, **27**: 119–31.

Sitsen, A.E. (1931) Beitrag zur Kenntnis des Sinus frontalis. *Anthropologischer Anzeiger*, **7**: 208–35.

Sjovold, T. (1984) A report on the heritability of some cranial measurements and non-metric traits. In: G.N. Van Vark & W.W. Howells (eds) *Multivariate Statistics in Physical Anthropology*, pp. 223–46. The Netherlands: D.Reidel.

Smith, A.B. (1993) Terminal Palaeolithic industries of Sahara: a discussion of new data. In: L. Krzyzaniak, M. Kobusiewicz & J. Alexander (eds) *Environmental change and Human Culture in the Nile Basin and Northern Africa until the second millennium* BC, pp. 69–75. Poznan: Museum Archeologiczne W Poznaniu.

Smith, F.H. (1976) *The Neanderthal Remains from Krapina. A Descriptive and Comparative Study*. Report of Investigations, University of Tennessee, 15, 359 pp.

Smith, F.H. (1982) Upper Pleistocene hominid evolution in South-central Europe: a review of the evidence and analysis of trends. *Current Anthropology*, **23**: 667–703.

Smith, F.H. (1984) Fossil hominids from the Upper Pleistocene of central Europe and the origin of modern Europeans. In: F.H. Smith & F. Spencer (eds) *The Origin of Modern Humans: A World Survey of the Fossil Evidence*, pp. 137–209. New York: Alan R. Liss.

Smith, F.H. (1985) Continuity and change in the origin of modern *Homo sapiens*. *Zeitschrift für Morphologie und Anthropologie*, **75**: 197–222.

Smith, F.H. (1992) The role of continuity in modern human origins. In: G. Bräuer & F.H. Smith (eds) *Continuity or Replacement? Controversies in* Homo sapiens *Evolution*, pp. 145–56. Rotterdam: Balkema.

Smith, F.H. & Paquette, S.P. (1989) The adaptive basis of Neanderthal facial form, with some throughts on the nature of modern human origins. In: E. Trinkaus (ed.) *The Emergence of Modern Humans*, pp. 181–210. Cambridge: Cambridge University Press.

Smith, F.H. & Spencer, F. (Eds) (1984) *The Origin of Modern Humans: A World Survey of the Fossil Evidence*. New York: Alan R. Liss.

Smith, F.H. & Trinkaus, E. (1992) Modern human origins in Central Europe: a case of continuity. In: J.J. Hublin & A.M. Tillier (eds) *Aux origines d' Homo sapien*, pp. 251–90. Paris: Presses Universitaires de France.

Smith, F.H., Simek, J.F. & Harrill, M.S. (1989) Geographic variation in supraorbital torus reduction during the later Pleistocene (*c.* 80,000–15,000 BP). In: P. Mellars & C.B. Stringer (eds) *The Human Revolution*, pp. 172–93. Edinburgh: Edinburgh University Press.

Smith, M.A. (1989) The case for a resident human population in the central Australian ranges during full glacial aridity. *Archaeology in Oceania*, **24**: 93–105.

Smith, M.A. & Sharp, N.D. (1993) Pleistocene sites in Australia, New Guinea and Island Melanesia: Geographic and temporal structure of the archaeological record. In: M.A. Smith, M. Spriggs & B. Fankhauser (eds) *Sahul in Review: Pleistocene Archaeology in Australian, New Guinea and Island Melanesia*, pp. 37–59. Canberra: Department of Prehistory, Research School of Pacific Studies, ANU. *Occasional Papers in Prehistory*, No. 24.

Smith, M.A. Spriggs, M. & Fankhauser, B. (1993) (Eds) *Sahul in Review: Pleistocene Archaeology in Australian, New Guinea and Island Melanesia Occasional Papers in Prehistory*, No. 24. Canberra: Department of Prehistory, Research School of Pacific Studies, ANU.

Smith, P. & Arensburg, B. (1977) A Mousterian skeleton from Kebara Cave. *Eretz Israel*, **13**: 164–76.

Smith, P.E.L. (1986) *Paleolithic archeology in Iran:* American Institute of Iranian Studies, Monographs No. 1. Philadelphia: The University Museum.

Soejono, R.P. (1982) New data on the Palaeolithic industry of Indonesia. In: M.A. de Lumley (ed.) *L'Homo erectus et la place de l'homme de Tautavel parmi les hominides fossiles*. 1er Congrés International de Paléontologie Humaine, Nice, pp. 578–92.

Solecki, R.S. (1963) Prehistory in the Shanidar Valley, northern Iraq. *Science*, **139**: 179–93.

Sondaar, P.Y. (1984) Faunal evolution and the mammalian biostratigraphy of Java. *Courier Forschungsinstitut Senckenberg*, **69**: 219–37.

Stanley, S.M. (1979) *Macroevolution: Patterns and Processes*. San Francisco: W.H. Freeman.

Steele, D.G. & Powell, J.F. (1992) Peopling of the Americas: Paleobiologcial evidence. *Human Biology*, **64**: 303–36.

Stewart, T.D. (1973) *The People of America*. New York: Scribner.

Stewart, T.D. & Newman, N.T. (1951) An historical resumé of the concept of differences in Indian types. *American Anthropologist*, **53**: 19–36.

Stine, O.C., Dover, G.J., Zhu, D. & Smith, K.D. (1992) The evolution of two west African populations. *Journal of Molecular Evolution*, **34**: 336–44.

Stoneking, M. & Cann, R. (1989) African origin of human mitochondrial DNA. In: P. Mellars & C.B. Stringer (eds.) *The Human Revolution*, pp. 17–30. Edinburgh: Edinburgh University Press.

Stoneking, M., Jorde, L.B., Bhatia, K. & Wilson, A. (1990) Geographic variation in human mitochondrial DNA from Papua New Guinea. *Genetics*, **124**: 717–33.

Stoneking, M., Sherry, S.T., Redd, A.J. & Vigilant, L. (1992) New approaches to dating suggest a recent age for the human mtDNA ancestor. *Philosophical Transactions of the Royal Society of London, Series B*, **337**: 167–75.

Storm, P. (1990) Mesolithic and Neolithic Sites from Java: Human Remains, Artifacts and the Subrecent Fauna. PhD Thesis, Department of Prehistory, University of Leiden.

Storm, P. and Nelson, A.J. (1992) The many faces of Wajak Man. *Archaeology in Oceania*, **27**: 37–46.

Strauss, L.G. (1989) Age of the modern Europeans. *Nature*, **342**: 476–7.

Strauss, L.G. (1994) Upper Paleolithic origins and radiocarbon calibration: More new evidence from Spain. *Evolutionary Anthropology*, **2–3**: 195–8.

Street, F.A. & Grove, A.T. (1976) Environmental and climatic implications of late Quaternary lake level fluctuations in Africa. *Nature*, **261**: 385–90.

Street, F.A. & Grove, A.T. (1979) Global maps of lake-level fluctuations since 30,000 B.P. *Quaternary Research*, **12**: 83–118.

Stringer, C.B. (1974) Population relationships of later Pleistocene hominids: A mutlivariate study of available crania. *Journal of Archaeological Sciences*, **1**: 317–42.

Stringer, C.B. (1978) Some problems in Middle and Upper Pleistocene hominid relationships. In: D. Chivers & K. Joysey (eds) *Recent Advances in Primatology, vol. 3*, pp. 395–418. London: Academic Press.

Stringer, C.B. (1984) Human evolution and biological adaptation in the Pleistocene. In: R. Foley (ed.) *Hominid Evolution and Community Ecology*, pp. 55–83. New York: Academic Press.

Stringer, C.B. (1985) Middle Pleistocene hominid variability and the origin of late Pleistocene humans, In: E. Delson (ed.) *Ancestors: The Hard Evidence*, pp. 289–95. New York: Alan Liss.

Stringer, C.B. (1988) The dates of Eden. *Nature*, **331**: 565–6.

Stringer, C.B. (1989a) The origin of modern humans: a comparison of the European and non-European evidence. In: P. Mellars & C.B. Stringer (eds) *The Human Revolution*, pp. 232–44. Edinburgh: Edinburgh University Press.

Stringer, C.B. (1989b) Documenting the origin of modern humans. In: E.Trinkaus (ed.) *The Emergence of Modern Humans*, pp. 67–96. Cambridge: Cambridge University Press

Stringer, C.B. (1992a) Replacement, continuity and the origin of *Homo sapiens*. In: G.Bräuer & F.H. Smith (eds.) *Continuity or Replacement: Controversies in*

Homo sapiens *Evolution*, pp.9–24. Rotterdam: Balkema.
Stringer, C.B. (1992b) *Homo erectus* et '*Homo sapiens* archaique' peut-on definir *Homo erectus*? In: J.J. Hublin & A.M. Tillier (eds.) *Aux origines d'Homo sapiens*, pp. 50–74. Paris: Presses Univeritaires de France.
Stringer, C.B. (1992c) Reconstructing recent human evolution. *Philosophical Transactions of the Royal Society of London, Series B*, **337**: 217–24.
Stringer, C.B. & Andrews, P. (1988) Genetic and fossil evidence for the origin of modern humans. *Science*, **239**: 1263–8.
Stringer, C.B., Hublin, J.J. & Vandermeersch, B. (1984) The origin of anatomically modern humans in Western Europe. In: F.H. Smith & F. Spencer (eds) *The Origin of Modern Humans: A World Survey of the Fossil Evidence*, pp. 51–135. New York: Alan R. Liss.
Stringer, C.B., Grün, R., Schwarcz, H.P. & Goldberg, P. (1989) ESR dates for the hominid burial site of Skhūl in Israel. *Nature*, **338**: 756–8.
Suzuki, H. & Hanihara, K. (Eds) (1982) *The Minatogawa Man: The Upper Pleistocene Man from the Island of Okinawa.* University Museum, University of Tokyo, Bulletin No. 19.
Suzuki, M. & Sakai, T. (1960) A familial study of torus palatinus and torus mandibularis. *American Journal of Physical Anthropology*, **18**: 263–372.
Svoboda, J. (1990) Moravia during the Upper Pleniglacial. In: O. Soffer & C.S. Gamble (eds) *The World at 18,000* BP: *High Latitudes*, pp. 193–203. London: Unwin Hyman.
Swisher III, C.C., Curtis, G.H., Jacob, T., Getty, A.G., Suprijo, A. & Widiasmoro (1994) Age of the earliest known hominids in Java, Indonesia. *Science*, **263**: 1118–21.
Szalay, F.S. (1981) Functional analysis and the practice of the phylogenetic method as reflected by some mammalian studies. *American Zoologist*, **21**: 37–45.
Szathmary, E. (1993) Genetics of aboriginal North Americans. *Evolutionary Anthropology*, **1**: 202–20.
Szilvassy, J. (1982) Zur variation, entwicklung und vererbung der stirnhohlen. *Annalen Naturhistorischen Museums in Wien*, **84**(A): 97–125.
Szilvassy, J. (1986) Eine neue methode zur intraserialen analyse von graberfeldern. *Mitteilungen der Antropologischen Gessellschaft in Wien*, **7**: 49–62.
Tagaya, A. & Katayama, K. (1988) Craniometric variation in Polynesians: multivariate positioning of male crania from Mangaia, Cook Islands. *Man and Culture in Oceania*, **4**: 75–89.
Taylor, R.E., Payen, P.A., Proir, C.A., Slota, P.J., Gillespie, R., Gowlett, J.A.J., Hedges, R.E.M., Jull, A.J.T., Zabel, T.H., Donahue, D.J. & Berger, R. (1985) Major revisions in the Pleistocene Age assignments for North American human skeletons by C-14 accelerator mass spectrometry: None older that 11,000 C-14 years BP. *American Antiquity*, **50**: 136–40.
Tchernov, E. (1992a) Biochronology, paleoecology and dispersal events of hominids in the southern Levant. In: T. Akazawa, K. Aoki & T. Kimura (eds) *The Evolution and Dispersal of Modern Humans in Asia*, pp. 149–88. Tokyo: Hokusen-sha Publ Co.
Tchernov, E. (1992b) Eurasian-African biotic exchanges through the Levantine corridor during the Neogene and Quaternary. *Courier Forschungsinstitut Senckenberg*, **153**: 103–23.

Tchernov, E. (1992c) Dispersal: a suggestion for a common usage of this term. *Courier Forschungsinstitut Senckenberg*, **153**: 21–5.

Templeton, A.R. (1992) Human origins and analysis of mitochondrial DNA sequences. *Science*, **255**: 737.

Thoma, A. (1973) New evidence for the polycentric evolution of *Homo sapiens*. *Journal of Human Evolution*, **2**: 529–36.

Thorne, A.G. (1977) Separation or reconciliation? Biological clues to the development of Australian society. In: J. Allen, J. Golson & R. Jones (eds) *Sunda and Sahul: Prehistoric Studies in Southeast Asia, Melanesia and Australia*, pp. 187–204. London: Academic Press.

Thorne, A.G. (1980a) The longest link: Human evolution in Southeast Asia and the settlement of Australia. In: J.J. Fox, R.G. Garnaut, P.J. McCawley & J.A.C. Mackie (eds) *Indonesia: Australian Perspectives*, pp. 35–43 Canberra: Research School of Pacific Studies.

Thorne, A.G. (1980b) The arrival of man in Australia. In: A. Sherratt (ed.) *The Cambridge Encyclopedia of Archaeology*, pp. 96–100. Cambridge: Cambridge University Press.

Thorne, A.G. (1981) The centre and the edge: The significance of Australian hominids to African paleoanthropology. In: R.E. Leakey & B.A. Ogot (eds) Proceedings of the 8th Panafrican Congress.

Thorne, A.G. (1984) Australia's human origins – how many sources? *American Journal of Physical Anthropology*, **63**: 227.

Thorne, A.G. & Wolpoff, M.H. (1981) Regional continuity in Australasian Pleistocene hominid evolution. *American Journal of Physical Anthropology*, **55**: 337–49.

Thorne, A.G. & Wolpoff, M.H. (1991) Conflict over modern human origins. *Search*, **22**(5): 175–7.

Tianyuan, L. & Etler, D.A. (1992) New Middle Pleistocene hominid crania from Yunxian in China. *Nature*, **357**: 404–7.

Tillet, T. (1983) *Paléolithique du bassin Tchadien septentrional (Niger-Tchad)*. Paris: Centre National de la Recherche Scientifique.

Tishkoff, S.A., Dietzsch, E., Speed, W., Pakstis, A.J., Cheung, K., Kidd, J., Bonné-Tamir, B., Santachiara-Benerecetti, A.S., Moral, P., Watson, E., Krings, M., Pääbo, S., Risch, N., Jenkins, T. & Kidd, K.K. Global patterns of linkage disequilibrium at the CD4 locus support a recent African origin of modern humans. (Unpublished data.)

Tobias, P.V. (1972) Recent human biological studies in Southern Africa,, with special reference to Negroes and Khoisans. *Transactions of the Royal Society of South Africa*, **40**: 109–33.

Tokunaga, S., Oshima, H., Polhaupessy, A.A. & Ito, Y. (1985) A palynological study of the Pucangan and Kabuh Formations in the Sangiran area. *Geological Research, Dev. Centre, Special Publications*, **4**: 199–219.

Torgersen, J.H. (1951) Asymmetry and skeletal maturation. *Acta Radiologica*, **36**: 521–3.

Torroni, A., Schurr, T.G., Cabell, M.F., Brown, M.D., Neel, J.V., Larsen, M., Smith, D.G., Vullo, C.M. & Wallace, D.C. (1993a) Asian affinities and continental radiation of the four founding native American mtDNAs. *American*

Journal of Human Genetics, **53**: 563–90.

Torroni, A., Sukernik, R.I., Schurr, T.G., Starikovskaya, Y.B., Cabell, M.F., Crawford, M.H., Comuzzie, A.G. & Wallace, D.C. (1993b) mtDNA variation of aboriginal Siberians reveals distinct genetic affinities with native Americans. *American Journal of Human Genetics*, **53**: 591–608.

Torroni, A., Neel, J.V., Barrantes, R., Schurr, T.G. & Wallace, D.C. (1994) Mitochondrial DNA 'clock' for the Amerinds and its implications for timing their entry into North America. *Proceedings of the National Academy of Sciences USA*, **91**: 1158–62.

Trinkaus, E. (1977) A functional interpretation of the axillary border of the Neanderthal scapula. *Journal of Human Evolution*, **6**: 231–4.

Trinkaus, E. (1982) Evolutionary trends in the Shanidar Neanderthal sample. *American Journal of Physical Anthropology*, **57**: 237.

Trinkaus, E. (1983a) Neanderthal postcrania and the adaptive shift to modern humans. In: E.Trinkaus (ed) *The Mousterian Legacy: Human Biocultural Change in the Upper Pleistocene*. BAR International Series, **164**: 165–200.

Trinkaus, E. (1983b) *The Shanidar Neanderthals*. New York: Academic Press.

Trinkaus, E. (1984) Western Asia. In: F.H. Smith & F. Spencer (eds) *The Origin of Modern Humans: A World Survey of the Fossil Evidence*, pp. 251–193. New York: Alan R. Liss.

Trinkaus, E. (1986) The Neanderthals and modern human origins. *Annals Revue of Anthropology*, **15**: 193–218.

Trinkaus, E. (1987) The Neanderthal face: evolutionary and functional perspectives on a recent hominid face. *Journal of Human Evolution*, **16**: 429–43.

Trinkaus, E. (1988) The evolutionary origins of the Neanderthals or, why were there Neanderthals? *L'Anatomie*, **3**: 11–29.

Trinkaus, E. (1989a) *The Emergence of Modern Humans: Biocultural Adaptations in the Later Pleistocene*. Cambridge: Cambridge University Press.

Trinkaus, E. (1989b) The Upper Pleistocene transition. In: E.Trinkaus (ed.) *The Emergence of Modern Humans: Biocultural Adaptations in the Later Pleistocene*, pp. 42–66. Cambridge: Cambridge University Press.

Troitzky, W. (1928) Zur frage der formbildung des schadelaches. *Zeitschrift für Morphologie und Anthropologie*, **30**: 504–34.

Turbón, D., Lalueza, A., Perez-Perez, A., Prats, E., Moreno, P. & Pons, P. (1994) Absence of the 9 bp mtDNA region V deletion in ancient remains of aborigines from Tierra del Fuego. *Ancient DNA Newsletter*, **2**: 24–6.

Turner, C.G. II (1971) Three-rooted mandibular first permanent molars and the question of American Indian origins. *American Journal of Physical Anthropology*, **34**: 229–41.

Turner, C.G. II (1976) Dental evidence on the origins of the Ainu and Japanese. *Science*, **193**: 911–13.

Turner, C.G. II (1979) Dental anthropological indications of agriculture among the Jomon people of central Japan. X. Peopling of the Pacific. *American Journal of Physical Anthropology*, **51**: 619–35.

Turner, C.G. II (1983) Sinodonty and Sundadonty. In: R.S. Vasilievsky (ed.) *Late Pleistocene and Early Holocene Cultural Connections of Asia and America*, pp. 72–6. Novosibirsk: USSR Academy of Science, Siberian Branch.

Turner, C.G. II (1985a) The dental search for Native American origins. In: R. Kirk & E. Szathmary (eds) *Out of Asia. Peopling of the Americas and the Pacific*, pp. 31–78. Canberra: *Journal of Pacific History*.

Turner, C.G. II (1985b) Comment on 'The supraorbital torus: A most remarkable peculiarity' by M.D. Russell. *Current Anthropology*, **26**: 355.

Turner, C.G. II (1986a) The first Americans: the dental evidence. *National Geographical Research*, **2**: 37–46.

Turner, C.G. II (1986b) Dentochronological separation estimates for Pacific Rim populations. *Science*, **232**: 1140–2.

Turner, C.G. II (1987) Late Pleistocene and Holocene population history of East Asia based on dental variation. *American Journal of Physical Anthropology*, **73**: 305–21.

Turner, C.G. II (1989) Teeth and prehistory in Asia. *Science*, **269**: 88–96.

Turner, C.G. II (1990a) Major features of Sundadonty and Sinodonty, including suggestions about East Asian microevolution, population history, and late Pleistocene relationships with Australian Aboriginals. *American Journal of Physical Anthropology*, **82**: 295–317.

Turner, C.G. II (1990b) Origins and affinity of the people of Guam: A dental anthropological assessment. *Micronesia (Suppl.)*, **2**: 403–16.

Turner, C.G. II (1992) Microevolution of East Asian and European Populations: A Dental Perspective. In: T. Akazawa, K. Aoki & T. Kimura (eds) *The Evolution and Dispersal of Modern Humans in Asia*, pp. 414–38. Tokyo: Hokusen-sha.

Turner, C.G. II & Scott, G.R. (1977) Dentition of Easter Islanders. In: A.A. Dahlberg & T.M. Graber (eds) *Orofacial Growth and Development*, pp. 229–49. Paris: Mouton Publishers.

Turner, C.G. II & Swindler, D.R. (1978) The dentition of New Britain West Nakanai Melanesians. *American Journal of Physical Anthropology*, **49**: 361–72.

Turner, W. (1901) The craniology, racial affinities and descent of the Aborigines of Tasmania. *Transactions of the Royal Society of Edinburgh*, **46**: 365–403.

Valladas, H., Joron, J.L., Valladas, G., Arensburg, B., Bar-Yosef, O., Belfer-Cohen, A., Goldberg, P., Laville, H., Meignen, L., Rak, Y., Tchernov, E., Tillier, A.-M. & Vandermeersch, B. (1987) Thermoluminescence dates for the Neanderthal burial site at Kebara in Israel. *Nature*, **330**: 159–60.

Valladas, H., Reyss, J., Joron, J.L., Valladas, G., Bar-Yosef, O. & Vandermeersch, B. (1988) Thermoluminescence dating of the Mousterian 'Proto-Cro-Magnon' remains from Israel and the origin of modern man. *Nature*, **313**: 614–16.

Valloch, K. (1976) Die alsteinzeitliche Fundstele in Brno-Bohunice. *Studie Archeologickeho ustavu Ceskoslovenske Akademie Ved v Brne*, **4**(1): 3–120.

Vallois, H.V. (1969) Les hommes de Cro-Magnon et les Guanches: les faits acquis et les hypotheses. *Anuario de Estudios Atlanticos*, **15**: 97–118.

Vallois, H.V. (1972) Le gisement et le squelette de Saint-Germain-la-Rivière. *Archives de l'Institute de Paléontologie Humain*, **34**: 47–115.

Vallois, H.V. & Billy, G. (1965) Nouvelles recherches sur les hommes fossiles de l'abri de Cro-Magnon. *L'Anthropologie*, **69**: 47–74, and 249–72.

Van der Broek, A. J. P. (1945) On exostoses in the human skull. *Acta Neerlandica Morphologiae Normalis et Pathologicae*, **5**: 95–118.

Van Valen, L.M. (1986) Speciation and our own species. *Nature*, **322**: 412.

Van Vark, G.N. (1984) On the determination of hominid affinities. In: G.N. Van

Vark & W.W. Howells (eds) *Multivariate Statistical Methods in Physical Anthropology*, pp. 323–49. Dordrecht: Reidel.

Van Vark, G.N. (1986) More on the classification of the Border Cave I skull. *5th Congress of the European Anthropological Association*, Lisboa.

Van Vark, G.N. (1990) A study of European Upper Paleolithic crania. *Fysisch-Antropologische Mededelingen*, **2**: 7–15.

Van Vark, G.N., Bilsborough, A. & Dijkema, J. (1989) A further study of the morphological affinities of the Border Cave 1 cranium with special reference to the origin of modern man. *Journal of Anthropology and Prehistory*, **100**: 43–56.

Van Vark, G.N., Bilsborough, A. & Henke, W. (1992) Affinities of European Upper Palaeolithic *Homo sapiens* and later human evolution. *Journal of Human Evolution*, **23**: 401–17.

Van Zinderen Bakker, E.M. (1982) African palaeoclimates 18,000 years BP. *Palaeoecology of Africa*, **15**: 77–99.

Vandermeersch, B. (1981) Les premiers *Homo sapiens* au proche-Orient. In: D. Ferembach (ed.) *Les processus de l'Homminisation*, pp. 19–26. Paris: CNRS.

Vandermeersch, B. (1985) The origin of the Neanderthals In: E. Delson (ed.) *Ancestors: The Hard Evidence*, pp. 306–9. New York: Alan R. Liss.

Vandermeersch, B. (1989) The evolution of modern humans: recent evidence from Southwest Asia. In: P. Mellars & C.B. Stringer (eds.) *The Human Revolution*, pp. 155–64. Edinburgh: Edinburgh University Press.

Vecchi, F. (1968) Sesso e variazioni di caratteri discontinui del cranio. *Rivista di Antropologia*, **55**: 283–90.

Veeh, H. & Veevers, J.J. (1970) Sea-level at −175 m off the Great Barrier Reef, 13,600 to 17,000 years ago. *Nature*, **226**: 536–7.

Veth, P.M. (1993) *Islands in the Interior*. Ann Arbor, MI: International Monographs in Prehistory.

Vigilant, L., Pennington, R., Harpending, H., Kocher, T.D. & Wilson, A. (1989) Mitochondrial DNA sequences in single hairs from a southern African population. *Proceedings of the National Academy of Science*, **86**: 9350–4.

Vigilant, L., Stoneking, M., Harpending, H., Hawkes, K. & Wilson, A. (1991) African populations and the evolution of human mitochondrial DNA. *Science*, **253**: 1503–7.

Vlcek, E. (1970) Relations morphologiques des types humains fossiles de Brno et Cro-Magnon au Pléistocène supérieur d'Europe. In: *L'Homme de Cro-Magnon 1868–1968*, pp. 59–72. C.R.A.P.E.

Vogel, J. & Beaumont, P.B. (1972) Revised radio carbon chronology for the Stone Age in southern Africa. *Nature*, **237**: 50–1.

Vogel, J. & Waterbolk, H. (1972) Groningen radiocarbon dates X. *Radiocarbon*, **14**: 6–10.

Von Koenigswald, G.H.R. & Weidenreich, F. (1939) The relationship between *Pithecanthropus* and *Sinanthropus*. *Nature*, **144**: 926–9.

Waddle, D.M. (1994) Matrix correlation tests support a single origin for modern humans. *Nature*, **368**: 452–545.

Wainscoat, J.S., Hill, A.V.S., Boyce, A.L., Flint, J., Hernandez, M., Thein, S.L., Old, J.M., Lynch, J.R., Falusi, A.G., Weatherall, D.J. & Clegg, J.B. (1986) Evolutionary relationships of human populations from an analysis of nuclear DNA polymorphisms. *Nature*, **319**: 491–3.

Waldeyer, W. (1880) Bemerkungen üer die squama ossis occipitis mit besonderer Berücksichtigung des 'Torus Occipitalis'. *Archaeologie für Anthropologiea*, **12**: 453–61.
Walker, D. (1973) Highland vegetation. *Australian Natural History*, **17**(2): 410–19.
Walter, H. (1983) Population genetical investigations on Indian tribals: Preliminary report. *Anthropological Anz*, **41**: 119–28.
Ward, R.H., Frazier, B.L., Dew-Jager, K. & Pääbo, S. (1991) Extensive mitochondrial diversity within a single Amerindian tribe. *Proceedings of the National Academy of Sciences, USA*, **88**: 8720–4.
Washburn, S.L. (1947) The relation of the temporal muscle to the form of the skull. *Anatomy Records*, **99:** 239–48.
Washburn, S.L. (1948) Sexual differences in the pubic bone. *American Journal of Physical Anthropology*, **6**: 199–208.
Waugh, L.M. (1930) A study of the nutrition and teeth of the Eskimos of North Bering Sea and Arctic Alaska. *Journal of Dental Research*, **10**: 387–93.
Webb, S.G. (1989) *The Willandra Lakes Hominids*. Canberra: Department of Prehistory, Australian National University.
Weidenreich, F. (1938) *Pithecanthropus* and *Sinanthropus*. *Nature*, **141**: 376–9.
Weidenreich, F. (1939) Six lectures on *Sinanthropus pekinensis* and related problems. *Bulletin of the Geological Society of China*, **19**: 1–110.
Weidenreich, F. (1940) The torus occipitalis and related structures and their transformations in the course of human evolution. *Bulletin of the Geological Society of China*, **19**(4): 479–544.
Weidenreich, F. (1943a) The skull of *Sinanthropus pekinensis*: a comparative study of a primitive hominid skull. *Palaeontologia Sinica, n.s. D.*, No. 10.
Weidenreich, F. (1943b) The 'Neanderthal man' and the ancestors of *Homo sapiens*. *American Anthropologist*, **45**: 39–48.
Weidenreich, F. (1946) *Apes, Giants and Man*. Chicago: Chicago University Press.
Weidenreich, F. (1951) Morphology of Solo man. *Anthropological Papers of the American Museum of Natural History*, **43**: 205–90.
Weiner, J.S. (1971) *Man's Natural History*. London: Weidenfeld and Nicolson.
Weinert, H. (1936) Der urmenschenschadel von Steinheim. *Zeitschrift für Morphologie und Anthropologie*, **35**: 63–518.
Welcker, H. (1862) *Untersuchungen uber wachsthum und bau des menschlichen schadels*. Leipzig: Engelmann.
Wendorf, F., Schild, R., Siad, R., Haynes, C.V., Gautier, A. & Kobusiewicz, M. (1976) The prehistory of the Egyptian Sahara. *Science*, **193**: 103–14.
Wendorf, F., Close, A.E., Schild, R., Gautier, A., Schwarcz, H.P., Miller, G.H., Kowalski, K., Krolik, H., Bluszcz, A., Robins, D. & Grün, R. (1990) Le dernier interglaciaire dans le Sahara oriental. *L'Anthropologie*, **94**: 361–91.
Wendorf, F., Close, A., Schild, R., Gautier, A., Schwarcz, H.P., Miller, G.H., Kowalski, K., Krolik, H., Bluszcz, A., Robinson, D., Grun, R. & McKinney, C. (1991) Chronology and stratigraphy of the Middle Palaeolithic at Bir Tarfawi, Egypt. In: J.D. Clark (ed.) *Cultural Beginnings: Approaches at Understanding Early Hominid Life-Ways in the East African Savanna Mosaic*, pp. 197–208. Bonn: Rudolf Habelt.
Wendorf, F., Schild, R. & Close, A.E. (1993) Middle Palaeolithic occupations at Bir

Tarfawi and Bir Sahara East, Western Desert of Egypt. In: L. Krzyzaniak, M. Kobusiewicz & J. Alexander (eds) *Environmental Change and Human Culture in the Nile Basin and Northern Africa Until the Second Millennium* BC, pp. 103–12. Poznan: Museum Archeologiczne W Poznaniu.

White, J.P., Crook, K.A. & Ruxton, B.P. (1970) Kosipe: A late Pleistocene site in the Papuan Highlands. *Proceedings of the Prehistoric Society*, **36**: 152–70.

White, M.J.D. (1978) *Modes of Speciation*. San Francisco: W.H. Freeman.

White, R. (1989a) Production complexity and standardization in early Aurignacian bead and pendant manufacture: evolutionary implications. In: P. Mellars & C.B. Stringer (eds) *The Human Revolution*, pp. 366–90. Edinburgh: Edinburgh University Press.

White, R. (1989b) Toward a contextual understanding of the earliest body ornaments. In: E. Trinkaus (ed.) *The Emergence of Modern Humans*, pp. 211–31. Cambridge: Cambridge University Press.

Whitmore, T.C. (1975) *Tropical Rainforests of the Far East*. Oxford: Clarendon Press.

Whyte, R.O. (Ed.) (1984) *The Evolution of the East Asian Environment*. Hong Kong: University of Hong Kong Press.

Wickler, S. & Spriggs, M. (1988) Pleistocene occupation of the Solomon Islands, Melanesia. *Antiquity*, **62**: 703–6.

Williams, M.A. (1975) Late Pleistocene tropical aridity: synchronous in both hemispheres? *Nature*, **253**: 617–18.

Williams, M.A.J. (1985) Pleistocene aridity in tropical Africa, Australian and Asia. In: I. Douglas & T. Spencer (eds) *Environmental Change and Tropical Geomorphology*, pp. 219–33. Boston: Allen & Unwin.

Williams, P.W., Mc Dougal, I. & Powell, J.M. (1972) Aspects of the Quaternary geology of the Tari-Koroba area, Papua. *Journal of the Geological Society of Australia*, **18**: 333–47.

Wilson, A.C. & Cann, R. (1992) The recent African genesis of humans. *Scientific American*, **266**: 22–7.

Wilson, A.C., Cann, R.L., Carr, S,M., George, M., Gyllensten, U.B. *et al.* (1985) Mitochondrial DNA and two perspectives on evolutionary genetics. *Biological Journal of the Linnean Society*, **26**: 375–400.

Wolpoff, M.H. (1980a) *Paleoanthropology*. New York: Knopf.

Wolpoff, M.H. (1980b) Cranial remains of Middle Pleistocene European hominids. *Journal of Human Evolution*, **9**: 339–58.

Wolpoff, M.H. (1982) The Arago dental sample in the context of hominid dental evolution. *1er Congres International de Palaeontologie Humaine*, Nice, pp. 389–410.

Wolpoff, M.H. (1985) On explaining the supraorbital torus. *Current Anthropology*, **26**: 522.

Wolpoff, M.H. (1986a) Describing anatomically modern *Homo sapiens*: a distinction without a definable difference. In: V.V. Novotny & A. Mizerova (eds) *Fossil Man – New Facts, New Ideas. Papers in Honour of Jan Jelinek's Life Anniversary*, pp. 41–53. Brno: Anthropos.

Wolpoff, M.H. (1986b) Stasis in the interpretation of evolution in *Homo erectus*: a reply to Rightmire. *Paleobiology*, **12**(3): 325–8.

Wolpoff, M.H. (1986c) More on Zhoukoudian. *Current Anthropology*, **27**(1): 45–6.

Wolpoff, M.H. (1989a) Multiregional evolution: the fossil alternative to Eden. In: P. Mellars & C.B. Stringer (eds.) *The Human Revolution*, pp. 62–108. Edinburgh: Edinburgh University Press.

Wolpoff, M.H. (1989b) The place of the Neanderthals in human evolution. In: E. Trinkaus (ed.) *The Emergence of Modern Humans*, pp. 97–141. Cambridge: Cambridge University Press.

Wolpoff, M.H. (1992a) Theories of modern human origins. In: G. Bräuer & F.H. Smith (eds) *Continuity or Replacement? Controversies in* Homo sapiens *Evolution*, pp. 25–63. Rotterdam: Balkema.

Wolpoff, M.H. (1992b) *Homo erectus* et les origines de la diversite humaine. In: J.J. Hublin & A.M. Tillier (eds) *Aux origines d'Homo sapiens*, pp. 97–155. Paris: Presses Universitaires de France.

Wolpoff, M.H. & Caspari, (1990) On Middle Paleolithic/Middle Stone Age hominid taxonomy. *Current Anthropology*, **31**: 394–5.

Wolpoff, M.H. & Thorne, A. (1991) The case against Eve. *New Scientist, 22nd June*: 37–41.

Wolpoff, M.H., Wu. X.Z. & Thorne, A.G. (1984) Modern *Homo sapiens* origins: a general theory of hominid evolution involving the fossil evidence from East Asia. In: F.H. Smith & F. Spencer (eds) *The Origin of Modern Humans: A World Survey of the Fossil Evidence*, pp. 411–83. New York: Alan R.Liss.

Woo, J.K. (1950) Torus palatinus. *American Journal of Physical Anthropology*, **8**: 81–111.

Workshop European Anthropologists (1980) Recommendations for age and sex diagnosis of skeletons. *Journal of Human Evolution*, **9**: 517–49.

Wright, R.V.S. (1992) Correlation between cranial form and geography in *Homo sapiens*: CRANID – a computer program for forensic and other applications. *Perspectives in Human Biology*, **2**: 128–34.

Wu, R. (1959) Human fossils found in Liukiang, Kwangsi, China. *Vertebrata PalAsiatica*, **3**: 109–18.

Wu, R. & Wu, X. (1984) Hominid fossils in China and their relation to those of neighbouring regions. In: R.O. Whyte (ed.) *The Evolution of the East Asian Environment: Palaeobotany, Palaeozoology and Palaeoanthropology*, pp. 787–95. Hong Kong: Centre for Asian Studies, University of Hong Kong.

Wu, X. (1961) Study on the Upper Cave man of Choukoutien. *Vertebrata PalAsiatica*, **3**: 181–211.

Wu, X. & Zhang, Z. (1985) *Homo sapiens* remains from late Palaeolithic and Neolithic China. In: W. Rukang & J.W. Olsen (eds) *Palaeoanthropology and Palaeolithic Archaeology in the People's Republic of China*, pp. 79–89. New York: Academic Press.

Yamaguchi, B. (1982) A review of the osteological characterisitcs of the Jomon population in prehistoric Japan. *Journal of the Anthropological Society of Nippon*, **90**: 77–90.

Yamaguchi, B. (1985) The incidence of minor non-metric cranial variants in the protohistoric human remains from eastern Japan. *Bulletin of the National Science Museum of Tokyo, Series D*, **11**: 13–24.

Yen, D.E. (1977) Hoabinian horticulture: The evidence and the questions from northwest Thailand. In: J. Allen, J. Golson & R. Jones (eds) *Sunda and Sahul*, pp. 567–600. London: Academic Press.

Zuraina, M. (1982) *The West Mouth, Niah, in the Prehistory of Southeast Asia.* Kuching (Sarawak): The Museum. *Sarawak Museum Journal* **31**, *Special Monograph, No. 3.*

Zuraina, M. & Tija, H.D. (1988) Kota Tampan, Perak. The geological and archeological evidence for a late Pleistocene site. *Journal of the Malaysian Branch of the Royal Asiatic Society*, **61**: 123–34.

Fossil and site index

Subject index